CHINESE ENERGY MARKETS

Also by Armelle Guizot

HEDGE FUND TECHNOLOGY

Chinese Energy Markets

Trading and Risk Management of Commodities and Renewables

ARMELLE GUIZOT

palgrave
macmillan

First published 2007 by
PALGRAVE MACMILLAN
Houndmills, Basingstoke, Hampshire RG21 6XS and
175 Fifth Avenue, New York, N.Y. 10010
Companies and representatives throughout the world

PALGRAVE MACMILLAN is the global academic imprint of the Palgrave Macmillan division of St. Martin's Press, LLC and of Palgrave Macmillan Ltd. Macmillan® is a registered trademark in the United States, United Kingdom and other countries. Palgrave is a registered trademark in the European Union and other countries.

ISBN-13: 978-0-230-55420-7 hardback
ISBN-10: 0-230-55420-2 hardback

This book is printed on paper suitable for recycling and made from fully managed and sustained forest sources. Logging, pulping and manufacturing processes are expected to conform to the environmental regulations of the country of origin.

A catalogue record for this book is available from the British Library.

Library of Congress Cataloging-in-Publication Data

Guizot, Armelle, 1972–
Chinese energy markets : trading and risk management of commodities and renewables / Armelle Guizot.
p. cm.
Includes index.
ISBN 0-230-55420-2 (alk. paper)
1. Energy policy–China. 2. Petroleum industry and trade–China.
3. Commodity exchanges–China. I. Title.
HD9502.C62G85 2007
333.790951–dc22 2007021643

10 9 8 7 6 5 4 3 2 1
16 15 14 13 12 11 10 09 08 07

Printed and bound in Great Britain by
Antony Rowe Ltd, Chippenham and Eastbourne

While best effort and attention have been taken in the preparation of this book, neither the publisher, the author nor any of the involved organizations, associations, institutions, and researchers accept responsibility for any errors it may convey, or unintentional coincidental similitude in the writing or in the information being provided, or for any losses howsoever arising from, or in reliance upon, its content. Additionally, the publisher and the author also reject all responsibilities for usage of information and communications potentially subject to possible interpretations, misunderstandings and allusions or used in the court of law under any government. The information, despite coming from wide research, may not be the subject of accusation of any type under any circumstances and is solely aimed to provide the reader with more transparency and knowledge about the topic. This statement shall protect both the author and the publisher from inappropriate, unforeseeable potentially accusatory statements and neither can accept any responsibility of liability for reliance by any person pertaining to this information.

The information and summaries contained in this book appear as a matter of record only and are provided for information purposes only. The summaries do not and should not be construed to constitute an investment recommendation, or an influence to alter economic policies, trading or recommend a particular country system over another. It shall not constitute an offer to sell or a solicitation of an offer to buy any of the products or companies or corporations' shares or any form on the basis of any contract or financial or liable commitment whatsoever. The summaries are based on information and data received from the subjects of each and from other third party sources and no guarantee of completeness or accuracy can be made nor is any representation or warranty made in this regard. We understand that this book does not contain a bias or subjective comments with respect to products and corporations it presents and thus those may not be sold into or that their sales may be restricted or influenced in particular markets; these are information formulated by various references, researchers, administrations, agencies and associations, and the accuracy and updates of their assertions have not been verified nor validated. They shall, under any circumstances, constitute a representation of true facts, even if they happen to be, and be used as reference basis for further digressions without thorough due diligence verifications. This book and its content has for sole purpose to inform and educate the reader about the current status of the Chinese energy markets in general.

Contents

List of Figures

List of Tables

List of Abbreviations

ARCO	Energy company, part of BP, www.arco.com
b/d	Barrels per day
bcm	Billion cubic metres
boe	Barrels of oil equivalent
BOO	Build-operate-own
BOOT	Build-own-operate-transfer
BOT	Build-operate-transfer
Btu	British thermal unit
CAPCO	Castle Peak Power Company Limited
CATARC	China Automotive Technology and Research Center
CCGT	Combined-cycle gas turbine
CCPN	Central China Power Network
CDM	Clean Development Mechanism (under the Kyoto Protocol)
CDU	Crude distillation unit
CESA	Clean Energy State Alliance
CHP	Combined production of heat and power
CNG	Compressed natural gas
CNOOC	China National Offshore Oil Corporation
CNPC	China National Petroleum Corporation
CO_2	Carbon dioxide
CPCIA	China Petroleum and Chemical Industry Association
CRED	Chinese Center for Renewable Developments
CREIA	China Renewable Energy Industry Association
CRS	Center for Resource Solutions
CSEP	China Sustainable Energy Program
CWP	Combined water and power
CYPN	Sichuan and Chongqing Power Network
DIS	Deferred investment scenario
DSM	Demand side management
E&P	Exploration and production

ECPN	East(ern) China Power Network
ECU	European currency unit
EEIJ/IEEJ	Institute of Electrical Engineers of Japan
EOR	Enhanced oil recovery
EPC	Engineering, procurement and construction
EU	European Union
FDI	Foreign direct investment
FSU	Former Soviet Union
GARP	Global Association of Risk Professionals
GCC	Gulf Co-operation Council
GCS	Green Certificate System
GDP	Gross domestic product
GHG	Greenhouse gas
GPS	Generation Performance Standards
Gt	Gigatonne (1 tonne $\times 10^9$)
GTL	Gas-to-liquids
GW	Gigawatt (1 watt $\times 10^9$)
GWh	Gigawatt hour or 1 watt * 10^9 hour
HNPN	Hainan Power Grid
IEA	International Energy Agency
IEEJ	Institute of Energy Economics of Japan
IEO	*International Energy Outlook*, annual publication by the IEA
IFC	International Finance Corporation
IGCC	Integrated Gasification Combined Cycle
IMF	International Monetary Fund
IOC	International Oil Company
IPP	Independent Power Producer
IWPP	Independent water and power producer
kb/d	Thousand barrels per day
Kgce	Kilogram coal equivalent
kJ	KiloJoule
kt	kilotonne
KUFPEK	Kuwait Company for energy exploration
kW	kilowatt
kWh	kilowatt-hour
LBNL	Lawrence Berkeley National Laboratory
LNG	Liquefied natural gas
LPG	Liquefied petroleum gas
mb/d	Million barrels per day
MBtu	Million British thermal units
mcm	Million cubic metres
Mm^2	10^6
MMS	Mandatory market share
MOF	Ministry of Finance

MOST	Ministry of Science and Technology
MSC	multiple service contract
Mt	Million (* 10^6) tons
Mtce	Million (* 10^6) tons coal equivalency
Mtoe	Million tonnes/tons of oil equivalent
MW	Megawatt (1 watt * 10^6)
MWe	MegaWatt electrical
MWh	Megawatt-hour
NAREIT	National Association of Real Estate Investment Trusts
NCPN	North(ern) China Power Network
NDRC	National Development and Reform Commission/Chinese national energy office
NELG	National Energy Leading Group
NEPA	National Environmental Protection Agency
NEPN	Northeast Power Network
NESP	China National Energy Strategy and Policy
NGL	Natural gas liquid
NOC	National Oil Company
NREL	National Renewable Energy Laboratory
NWPN	Northwest Power Network
OAPEC	Organization of Arab Petroleum Exporting Countries
OCGT	Open-cycle gas turbine
oe	Oil equivalent
OECD	Organization for Economic Co-operation and Development
OGCR	Office of Government and Community Relations
OPEC	Organization of Petroleum Exporting Countries
PBF	Public Benefits Fund
PPP	Purchasing power parity; the rate of currency conversion that equalizes the purchasing power of different currencies. PPPs compare costs in different currencies of a fixed basket of traded and non-traded goods and services and yield a widely-based measure of standard of living
ppm	Parts per million
PR	Proved reserves
PSA	Production-sharing agreement
PV	Photovoltaic or Power Voltage
REDP	Renewable Energy Development Project
REEEP	Renewable Energy and Energy Efficiency Partnership
RMB	Renminbi or Yuan Chinese currency
ROPED	Reijing Research Institute of Petroleum Exploration and Development
RPS	Renewable Portfolio System
SCPN	South China Power Network
SECSC	Shanghai Energy Conservation Supervision Center
SEPA	State Environmental Protection Administration

SERC	State Electricity Regulatory Commission
SETC	State Economic and Trade Commission
SGCC	State Grid Corporation of China
SO_2	Sulfur dioxide
SOE	State-owned enterprise
SPC	State Power Corporation
SSTC	State Science and Technology Commission
tce	Tons coal equivalency
tcf	Thousand cubic feet
tcm	Trillion cubic metres
TFC	Total final consumption
toe	Tonnes of oil equivalent
TPES	Total primary energy supply
TWh	Terawatt (* 10^{12}) hours
UNDESA	United Nations Department of Economic and Social Affairs
UNDP	United Nations Development Program
UNFCCC	United Nations Framework Convention on Climate Change
USGS	Chinese oil and gas research institute
USGS	United States Geological Survey
VAT	Value added tax
WAPS	World Alternative Policy Scenario
WEM	World Energy Model
W_p	Peak Watt
WTI	Western Texas Intermediate
WTO	World Trade Organization

EQUIVALENTS

10^1 deca (da)
10^{-1} deci (d)
10^{-12} pico (p)
10^{12} tera (T)
10^{-15} femto (f)
10^{15} peta (P)
10^{-18} atto (a)
10^{18} exa (E)
10^{-2} centi (c)
10^2 hecto (h)
10^3 kilo (k)
10^{-3} milli (m)
10^6 mega (M)
10^{-6} micro (μ)
10^9 giga (G)
10^{-9} nano (n)

Introduction

China's emerging middle class requires preparation and security for strategic commodities reserves. One way to plan for emerging needs is to secure more energy partnerships throughout the world and to invest in more commodities projects with, for instance, the foreign exchange reserves surplus of about $1 trillion equivalent as of end of 2006. Very much like the growing foreign currency reserves, the accumulation of Chinese commodities reserves has been quiet and remarkable, despite consumption overriding supply. One could believe that the commodities reserves that are of real tangible value may serve as collateral assets for the bubble-like foreign exchange currency reserves of 2006 mostly financed by American Treasury bills and growing home equity loans that began to deplete in November 2005. This means that the current value of the renminbi is also evaporating, in concert with the gradual slow down and fall of matured real estate markets. While the Chinese yuan is appreciating in the face of other currencies, the investments driving the Chinese economy are no longer going to return as much as was the case prior to the yuan's re-evaluations. They will gradually become of equal value to those of matured economies. This would also imply that the valuations of Chinese companies will rise apropos their peers as they comprise manufacturing hubs with less unemployment and more capital creation, while the service consuming economies will become impoverished and dependent on this wealth creation. China creates manufacturing centers and India intellectual capitalism at lower cost, while technologies replace service economies. China's dilemma is that while it concentrates its financial reserves in China, it has subcontracted geopolitical alliances throughout the world to secure commodities imports for the next decades, primarily because China has been a net importer of oil since 1993. All contracts are still negotiated in dollars despite the fact that the main geopolitical zones at risk are far from the dollar and the yuan. The prices of those contracts are mainly linked to the relative value of the dollar against the yuan and thus, upon the arrival of financial derivatives in the Chinese markets, the real 'dollar' value of these tangible assets will change, but not for the better. The commodities reserves may serve as tangible collateral assets to support and hedge the actual value of the Chinese foreign exchange currency reserves evaluated at US$1.2 trillion as of the first quarter of 2007. Yet, with the rising consumption and living standards of the 1.5 billion Chinese, the current commodities reserves will soon appear insufficient.

As of 2005, Chinese proven oil reserves amounted to 18.3 billion barrels. Chinese oil production in 2005 amounted to 3.62 million barrels per day of which 3.49 million barrels were crude oil. The Energy Information Administration reported that Chinese oil production in mid-2006 was about 3.8 million barrels a day. They also reported the following statistics: Chinese oil consumption is around 6.53 million barrel per day; crude oil refining capacity amounts to 4.65 million barrels per day; 2005 natural gas reserves totalled 53.3 trillion cubic feet (Tcf) while production creates a yearly 1.21 Tcf.

Chinese energy companies have bought stakes in many global geopolitical strategic areas to diversify partnerships and acquire economic means for negotiating in global affairs for the future. Until 2005, Chinese energy companies remained old fashioned and suffered from a lack of innovation, technologies, synergies, scientific diversification and international partnerships. Due to their lengthy isolation from world affairs, they have been immune from speculative financial engineering and inflationary trading. While the Chinese energy markets' infrastructures and operations remain highly traditional and unsophisticated, the financial statements of the companies have also been immured from financial engineering and speculative fictitious valuations. So while they are producing excess tangible reserves, they are also behind in commodities financial engineering and risk management partnerships and trading. The development of financial engineering instruments such as derivatives had just started to appear on the Chinese market in 2005, but many new reserves are also being explored in Africa, South America, Northern Europe, and Russia. However, the new world order creates many challenges as to the cost of commodities transportation from point to point. It is obvious that the global competition for natural resources will soon intensify. And those who can will use their accumulated built-in reserves to control policies, possibly dangerously. How will this be seen in the real world? The fallacy of replacing fictitious financial derivatives by tangible real value assets will inevitably create tensions in global futures markets. Geopolitical insecurity, rising competition, and insufficient operational risk management within companies enables us to understand that it may be more costly in the long term to import from distant areas than to invest in nearby production centers. While this does not hold in the short term, it is a certainty over the longer term. The rising global economic competition between states also engenders insufficient risk management, transparency, and corporate governance among international corporations. Now, while we have the means to invest in new technology and commodities production, global oil companies are going out of their way to promote alternative solutions or to invest in neighboring production centers (for example, the United Kingdom and Norway in 2006). However, this book will also demonstrate that the diversification to alternative energy solutions is just as expensive, if not more so, than traditional solutions, in that they are inefficient and inadequate for 1.5 billion Chinese citizens and are not enough to respond to the demands of the rising Chinese middle class. The book also discusses operational risks and corporate governance in Chinese energy companies, which, as a stand alone solution, will not be able to respond to the rising consumption of the new Asian democracies' middle classes.

CHAPTER 1

The History of Chinese Energy Policy

1.1 THE HISTORY OF CHINESE COMMODITIES MARKETS

The history of Chinese energy policy is relatively short. It is less than 60 years since the first discovery of exploitable energy resources on Chinese soil. However, in an era of advanced technologies and global competition, the Chinese government has been improving its infrastructures and focusing on expanding its energy policies and the means to secure its own reserves since the beginning of the twenty-first century (Figure 1.1).

American Mobil, Texaco and British Asia companies used to hold the vast majority of global market shares prior to 1949; the total global output of crude oil in 1948 was only 89,000 tons. In September 1949, the People's Liberation Army took control of the Yumen Oil Field in Gansu Province, and this event marks the start of the Chinese government setting its own policies on petroleum production and the accumulation of reserves. In the aftermath of World War II, governments throughout the world were designing political and economic plans and policies according to industrial development and the modernization of transportation. The Chinese government similarly became aware of energy resources and considered them to be an important strategic resource. Thus, like its global counterparts, China developed its own policies to implement a national petroleum industry in the post-war era.

The first important Chinese onshore petroleum enterprise was initiated in the Daqing oilfield in northeast China in 1959. This is one of the major oilfields in China and is still used as the main reference basis in the pricing of oil for China. Around the Daqing oil field, other drilling enterprises started. Despite the focus on one geographical region, this was considered sufficient to sustain national consumption through to the late 1970s. However, with

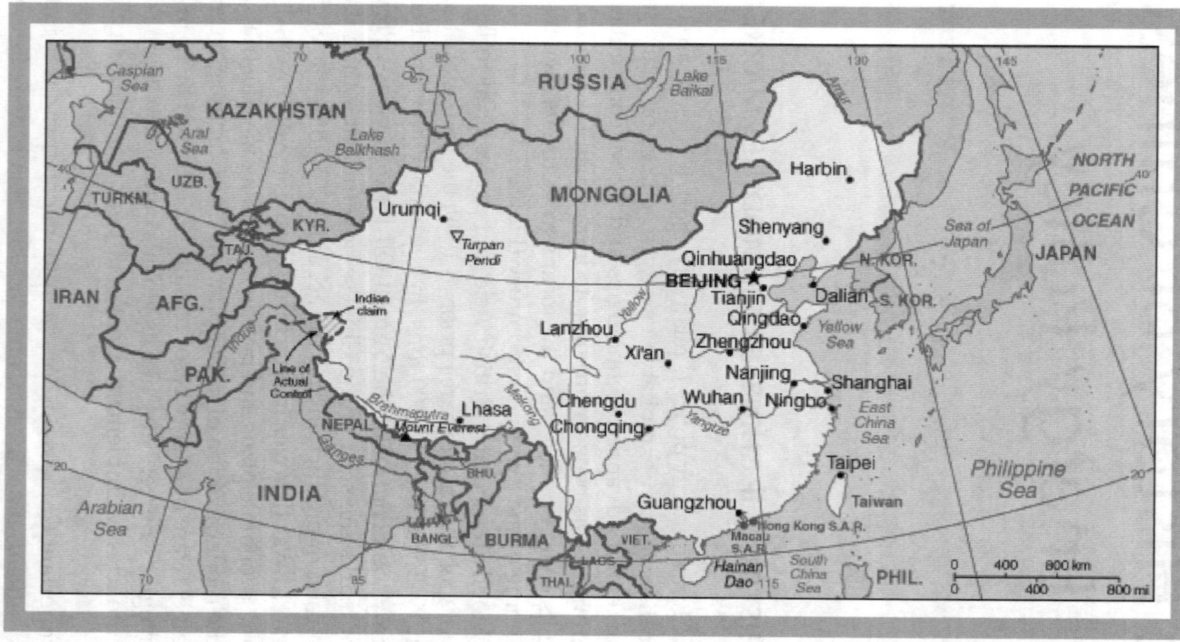

Figure 1.1 Map of China
Source: CIA World Factbook.

industrialization and the rapid development of cars and other vehicles, consumption and demand increased rapidly. Chinese production in key onshore fields declined at the same time as oil began to be used as an alternative source of energy for heating and the increase in consumption became exponential.

With the development of new technology, China's offshore area and far western onshore region proved to hold substantial resources and reserves. And this is when China began to adopt foreign technology, expertise and capital in the drilling of its neighboring reserves. The Chinese Ministry of Oil Industry initiated its first political conference in February 1956 to promote oil policies. Initially, Chinese policy aimed to develop its internal capacities, to prospect and produce petroleum reserves from the large basins. With new technologies and capabilities, new oil sources were discovered, including Karamay in the Xinjiang Uygur Autonomous Region and Lenghu in Qinghai Province. The Chinese enterprises involved in commodities markets not only strove continuously for new global partnerships, but they also invested in new technological developments, advanced methodologies and products. These tactics embarked China on a voyage towards new 'unknown' markets in the global search for oil, and affected the more serious design of internal strategic policy.

China's output of crude oil hit a record high of 1 million tons in 1956. In 1958, The Chinese Ministry of Oil Industry established three petroleum branches – in Songliao, North China, and East China regions – to monitor and develop Chinese energy policy. Songliao Basin is known as one of the main resources and as the bridge to more oil in the eastward part of the region. In the early 1960s, to further develop the industry, a production campaign was launched, marking a significant turning point. China became self-sufficient and independent in terms of oil policies in 1965. Developments were expanded to the Bohai Sea which is still a main resource center for CNOOC. During the mid-1960s new reserves, such as Shengli, Dagang and Liaohe, were also found. These new discoveries enabled Chinese crude oil to reach a production record of 100 million tons in 1978, equaling other peer international main producers.

Further reforms and new policies were initiated in the 1970s with the third development plan for oil and natural gas research. The search for oil extended to the Tarim Basin, the Junggar Basin and the Turpan Hami Basin toward the western region of China, the continental shelf in the Hunghai Sea, and the East China Sea in the eastern region as well as the offshore areas of the South China Sea. From 1979 to 1996, Chinese proven oil reserves amounted to 1.27 times that of thirty years earlier and 5.5 times that of natural gas. Chinese annual crude oil output increased by 51.5 per cent in comparison to its early historical start of 1950 while natural gas production rose by at least 46.5 per cent from the mid-century mark. The study of Chinese energy markets demonstrates that the geopolitical allocation of energy production will be complemented and subordinated by new resources (alternatives), strategies (geopolitical partnerships), and means (financial engineering and infrastructures). This will also be reflected in the pricing of oil types across various geopolitical zones. It is expected that commodities pricing structures will be subject to wider pricing spreads or differences and more opportunities for the exploitation of financial inefficiencies and arbitrage trading. The uniqueness of this analysis is the taking into account of the oil pricing trends of all geopolitical points and integrating various research forecasts to project futuristic pricing variations.

In 1996 Chinese oil output accumulated approximately 157 million tons which rose to some 161 million tons in 1998. Chinese natural gas accumulated some 20.1 billion cubic meters by the end of the century. In 1998, the Chinese government expanded its oil and natural gas prospecting and development at around twenty different sites throughout the country. The twenty newly created petrochemical industrial enterprises were created and financed with a crude oil capacity totaling about 210 million tons per year. Consequently, Chinese oil and gas production have increased by 19.1 per cent since the middle of the last century, or since the inception of petroleum discovery and policy-making in China in 1949.

The first historical cooperation in the global competition for oil between the United States and China occurred in 1922.

From 1914 to 1921, the US Mobil Oil Company drilled samples to survey for oil in China. They found six basins in middle-west China and were among the first to assert that the Chinese market had substantial potential oil reserves. At that time, however, they could only retrieve 175 million tons. From 1922 to 1994 various entities, corporations, associations and government agencies drilled and tested for oil in China at least 15 times. Technological progress advanced the skills of these entities in their search for oil. For instance, in 1935 China Mine Summary found about 175 million tons. Two years later, Xie Jiarong forecast that Chinese resources amounting to 182 million tons surrounded the middle-west basins. Yumen and Changqing's reserves – in addition to the middle-west basins – inspired further predictions by 1944 that Chinese reserves totaled approximately 521 million tons. With more investment and improvements in machinery, the Ministry of Fuel Industry concluded in 1954 that the ten Chinese oil resources already being exploited could provide 1.5 billion tons. In the 1960s, Chinese oil corporations expanded the search and from 1959 to 1963 numerous substantial findings amounted to potential reserves worth about $10 billion in 1964, through about 20 basins. And so the Shengli and Dagang oil fields were also added to the cumulative rising wealth of Chinese natural resources between 1965 and 1970. Only a year later, in 1971, Chinese oil resources totaled 20 billion tons. During the following five years, the territorial search expanded towards the eastern region of China, along the sea coast, and further developments were achieved in Liaohe, Zhongyuan,

Huabei, Henan, Jianghan, Subei, Erlian and Baise. The development of marine and ocean drilling produced resources enough to meet Chinese demand until 1993. The development of eastern production centers was also increasingly substantial as more city centers developed alongside sea ports and in neighboring regions. The eastern regions remained untouched for much longer as it is geologically more difficult to drill and install infrastructural machinery in mountainous landscapes. While developments improved along the eastern coasts, new areas of interest appeared along with advanced progressive drilling methods and new resources continued to be added, with the rich-soiled Xinjiang, Qinghai and Changqing in the central region of China. Thus by 1997, China benefited from 94.9 billion tons of petroleum resources. The new resources were estimated to comprise 292 million tons of the total – more than ten times the estimation three years previously. Taking all these new data into account, the macro total petroleum resources in China in 1997 ought to be estimated at 105 billion tons.

By this time there was a new sense of urgency in the discovery of country-wide resources as China's human resources and energy consumption were outrunning its production abilities. Effectively, in the four years to 1993, China had undergone a net importing balance, causing it substantially to change its geopolitical strategies. Yet it still remained quietly uninvolved in global financial markets. Chinese government policy did not realize at that time the tremendous capacities and consequences of the new technological era. China still believed it could isolate its people and its resources and live substantially cut off from neighbors and global competitors. Technology forced it to realize otherwise. It is with the rise of the technological bubble that the Chinese government had: (1) to accelerate its capacities to balance its internal markets in terms of demand and supply (for financial and commodities markets); and (2) to increase its abilities to meet global standards to better integrate in specific ways its own challenges and needs. Following this breakthrough, numerous agencies and research centers developed at a phenomenal pace. For instance, the China Energy Society and geologists Guan Shicong and Gin Tongluo evaluated Chinese reserves at about 45 to 66 billion tons from 1979 to 1985. In addition to this agency, in 1984 the Reijing Research Institute of Petroleum Exploration and Development (ROPED) increased its drilling activities and discovered 114 new basins. Between 1984 and 1987 the expected total resources amounted to 78.75 billion tons. Other agencies, such as the Science and Technology Bureau of CNPC, also started to initiate large scale oil programs in order to meet common standards and Chinese rising demand. Thus by 1994 the Chinese reserves were up to 15.3 billion tons. At the beginning of the 1990s, other associations and research institutes added 81.47 billion tons in 1991 and 83.45 billion tons in 1993. With governmental subventions and additional efforts to meet growing demand, total oil resources rose. Daqing's reserves have grown by six times its original capacity and still met Chinese demand in 2006. (The pricing of

Daqing oil is in correlation with global markets and will be discussed in the next chapters.) The Chinese authorities have also enhanced discoveries in the Shengli, Dagang, Huabei, and Zhongyuan oil fields, partly due to the global petroleum oil disputes in the early 1970s.

Research analysis discovered that the Chinese authorities could meet increasing consumption demand if they were able to evaluate the approximate maturity of their maximal reserves and if they were also able to recognize the degree of hydrocarbon distribution in each basin. Thus, more advanced research led to new forms of energy such as Paleozoic and Mesozoic, which were found in Qaidam Basin, with 2.86 billion tons. Consequently, oil resources increased from 1.24 billion tons in 1993 to the current 4.1 billion tons. New basins such as Junggar, Ordos and Qiangtang allowed China to add a total of new sources amounting to 5.5 billion tons. China's main oil corporations also added East China's Songliaa and Bohai Bay basins. Almost 3.5 billion tons of resources were found as far as 3500 meters underground. New advanced drilling technologies and deep drilling machinery enabled Chinese corporations to continue the discovery of new energy and resources. Thus Santangku, Kumishi and Duihuang basins were added to the growing list of Chinese domestic strategic resources. However, new discoveries are relatively small compared with historical ones, such as Daqing, and these new sources, such as Yanqi Basin only account for 27 million tons. Nonetheless, by the beginning of 1997, the discovered reserve was far greater than the resources estimated previously.

Since China joined the World Trade Organization in 2001, Chinese companies have entered into global partnerships and deals to standardize its policies with regards to energy and commodities in order to satisfy its domestic needs and to adjust to global competitive demands. Institutional reform of the existing oil companies and enterprises is also needed to compete with global peers. Chinese companies' income from sales of oil products are estimated to have doubled between 1995 and 2005, reaching 250 or 300 billion yuan. The productivity of the sector's labor force is also estimated to have doubled that of 1995, reaching 200 tons or more in per capita production of oil and gas in less than ten years.

Chinese corporations were created by the government. In response to rising global demand and technological progress, China created CNOOC in 1982 to survey offshore territories for oil. With the project in mind of globalizing Chinese corporations via offshore partnerships and projects, the Chinese authorities multiplied foreign alliances. Thus, since CNOOC's inception, China has experienced rapid economic growth, which has generated a strong increase in demand for petroleum and other primary energy sources. For instance, China's gross domestic product (GDP) increased throughout the 1990s at an exponential real annual rate of 10.4 per cent transforming China into the fastest growing economy in the world, primarily benefiting from its underpaid labor and pegged yuan against other currencies, and inducing many foreign firms to inject capital into various industries. At this time the world's mature economies transformed their production centers into service hubs, mostly easily substituted by technology and machinery, while China dominated the tangible capital wealth of the world by holding a monopolistic 50 per cent of the global manufacturing market. Guizot (2007) observed that China holds phenomenal economic records in most industries since the early 1990s and that it is applying economic tricks such as special tariffs and taxing incentives to attract foreign investors and retain them in high-technology and advanced industries to participate in and promote Chinese world leadership in these fields. China is also securing a brain drain by hiring intelligence to maintain its geopolitical competitive leadership. It is also promoting the offshoring of its manufacturing capital to Africa to leverage its own labor force into new sectors such as science, technology, or engineering. And while creating a new economic leadership, China is converting its gargantuan foreign exchange reserves into commodities contracts throughout the world. In the 1990s, petroleum consumption increased at a compound annual rate of 6.3 per cent and China's oil production grew at a compound annual rate of 1.7 per cent. Since 1993, China has been a net importer of oil which influences the increasing importance placed on the further development and diversification of offshore crude oil production activities. Particularly among China's three main oil and gas companies: China National Petroleum Corporation (CNPC), China National Petrochemical Corporation (Sinopec) and CNOOC.

Chinese commodities markets' data according to the Oil and Gas Committee's Global Reserves Committee's mapping terminology and definitions indicate that total oil reserves amount to 94 billion tons, with a final recoverable total of 14 billion tons; proven reserves account for 5.28 billion tons; Chinese oil proven rate is 37.5 per cent; accumulated recoverable total is 3.03 billion tons; total in reserves at immediate disposal amounts to 2.25 billion tons; and finally, output in 1997 totaled 160.74 million tons. China also uses natural gas as a complementary source of energy, of which it has total reserves of 3.8 billion cubic meters; final recoverable of 1.05 billion cubic meters; proven reserves of 95 million cubic meters; a proven rate of 9 per cent; accumulated recoverable totaling 22.8 million cubic meters; still to be used reserves of 72.2 million cubic meters; and finally an average annual output since 1997 of 2.27 million cubic meters. The insufficient natural resources cannot be proportionally shared among entities. Since June 2006, world oil prices have decreased by approximately $10 a barrel. Between June and December 2006, the pricing of Chinese oil companies' stocks rose by about 30 per cent. The rise brought them into line with the valuations of competitive international peers, so that the foreign exchange gradual appreciation of the yuan to July 2005 does not hold back the progress of Chinese companies in comparison with their international energy competitors.

1.2 AFTER CHINA BECAME A NET IMPORTER OF OIL IN 1993

By the end of 1994, based on 20 billion tons of proven and controlled reserves in 150 basins all over the country, China National Petroleum Corporation (CNPC) reported that the petroleum resources to be discovered in China amount to 74 billion tons. The total petroleum resources in China are 94 billion tons. China became a net oil importer in 1993, and dependence on foreign oil has been increasing exponentially since then (Figure 1.2). In 1999, 22 per cent of total oil consumption came from net imported oil, a figure that is expected to reach 40 per cent by 2010. According to the *Chinabidding and Xinhua Journal* (21 July 2006) and the China Petroleum and Chemical Industry Association (CPCIA), between July 2005 and July 2006 China produced 91.664 million tons of crude oil, a rise of 2.1 per cent over the corresponding period in 2004–05. The output of refined oil products was 84.822 million tons, up 5.6 per cent year on year.

China started to build national oil reserve bases in 2004. The first four are in Zhenhai and Daishan in Zhejiang Province, Huangdao in East China's Shandong Province, and Dalian in the northeastern Liaoning Province. The bases are included in the government's social and economic development plan for 2006–10. According to NDRC, the National Energy Office, they were completed in 2005 and have a total storage capacity of between 10 and 12 million tons. China worked out a three-phase blueprint for building the national petroleum reserves to cope with the global oil price fluctuation. China aims to increase the oil reserves by 12 million tons in the first phase,

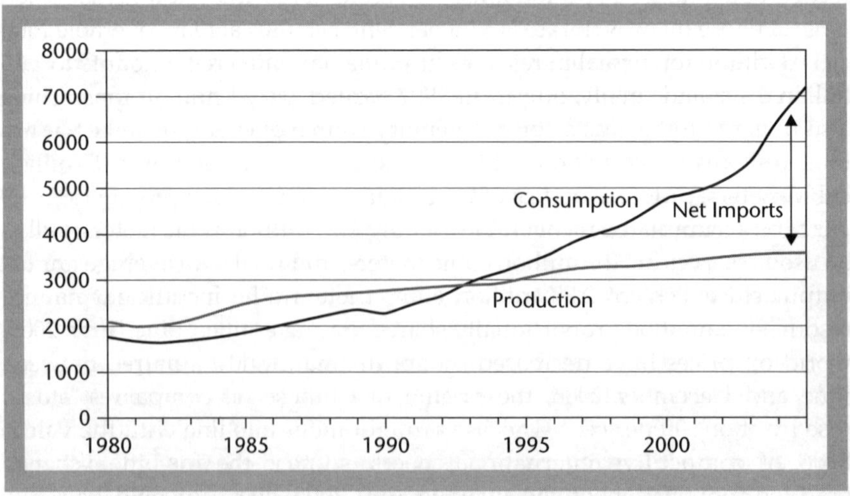

Figure 1.2 China's oil production and consumption, 1980–2005 (thousand b/d)

and by 28 million tons of petroleum in the second and third phases. Currently, China imports about 100 million tons of oil a year, and the gradual growth in stored oil does not yet affect international oil pricing curves. What does affect oil prices is financial engineering speculations on oil trading. The price of oil has an embedded premium – especially since 2000 – due to the rise of global financial engineering speculations. Since the downfall of Enron, more and more financial institutions have got into the lucrative business of commodities and energy trading. Global economies are currently in a preliminary, intermediary phase of peaceful economic contestation, during which states are competing to gain substantial commercial corporate contracts, to secure strategic energy reserves, to ally with geopolitical partnerships and to maximize economic efficiencies. Compared with its huge imports, China's strategic oil reserves remain globally irrelevant. The petroleum amount reserved by the China National Petroleum Corporation and the China National Petrochemical Corporation, two major oil giants in China, is only equal to 21 days of China's crude oil imports. In comparison, the United States' petroleum strategic reserve is equivalent to 158 days of crude oil imports, Japan's is 161 days and Germany's 117 days.

According to the statistics of the General Administration of Customs, China's net imports of crude oil were 70.33 million tons in the first six months of 2005, up 17.6 per cent year on year, and those of refined oil products were 12.03 million tons, with a growth of 48.3 per cent. According to the *Xinhua China Securities Journal* of July 2006, the operational draft for a new national petroleum reserve center has been submitted to the central government.

China is thus becoming more vulnerable to the uncertainty of global oil markets and to political volatility in oil producing areas of the world. The import dependency was 34 per cent in 2001 and will reach 50 per cent by 2020. Until 2005, China was not part of international statistical reports in the financial markets. Yet, financial markets have played an increasingly important role in the pricing of oil globally. One example of oil pricing distortions was that initiated by Enron. There is a fictitious pricing premium embedded in the recent inflationary pricing of oil due to financial engineering speculation, the over-usage of derivatives and the inability to provide accurate collateralized asset valuations for commodities trading operations in financial banks. Another reason for the increase in pricing structure was disparate global regulation, which forced institutional entities to trade on market inefficiencies. Hedge funds also take advantage of geopolitical regulatory imbalances to arbitrage systemic valuations. For instance, Chinese foreign exchange valuations are not included in the Foreign Exchange and Derivatives Market Activity published by the Bank of International Settlements in 2004. Before 2006, all oil prices correlated with each other and there was little disparity between geopolitical points of production. Global oil pricing curves tripled in relation to one another from 2001 to October 2006 and doubled from 2001 and June 2006. Effectively, average oil prices decreased by approximately

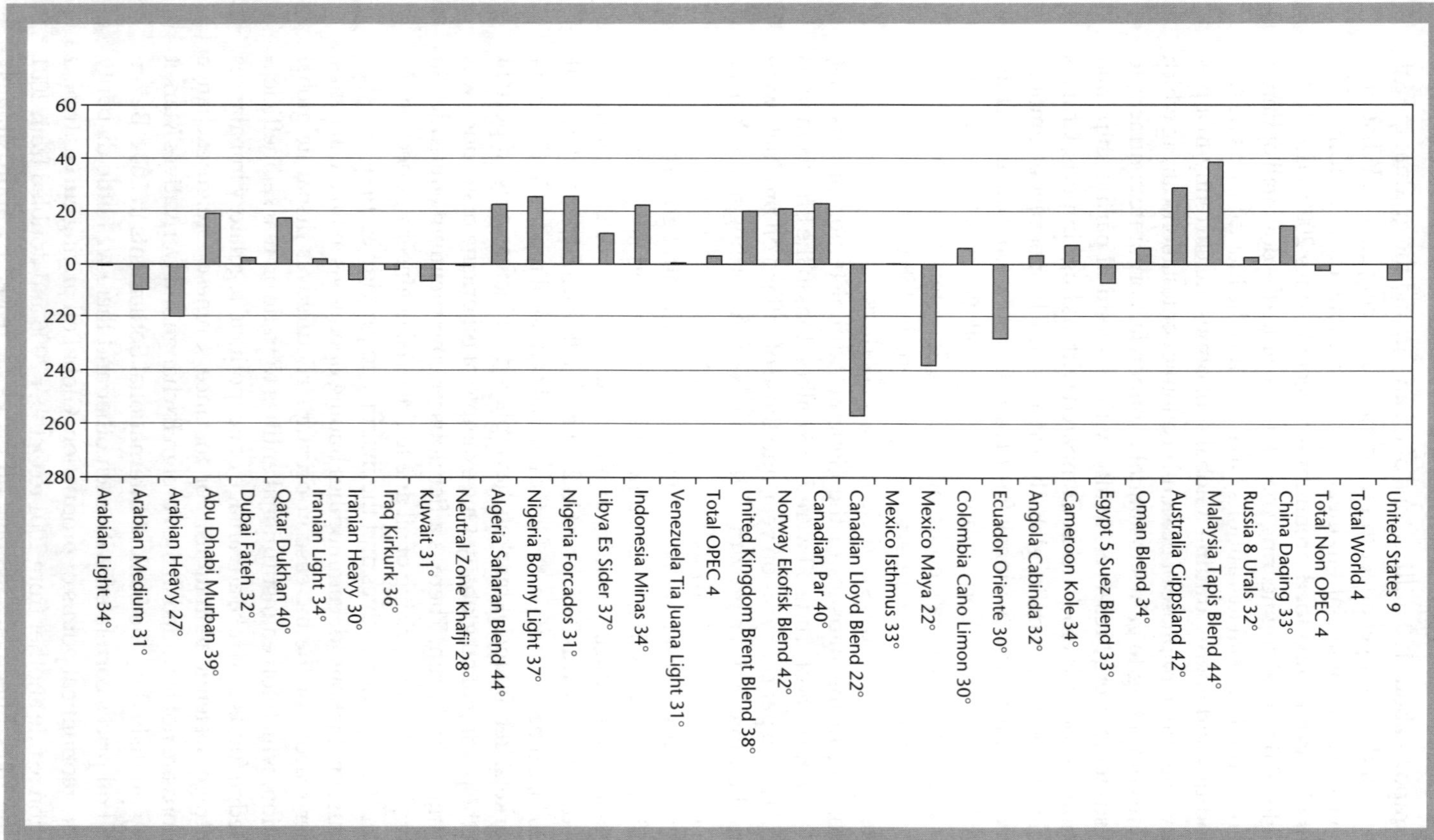

Figure 1.3 Estimated oil price spreads of given countries and world reference basket price $240/barrel, 2030
Note: Assuming no major drastic changes in living standards
Source: Author's estimates (2006).

$10 per barrel from June to October 2006. From 2007, the pricing structures of oil curves between geopolitical zones will diverge from traditional correlations and will incur widening of pricing spreads due to transportation costs and financial fluctuations (Figure 1.3).

Greater insecurities are also likely to develop between geopolitical areas. These cost variations and foreign exchange valuations will be directly correlated to trading deficits versus surpluses. But with growing time gaps, oil price spreads between given countries are going to widen and will have a consequential economical impact on individual countries' gross domestic product, consumer pricing index, and purchasing powers, which will also affect living standards. The author proposes an economic model to evaluate the price of oil in the future in the following chapters. The more partnerships and tangible production centers are initiated now – in 2006 – the more security in the future for that particular country. Commodities will be the weapon of choice in the future in negotiating vetos on important decisions in global matters and in securing specific trading partners. Oil prices will also directly affect the value of foreign exchange and the cost of credit. Estimates show that as of 2006, the values of tangible markets do not reflect their actual real values due to the lagging effect of the events of 2001 and the numerous accounting scandals of 2002. Since then, financial markets have been subject to much greater speculation by large financial (and non-financial) institutions which have made the accurate pricing of oil difficult. Commodities prices are the most volatile and difficult to explain as the prices are highly correlated to transportation and delivery conditions. Commodities prices have also been subject to conspicuous manipulation and unhelpful influences that have reduced actual economic variables. These effects have become more obvious since 2000. Markets are inefficient and highly used and eroded by large operational risks and gaps within corporation's databases and technological frameworks. The pricing of oil depends on the main producing countries (Organization of Petroleum Exporting Countries, OPEC) within the global pricing structure, regardless of any single powerful entity.

CHAPTER 2

Oil Availability: Supply and Production

In this book, the reader will be able to review pricing for about fifty years of real historical data with regard to oil production and supply. World oil reserves trended upwards from 1950 to the end of 2006. While production reserves still remain substantial in sustaining actual timing phases, research reveals that production capacities are being outrun by consumption and exponential over-speculation within financial market trading institutions speculating with commodities pricing structures, particularly since 2001. Figure 2.1 shows the energy supply structure of the world in 2001.

We do not know how or what alternatives may be used to sustain new global needs beyond the second part of this new century. *Oil and Gas Journal* estimates that proved oil reserves are approximately 15 billion barrels, about 1 per cent higher than the 2005 estimate for January 2006, while according to www.globalfirepower.com, the Chinese proved oil reserves as of 2005 stood at 18.3 billion barrels. China's proven reserves thus represent approximately one-tenth those of Canada. According to the US Energy Information Administration (EIA), China is the world's second largest consumer of oil behind the United States. In 2006, the EIA reported that China consumed 7.4 million barrels of oil per day, while its production was only 3.8 million barrels per day. This places China as the third largest net importer of oil after the US and Japan.

According to the *China Daily* (14 June 2005) Xinjiang Uygur Autonomous Region plans to overtake northeast China's Daqing to become the country's largest oil production area in the next few years. Xinjiang, currently China's third largest oil producing area, plans to more than double its oil production to 50 million tons in the near future from 22 million tons in 2005. According to the *China Daily* (2006), Xinjiang is expected to pump 100 million tons of crude oil a year in the future, or nearly 60 per cent of China's

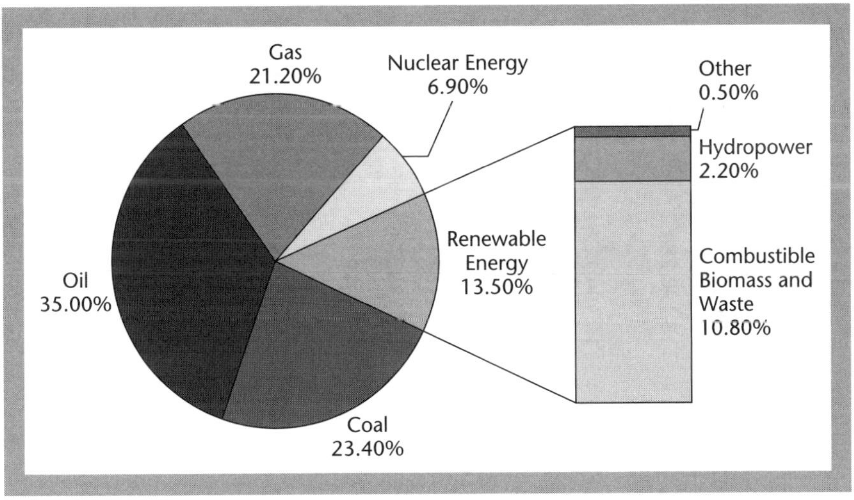

Figure 2.1 Energy supply structure of the world, 2001
Source: Chinese Energy Ministry Energy and Technology Symposium, 2003.

total in 2005. The targets, if realized, will greatly help China offset the slowing production of its older oilfields. Table 2.1 indicates the efficiency of the main production centers by provincial regions.

Total recoverable petroleum resources are derived from estimates of findings to be discovered and cumulative production. The total recoverable resources rise in function of those variables. The final recoverable resources are increasing and are directly correlated with increasing exploration in previously unexplored areas and the improved exploration maturity. Explorations are enhanced with new neighboring production centers. Production associations and experts provide an approximation of the production data from the recoverable resources. According to the Oil and Gas Research Institute, USGS, Chinese recoverable resources were approximately 11.9 and 9.702 billion tons respectively in 1988 and 1987. The Thirteenth World Petroleum Congress announced that the total recoverable oil resources in China were 10.6 billion tons in 1991. And the Fifteenth World Petroleum Congress announced that the recoverable oil resources in China were about 14 billion tons in 1997. This represents about 3.4 billion tons more than 1991. According to ROPED, recoverable oil resources in China in 1991 approximate about 11.8 billion tons. This figure rose to 12.7 billion tons in 1995. These statistics represent 4 billion tons more than the estimate of four years previously. The Ministry of Geology and Minerals estimated in 1993 that the total recoverable Chinese oil resources amounted to about 11.8 billion tons. The capacity increase from 1991 to 1995 represents about 900 million tons more than that of 1993. On a net cumulative basis and comparing the incremental increase,

Table 2.1 Net profits from petroleum and natural gas extraction, by region, as of May 2006 (thousand yuan)

Region	2006/05	2006/04	2006/03	2006/02	2005/12	2005/11
National total	55,472,968	43,331,281	36,776,847	26,767,573	118,190,284	114,114,237
Beijing	*	*	*	*	*	*
Tianjin	7773610	5876767	4669206	3712129	10,999,712	10,713,263
Hebei	1915529	1739660	1601849	960326	3912577	3997794
Shanxi	–	–	–	–	–	–
Inner Mongolia	240504	200051	207272	114729	365137	583778
Liaoning	1937202	1237548	1240731	761231	4110235	3819973
Jilin	2029913	1611102	1423718	1072390	4338270	4097029
Heilongjiang	14,437,381	9985862	9146074	6189743	31,532,681	31,395,382
Shanghai	*	*	*	*	*	*
Jiangsu	*	*	*	*	1457532	1394898
Zhejiang	–	–	–	–	–	–
Anhui	–	–	–	–	–	–
Fujian	–	–	–	–	–	–
Jiangxi	–	–	–	–	–	–
Shandong	6867135	5751449	4785577	3965349	17,487,015	15,992,737

Henan	1202756	1081233	848782	586117	3029981	2278937
Hubei	429585	342665	320036	267765	1193773	1330508
Hunan	–	–	–	–	–	–
Guangdong	1238081	1242323	1238096	903183	4846820	4844103
Guangxi	–	–	–	–	–	–
Hainan	59570	54179	70899	62990	*	199794
Chongqing	3877	7247	5600	3935	−16756	−14408
Sichuan	753287	565614	322714	369454	−248797	194279
Guizhou	–	–	–	–	–	–
Yunnan	*	*	*	*	*	*
Tibet	–	–	–	–	–	–
Shaanxi	4685900	3541680	2771828	1984520	11,401,441	11,591,420
Gansu	806587	675918	594268	371262	2214225	1877753
Qinghai	586670	482690	403030	317550	2392820	1556910
Ningxia	*	*	*	*	*	*
Xinjiang	10,009,548	8432891	6600402	4819017	18,872,652	18,154,344

(*Continued*)

Table 2.1 (Continued)

Region	2005/10	2005/09	2005/08	2005/07	2005/06	2005/05	2005/04
National total	107,987,046	94,978,398	83,111,874	68,674,865	56,308,212	44,272,630	33,731,742
Beijing	*	*	*	*	*	*	*
Tianjin	9783863	8647032	7517620	5633798	4720485	3827700	2884308
Hebei	4039981	3557517	3249323	2791726	2619480	1965464	1427952
Shanxi	–	–	–	–	–	–	–
Inner Mongolia	526529	420110	360808	294471	244563	167359	92773
Liaoning	3552724	3294707	2856616	2399381	1963431	1420700	1157012
Jilin	3914403	3351976	2727894	2438281	2042822	1415746	926007
Heilongjiang	30,768,414	27,288,747	23,906,894	20,281,392	16,657,996	11,082,219	9356138
Shanghai	*	*	*	*	*	*	*
Jiangsu	1295137	1190165	1058549	910611	788093	664934	533754
Zhejiang	–	–	–	–	–	–	–
Anhui	–	–	–	–	–	–	–
Fujian	–	–	–	–	–	–	–
Jiangxi	–	–	–	–	–	–	–
Shandong	14,776,750	11,821,767	10,320,861	8416831	6168024	5362682	3768524

Henan	2196816	1975965	1462592	1091601	947874	724021	454274
Hubei	1388710	1207240	1007085	863040	738971	569420	467094
Hunan	–	–	–	–	–	–	–
Guangdong	4603115	4116960	4253184	4014745	3707135	3289023	2884543
Guangxi	–	–	–	–	–	–	–
Hainan	169727	171675	149167	114438	88712	69356	62789
Chongqing	−11263	−9938	−8978	−5448	−6375	−4027	−3495
Sichuan	659594	587353	482530	416488	182755	457625	374854
Guizhou	–	–	–	–	–	–	–
Yunnan	*	*	*	*	*	*	*
Tibet	–	–	–	–	–	–	–
Shaanxi	9619017	8695874	7837757	5868634	4828279	4334891	3324423
Gansu	1803594	1650533	1394365	1155840	1046201	894461	640100
Qinghai	1632550	1594960	1480660	1253470	1118770	483180	391138
Ningxia	*	*	*	*	*	*	*
Xinjiang	17,128,340	15,288,092	13,000,082	10,671,372	8384656	7488506	4962378

Notes: *: no extractions for given month; – : not available/no production.

research shows that the new statistics are higher than the old figures. This rising net accumulation is due to the fact that most basins experience a low exploration maturity. Data rise with the improved exploration maturity. In addition, cumulative reserves increase due to advanced technological processes and new engineering methods of finding, extracting or conserving reserves. The recovery of low-porosity and low-permeability dams is also improved with progress in oilfield development technologies. Explored resources and recovery are also greater due to progress in drilling technology in oil exploration and development. The total Chinese petroleum recoverable resources may be projected to 20–23 billion tons. According to the International Energy Administration, China was the twelth main producing country in the world in 2006. With added valued international partnerships and new technology progresses since December 2001, China is likely to become a main producer and consumer. The International Energy Administration also reports that the largest increases in proved oil reserves is in Iran. Iranian oil reserves increased by 5 per cent, from 125.8 billion barrels in 2005 to 132.5 billion barrels in 2006. Saudi Arabia reports similar trends with reserves which increased by 4.9 billion barrels or 2 per cent in 2006. Kuwait's reserves also increased by 2.5 billion barrels, up 3 per cent from 2005. Venezuela's reserves also gained 2.5 billion barrels or 3 per cent. According to *Oil and Gas Journal,* Chad reported 1.5 billion barrels of proved oil reserves in 2006 and is believed to have great resources. Africa has significant reserves that are yet to be adequately financed and sponsored. According to Stephanie Hanson in 'Vying for West Africa's Oil' at www.cfr.org, 7 May 2007, 'Africa contains only 10 per cent of the world's proven oil reserves, but in an increasingly uncertain market, oil companies are eyeing crude from West Africa's Gulf of Guinea for reasons beyond sheer output.' In 2006, for example, Sinopec, China's state-owned energy company, bid $2.2 billion for two deepwater blocks off the Angolan coast among many other investment projects in other African countries. Esther Pan reports in 'China, Africa and Oil', published with www.efr.org on 26 January 2007 that Chinese corporations invested a total of $175 million in African countries, primarily on oil exploration projects and infrastructure. On 9 January 2007 state-owned Chinese energy company CNOOC said that it would buy a 45 per cent stake in an offshore oil field in Nigeria for $2.27 billion. Where Western markets failed to colonize emerging markets for natural resources, China contributes to 64 per cent of Sudan's oil exports to supply its new middle class consuming force.

Oil reserves remain primarily within the Middle Eastern region which owns almost half the world total (see figures 2.2 and 2.3 and tables 2.2–2.4).

In future, the pricing of the oil structure, will more likely depend on the contracts set between large consuming nations, such as China or India, and producing countries such as South America, Africa, the countries of the Middle East and Russia. In 2003, despite geopolitical tensions, those countries multiplied partnerships, commercial contracts and corporate cooperation.

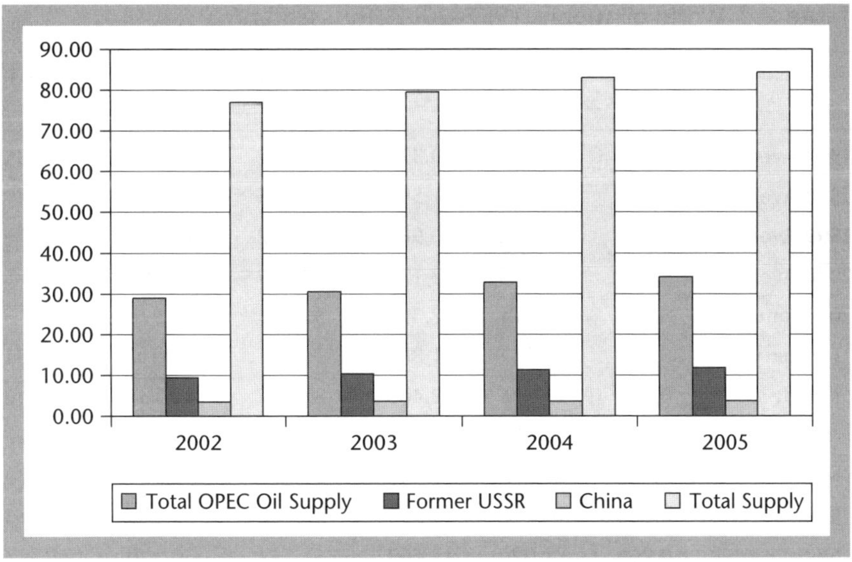

Figure 2.2 World oil supply, 2002–05 (million b/d)
Sources: OECD and EIA, http://www.eia.doe.gov/emeu/ipsr/appa.html.

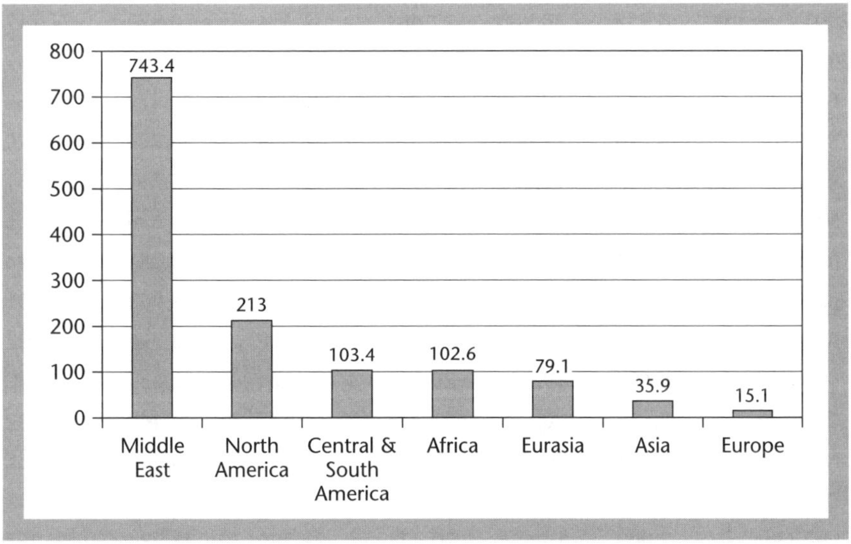

Figure 2.3 World proved oil reserves by region, 1 January 2006 (billions of barrels)
Source: 'Worldwide Look at Reserves and Production', *Oil and Gas Journal*, 103 (47) (19 December 2005): 24–5.

Table 2.2 World oil supply, 1997–2007 (thousand b/d)[a]

	United States[b]	Persian Gulf[c]	OAPEC[d]	OPEC-121[e]	OPEC-11[f]	World
1997 Average	9,461	19,578	20,372	30,442	29,729	74,185
1998 Average	9,278	20,895	21,594	31,615	30,881	75,679
1999 Average	8,993	20,232	20,961	30,512	29,767	74,879
2000 Average	9,058	21,520	22,175	32,329	31,583	77,793
2001 January	8,453	21,571	22,091	32,721	31,976	78,316
February	8,825	21,287	21,869	32,261	31,512	78,204
March	8,978	21,975	22,488	32,867	32,137	78,946
April	8,963	21,425	21,959	32,101	31,363	77,961
May	9,058	21,292	21,846	31,867	31,134	77,320
June	8,940	19,712	20,235	30,408	29,680	75,776
July	8,966	21,020	21,463	31,632	30,920	77,747
August	8,913	21,459	22,014	32,106	31,406	77,934
September	9,002	20,628	21,242	31,198	30,489	77,579
October	9,130	20,547	21,233	31,122	30,369	77,576
November	9,241	20,413	21,105	31,009	30,225	78,004
December	9,012	19,523	20,261	30,075	29,255	77,243
2001 Average	8,957	20,905	21,484	31,614	30,872	77,717
2002 January	9,014	19,429	20,122	29,615	28,705	76,537
February	9,077	19,492	20,213	29,622	28,672	76,662
March	9,116	19,644	20,373	29,834	28,894	76,470
April	9,144	18,551	19,313	28,671	27,731	75,951
May	9,312	19,246	20,067	29,536	28,606	76,645
June	9,154	18,976	19,765	29,246	28,351	76,344
July	8,973	19,573	20,375	29,833	28,963	76,751
August	9,151	19,323	20,101	29,708	28,798	76,592
September	8,686	19,881	20,668	30,524	29,639	77,170
October	8,553	20,618	21,444	31,343	30,448	78,582
November	8,921	20,790	21,608	31,415	30,600	78,861
December	8,902	20,814	21,645	29,319	28,499	76,902
2002 Average	9,000	19,698	20,478	29,890	28,994	76,957
2003 January	8,901	21,447	22,289	29,773	28,954	77,459
February	8,936	21,892	22,685	31,194	30,369	79,216
March	8,928	21,833	22,661	31,845	30,995	79,702

	United States[b]	Persian Gulf[c]	OAPEC[d]	OPEC-121[e]	OPEC-11[f]	World
April	8,781	20,794	21,745	31,115	30,205	78,567
May	8,746	20,718	21,675	31,263	30,343	78,568
June	8,677	19,964	20,890	30,575	29,655	77,866
July	8,550	20,118	21,018	30,806	29,886	78,675
August	8,719	20,597	21,487	31,340	30,420	79,237
September	8,971	20,857	21,755	31,700	30,780	80,127
October	8,774	21,556	22,449	32,390	31,470	81,097
November	8,769	21,655	22,493	32,434	31,514	81,446
December	8,830	22,205	22,959	33,126	32,146	82,784
2003 Average	8,797	21,134	22,006	31,466	30,563	79,565
2004 January	8,780	22,507	23,203	33,324	32,294	82,322
February	8,810	22,437	23,120	33,253	32,223	82,354
March	8,868	22,191	22,838	32,997	31,968	82,191
April	8,779	21,985	22,598	32,792	31,762	81,925
May	8,790	21,805	22,422	32,607	31,577	81,564
June	8,584	23,205	23,869	34,117	33,088	83,467
July	8,708	23,545	24,280	34,537	33,508	83,953
August	8,731	23,435	24,137	34,365	33,315	82,941
September	8,283	23,935	24,637	34,885	33,815	83,426
October	8,470	23,842	24,537	34,912	33,822	84,205
November	8,818	23,357	24,098	34,407	33,307	84,034
December	8,780	23,567	24,270	34,640	33,540	83,661
2004 Average	8,700	22,986	23,669	33,905	32,854	83,005
2005 January	8,687	23,643	24,473	34,974	33,864	83,953
February	8,792	23,713	24,530	35,099	33,979	84,326
March	8,854	23,763	24,617	35,277	34,131	84,452
April	8,880	23,923	24,772	35,438	34,280	84,896
May	8,917	23,733	24,573	35,335	34,157	85,313
June	8,727	23,848	24,601	35,481	34,304	84,871
July	8,353	24,058	24,809	32,742	34,523	84,482
August	8,399	24,023	24,789	32,782	34,418	84,776
September	7,060	24,283	25,091	36,119	34,711	84,072
October	7,365	23,898	24,760	35,760	34,392	84,094

(Continued)

Table 2.2 (Continued)

	United States[b]	Persian Gulf[c]	OAPEC[d]	OPEC-121[e]	OPEC-11[f]	World
November	7,918	23,798	24,660	35,703	34,295	84,865
December	7,930	23,703	24,588	35,618	34,200	84,657
2005 Average	8,322	23,866	24,689	35,529	34,272	84,564
2006 January	E 8,225	23,554	***24,450***	***35,349***	***33,921***	***84,417***
February	E 8,232	23,759	***24,699***	***35,399***	***33,981***	***84,412***
March	E 8,096	23,634	***24,645***	***35,267***	***33,839***	***83,924***
April	E 8,239	23,658	***24,685***	***35,293***	***33,865***	***84,246***
May	E 8,348	23,458	***24,495***	***34,966***	***33,638***	***84,188***
June	E 8,463	23,713	***24,661***	***35,300***	***34,007***	***84,081***
July	E 8,456	24,098	***25,078***	***35,697***	***34,230***	***85,427***
August	E 8,486	24,128	***25,106***	***35,823***	***34,355***	***85,186***
September	E 8,499	23,778	***24,801***	***35,446***	***34,000***	***84,720***
October	E 8,455	23,573	***24,601***	***35,269***	***33,885***	***84,934***
November	E 8,378	23,243	***24,193***	***34,894***	***33,435***	***84,366***
December	E 8,556	23,133	***24,103***	***34,813***	***33,323***	***84,200***
2006 Average	E 8,370	23,644	***24,626***	***35,293***	***33,873***	***84,511***
2007 January	E 8,462	22,945	***23,981***	***34,659***	***33,069***	***83,844***
February	PE 8,727	22,837	24,000	34,553	32,948	84,675
2007 2-Month Average	PE 8,588	22,894	23,990	34,609	33,012	84,239

Notes: a. 'Oil Supply' is defined as the production of crude oil (including lease condensate), natural gas plant liquids, and other liquids and refinery processing gain (loss). For definitions of these terms see: http://www.eia.doe.gov/emeu/ipsr/appc.html; b. US geographic coverage is the 50 states and the District of Columbia. Beginning in 1993, includes fuel ethanol blended into finished motor gasoline and oxygenate production from merchant MTBE plants. For definitions of fuel ethanol, oxygenates and merchant MTBE plants see: http://www.eia.doe.gov/emeu/ipsr/appc.html; c. The Persian Gulf countries are Bahrain, Iran, Iraq, Kuwait, Qatar, Saudi Arabia, and the United Arab Emirates. Production from the Kuwait-Saudi Arabia Neutral Zone is included in Persian Gulf production; d. OAPEC: Organization of Arab Petroleum Exporting Countries: Algeria, Iraq, Kuwait, Libya, Qatar, Saudi Arabia, and the UAE; e. OPEC-12: Organization of Petroleum Exporting Countries: Algeria, Angola, Indonesia, Iran, Iraq, Kuwait, Libya, Nigeria, Qatar, Saudi Arabia, the United Arab Emirates, and Venezuela; f. OPEC-11 does not include Angola.

E = Estimated data. RE = Revised estimated data. PE = Preliminary estimated data.

Revised data are in ***bold italic font.***

Monthly data are often preliminary and also may not average to the annual totals due to rounding.

For historical series see: http://www.eia.doe.gov/emeu/ipsr/t44.xls.

Source: http://www.eia.doe.gov/emeu/ipsr/source1.html.

Table 2.3 World oil reserves by country, 1 January 2006 (billion barrels)

Country	Oil reserves	Country	Oil reserves
Saudi Arabia	264.3	China	18.3
Canada	178.8	Qatar	15.2
Iran	132.5	Mexico	12.9
Iraq	115.0	Algeria	11.4
Kuwait	101.5	Brazil	11.2
UAE	97.8	Kazakhstan	9.0
Venezuela	79.7	Norway	7.7
Russia	60.0	Azerbaijan	7.0
Libya	39.1	India	5.8
Nigeria	35.9	Rest of world	68.1
United States	21.4	World total	1292.5

Source: 'Worldwide Look at Reserves and Production', *Oil and Gas Journal*, 103 (47) 19 December 2005: 24–45.

Let us review in particular the involvement of African countries in global oil markets. According to Esther Pan in 'China, Africa and Oil', reporting in January 2007:

Angola is the second-largest oil producer in sub-Saharan Africa after Nigeria. Angola is one of the main oil producers with an estimated productivity of 2 million barrels per day by 2008. Additionally, Angola disposes of offshore gas resources. Angola generated more than $20 billion in foreign direct investment since 2003 with energy supplies alone. Oil accounts for over 40 per cent of gross domestic product (GDP) and 90 per cent of government's revenues. The World Bank reports in 2005 that Angola represents half of China's oil imports from Africa. Equatorial Guinea has total proven oil reserves of about 1.28 billion barrels. Oil production averaged 371,700 barrels per day in 2004 which accounts for 90 per cent of the total national exports in 2003. Equatorial Guinea maximized productions at 350,000 barrels per day as of December 2004. Gabon disposes of proven oil reserves of roughly 2.5 billion barrels and produces about 230,000 barrels per day. This represents a decline of 37 per cent since its peak production levels in 1997. Exports of crude oil account for approximately 60 per cent of the government's budget and more than 40 per cent of GDP. Nigeria is the largest African oil producer to supply Western hemisphere. It produces 2.5 million barrels per day. Nigeria's proven oil reserves are 35.2 billion barrels but targets 40 billion barrels by 2010. Oil natural resources account for 80 per cent of total economic revenues. Sudan produces 500,000 barrels per day in 2005 in the northern region of Darfur. Sudan has proven reserves of some 563 million barrels of oil. Republic of Congo has 1.5 billion barrels in proven reserves and averaged 235,000 barrels of crude oil production per day in 2004. Oil accounts for about 80 per cent of the country's revenues in 2005 and 90 per cent of exports.

Table 2.4 Estimated world oil resources, 1995–2025[a] (billion barrels)

Region	Proved reserves	Reserve growth	Undiscovered	Total
OECD				
United States	21.4	76.0	63.0	180.4
Canada	176.5	12.5	32.6	223.9
Mexico	12.9	25.6	46.8	84.3
OECD Europe	15.1	20.0	35.9	71.0
Japan	0.1	0.1	0.3	0.5
Australia/New Zealand	1.5	2.7	5.9	10.1
Non-OECD				
Russia	60.0	106.2	115.3	291.5
Other non-OECD Europe/Eurasia	19.1	32.3	55.6	107.0
China	18.3	19.6	14.6	52.5
India	5.8	3.5	6.5	16.4
Other non-OECD Asia	10.3	14.6	23.9	48.8
Middle East	743.4	252.5	269.2	1,265.1
Africa	102.6	73.5	124.7	300.8
Central and South America	103.4	90.8	125.3	319.5
Total world	1,292.5	730.2	936.9	2,961.6
OPEC	901.7	395.6	400.5	1,697.8
Non-OPEC	390.9	334.6	538.4	1,263.9

Notes: a. The US Geological Survey's assessment extends only to 2025.
Reserves include crude oil (including lease condensates) and natural gas plant liquids.

Sources: Proved reserves as of 1 January 2006; *Oil and Gas Journal*, 103(47) (19 December 2005): 24–25. Reserve growth (total) and undiscovered: US Geological Survey, *World Petroleum Assessment* 2000, website http://pubs.usgs.gov/dds/dds-OBD. Estimates of regional reserve growth: Energy Information Administration, Office of Integrated Analysis and Forecasting.

However, while new resources are being discovered throughout the world, older producing centers are diminishing in capacity. For instance, Mexico has reduced production and is down by 1.7 billion barrels as compared to figures from the end of the twentieth century. Norway is also recording losses of 0.8 billion barrels, the United States shows production down by 0.5 billion barrels, and the United Kingdom by 0.5 billion barrels. These countries are experiencing sustained productivity at a decreasing rate. The Middle East and Canada record the largest production concentration with a production capacity of 71 per cent of the world's total proved oil reserves. According to the Canadian Association for Petroleum Producers, Canada is considered as a main reserve with 174.1 billion barrels. According to the *Oil and Gas Journal*, World Energy, OPEC and British Petroleum Statistical Review of World Energy, about 40 per cent of the oil reserves' producers are OPEC countries, which account for at least 65 per cent of the world's total reserves. In January 2006, the Chinese oil supply accounted for 18.3 billion barrels. The International Energy Oil Administration estimates that the 2030 global oil supply will exceed the 2003 level by 38 million barrels per day. Increases in production are expected for both OPEC and non-OPEC producers with about 38 per cent from OPEC.

In 2006, OPEC expects to produce 45.3 million barrels a day, and non-OPEC producers expect to produce 72.6 million barrels a day by 2030. Non-OPEC oil supply growth from 1993 to 2006 resulted in an OPEC market share substantially under its high of 52 per cent in 1973. And OPEC created 39 per cent of the world's oil supplies in 2003. About 62 per cent of the rise in petroleum demand over the next 25 years will be sustained by increased production from non-OPEC suppliers. Non-OPEC production in 2030 is projected to be almost 24 million barrels per day higher than it was in 2003. Earlier estimates of 2010 OPEC production capacity diminished in the 2005 projections. OPEC countries have not made significant changes in operational transparency infrastructures, and global oil prices will remain high for a while. Prices have remained high primarily as a result of the increased geopolitical tensions in the Middle East. OPEC adopted conservative policies to maintain high oil pricing, to gain control and to benefit from an added premium embedded in the barrel price. This added premium was particularly high from January to June 2006 as the barrel decreased by $10 from June to October 2006. The lack of policy reforms at OPEC also means that substantial re-engineering changes are needed to remedy global economic imbalances or inadequate dispatching of oil supplies to rising new major oil dependants. The lack of restructuring among OPEC leaders and within the OPEC organization also contributes to the illegitimate rise of world oil prices. And while OPEC could raise daily production to higher levels, the organization has purposely maintained low production levels from 2003 and predicted until 2030 to support global oil prices at sustainably high levels, around the $60 limit. The inflationary pressure put on oil prices also directly affects global economic imbalances with a rising American deficit that becomes increasingly expensive to finance and an abnormally high euro blocking exports and economic growth within the eurozone. OPEC suppliers aim to increase production capacity by only 4 million barrels per day between 2003 and 2030, while they have the ability to raise daily production significantly or alter the supply levels to change the state of the global economic landscape. Figures 2.4 and 2.5 provide more information about projected production estimates for OPEC and other production centers.

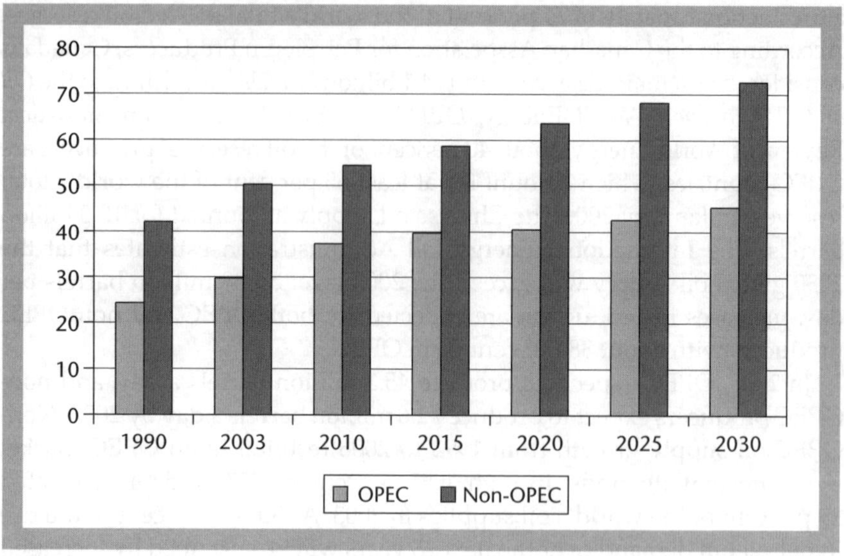

Figure 2.4 OPEC and non-OPEC total petroleum liquids production, 1990, 2003 and 2010–30 (million barrels a day)
Sources: 2003 Energy Information Administration EIA, International Energy Annual 2003, Annual Energy Review 2004 DOE, EIA 2000 Washington DC, August 2005, www.eia.doe.gov/emeu/; projection EIA, Annual Energy Outlook 2006 DOE EIA, 2006.

In 2003 oil pricing projections were in line with reality, while they significantly increased – by almost 50 per cent – from 2003 to 2006. The 2006 pricing structure had already reached levels that are similar to those projected for 2015–20. In June 2006 and late July 2007, oil prices flirted with the $78–$80 range which represented a level even higher than the 2030 forecasts. Global oil prices decreased by approximately $10 a barrel and came back to around $60 a barrel at the end of June 2006, despite rising oil consumption during summer 2006 and geopolitical tensions in Libya and Israel. This fits the projections of oil prices of 2015. Increased price volatility exacerbates the gaps in pricing structures between entities and geopolitical zones. It reduces the economic accuracy and creates further operational risks in the future pricing of energy.

There is a mismatch between what government agencies have reported to the general public and what the public has actually experienced. For example, INSEE (Institut National de la Statistique et des Études Économiques, www.insee.fr) reports that the differential gap in sentiments about inflation experienced by consumers versus that reported by government entities has significantly diverged since 2000 by at least 3 per cent. Eurostat and INSEE assert that inflation reported by government statistics in 2000 is between 1.5 to 2 per cent while consumers witness inflation on their wallets rising from 2 per cent in 1999, to 3 per cent in 2001, to 4 per cent in 2002, and to 5 per cent

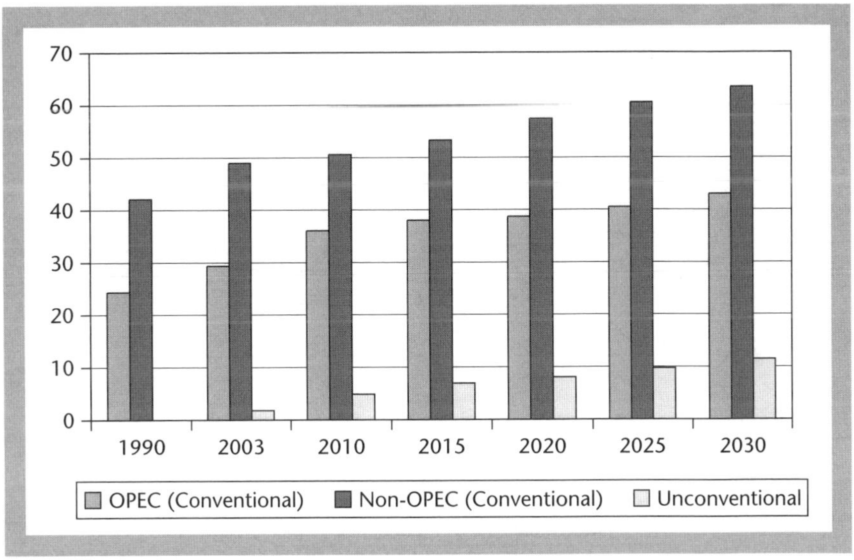

Figure 2.5 OPEC, non-OPEC and unconventional oil production, 1990, 2003 and 2010–30 (million barrels a day)
Sources: Energy Information Administration, 2003; International Energy Annual, 2003; May–June 2005: www.eia.doe.gov/iea; 2030: EIA system for the Analysis of Global Energy Markets, 2006.

in 2003 and 2004. Figure 2.6 reflects the disparities in oil pricing under different assumptions.

The historically high oil prices since 2001 and in the first quarter of 2006 are correlated to a global, inexplicable and unjustifiable gradual inflationary pressure on daily goods. For instance, *Le Figaro* magazine of 6 January 2007 includes an economic report about the operational difference between inflationary statistics reported by government and the actual perception of inflation by consumers. *Le Figaro* asserts that according to INSEE prices of daily goods rose significantly from 2001 to December 2006. For example, real estate rose by 90.22 per cent in Paris, rents 30 per cent, a loaf of bread 39.5 per cent, a cup of coffee by 44.73 per cent, cereal 95.9 per cent, tobacco 46.26 per cent, chicken 24 per cent, champagne 31.19 per cent, a pair of jeans 5 per cent, a movie 20.48 per cent, Paris subway passes (2 zones) 18.35 per cent, cars 15.8 per cent, while prices in a few sectors, such as electronics and entertainment, fell by 8 times. As shown below, the price of oil from June 2006 until December 2006 levels off around the reference levels of $60 a barrel. And the price of oil fell by 13 per cent in the first two weeks of January 2007 alone. While oil prices have exhibited significant (sometimes extreme) pricing volatility since 2001, the changes in oil pricing are minimally reflected at macro levels and have a relatively small impact on other economic variables.

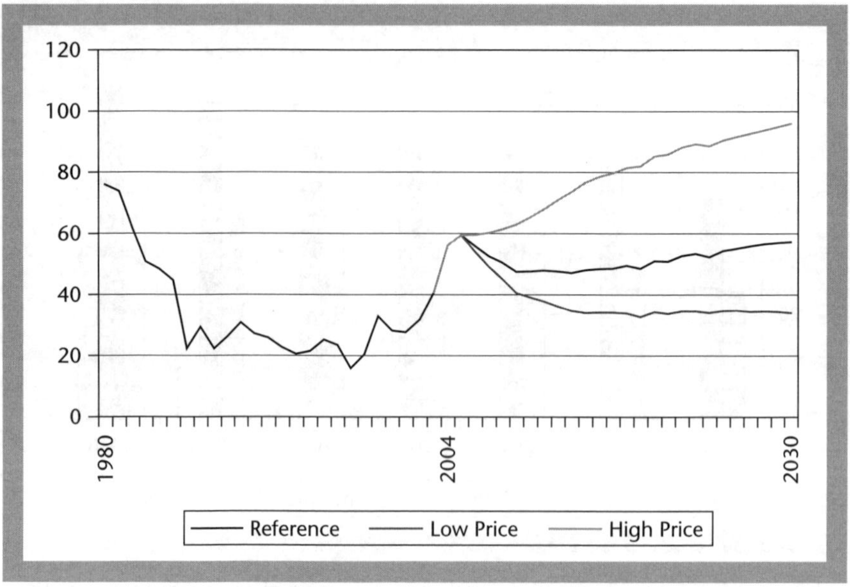

Figure 2.6 World oil prices in three cases, 1980–2006 ($ per barrel)
Sources: EIA, Annual Energy Review 2004; Annual Energy Outlook 2006.

Financial speculation exacerbated by large financial institutions and hedge funds creates more volatility in commodities prices in order to profit more significantly from derivatives trading. The following explains the worst, reference and best pricing scenarios as of 2006 given worst, unchanged and ameliorated external factors and assumptions, respectively.

Non-OPEC supply became increasingly diverse from the beginning of the 1970s. Similarly, non-OPEC oil production participated in the erosion of power of the main OPEC organization. OPEC participation in global production fell from 52 per cent in 1973 to 39 per cent in 2003 due to the increasingly wide distribution of production and dispersion to subsidiary centers in non-OPEC countries. Non-OPEC producers created cumulative strategic reserves and contingency resources to meet global demands over time. The North American market was one of the main producers from 1970 to the mid-1980s. The North Sea and Mexico gained more controls in the 1980s. Since 1990, newly emerging economies such as South America, West Africa, the non-OPEC Middle East, and China are accessing new market shares. Non-OPEC countries also partially control global oil prices until at least 2030 as they have sufficient estimated proven reserves to meet global demand until then assuming no major changes from 2006 infrastructures. Non-OPEC members expect to achieve such goals by producing 48.9 million barrels per day in 2003, rising to 72.6 million barrels per day in 2030.

With these estimates, the non-OPEC market is estimated to sustain 62 per cent of the world's oil supply. Estimates in 2003 were at 61 per cent. Production is also targeted to rise, with added unconventional alternative liquids such as oil sands and ultra-heavy oils with an estimated unconventional production rising to 5.7 million barrels per day in 2025, unconventional supplies reaching 9.7 million barrels daily in 2025 and 11.5 million barrels daily by 2030. Unconventional liquids production is expected to rise to 16.3 million barrels per day in 2025 and 21.1 million barrels per day in 2030. According to the International Energy Organization 2006 report, North Sea production and Norway raised production to 3.6 million barrels per day in 2006, but these are estimated to decline to 2.5 million daily barrels by 2030. The United Kingdom is also expected to produce about 2.2 million daily barrels by 2010 and then is forecast to decline to 1.4 daily million barrels by 2030.

The Daqing oilfield

The Daqing oilfield is located on a large plateau between the Songhua River and Nunjiang River in Heilongjiang Province. Daqing is the largest and considered to be the oldest oil production reserve in China. It is also considered to be one of the biggest oilfields in the world. It was discovered over fifty years ago and is still operating on a daily basis. Daqing was put into operation in May 1960, and produces 66 per cent of Chinese oil. Crude oil output reached 50.3 million tons in 1976, making it one of the ten largest oil fields in production at that time. Statistics show that Daqing's proven recoverable reserves exceed 5.3 billion tons, and it has produced 1.405 billion tons of crude oil, accounting for 47.2 per cent of the national total. Daqing oil is the main pricing reference for Chinese oil in global oil tables. It has been used as the reference base since the 1970s.

The Shengli oilfield

The Shengli oilfield is located in the Yellow River Delta in the north of Shandong Province and on the shore of the Bohai Sea. The Shengli oilfield started production in 1962 and is used as a main strategic domestic oil base to serve most of the eastern coast of China, including Shanghai and surrounding urban centers. Shengli's oil output in 1993 accounted for 20 per cent of China's total production. Shengli is the second largest oilfield in the country.

The Liaohe oilfield

The Liaohe oilfield is the third largest oilfield in China and is situated on the lower reaches of the Liaohe River. Liaohe started to produce oil in the

1960s. The proven recoverable reserves are over 90 million tons and its controlled reserves exceed 100 million tons.

The Dagang oilfield

The Dagang oilfield is located in the Bohai New Area about 50 km to the southeast of Tianjin city. Dagang is the only land-based, large-scale oil–gas field in the coastal regions. Dagang was discovered in 1966 and was found to occupy a total land area of 18,600 square km. It also has proven oil reserves of 725 million tons and natural gas reserves of 28.719 billion cubic meters.

The Huabei North China oilfield

The Huabei oilfield was the first buried oilfield to be discovered in China. It was found in 1975. Huabei is located in a plateau in the central section of Hebei Province.

The Jidong East Hebei oilfield

The Jidong oilfield is located in the east part of Hebei Province, adjacent to Tangshan city. It is an offshore oilfield which was first exploited in 1978.

The Zhongyuan oilfield

The Zhongyuan oilfield is located in the north around Shandong and south within Henan Province. Zhongyuan oilfield was discovered in 1975. The Zhongyuan oilfield has an annual output of about 4 million tons.

The Sichuan oil–gas field

Sichuan oilfield is located in the central section of Sichuan Province and it is the largest natural gas production base in China. Sichuan was first exploited in 1953. Its annual output of natural gas ranks first among the 21 similar fields in China.

The Datianchi gas field

The Datianchi gas field is located in the east of Sichuan Province and represents the third largest field in China. Datianchi gas field has three gas layers with a total length of 200 km and width of 20 km. Datianchi has proven reserves greater than 50 billion cubic meters. Datianchi's expected annual output was about 3 billion cubic meters in 2000.

The Changqing gas field

Changqing gas field is located in northwestern China's Erdous Basin. Changqing is the largest land-based gas field, and is considered among the largest gas fields in the world. Changqing started operations in 1995. In 1997, Changqing produced 3.3 million tons of crude oil. Changqing has proven natural gas reserves of 300 billion cubic meters.

The Tarim oilfield

Tarim oilfield was discovered by the China National Petroleum Corporation (CNPC) in April 1989. However CNPC did not start production until 1990. The proven reserves of oil–gas came to over 400 million tons, including 206.2 billion cubic meters of natural gas in 1996.

The Turpan-Hami oilfield

Turpan is located in the Turpan-Hami Basin. Turpan's field covers 53,000 square km. Turpan is divided into 14 oil–gas fields with the proven reserves being 220 million tons.

The Cainan oilfield

Cainan oilfield is in the Kurban-Tonggute Desert in the Junggar Basin. It was the first integrated oilfield to be found in a desert in China. It was discovered in 1991 and put into operation in 1995. The annual output has reached 1.483 million tons.

The offshore oil–gas fields in the Bohai Sea

The Suizhong 36-1 oilfield is located in Liaodong Bay, about 50 km to the west of Suizhong County in Liaoning Province. Bohai is a large offshore oilfield with a proven reserve of about 300 million tons. The field started its operations in 1995. Chengbei oilfield is in the southwest part of the Bohai Sea and was exploited in 1986. It is the first Chinese–Japanese joint venture. The Qinhuangdao 32-6 oilfield is located in the central part of the Bohai Sea. Qinhuangdao was also found in 1995 and has an estimated proven reserve of 200 million tons.

The offshore oil–gas fields in the East China Sea

Pinghu oil–gas field is located on the continential side of the East China Sea, and is approximately 420 km from Shanghai. It is the first combined oil–gas field with natural gas as its main product. The East China resort was exploited from November 1996 to the end of 1998. The Eastern China field has also been providing qualitative natural gas of 1.2 million cubic meters to serve the urban population along the eastern sea coast and in the Pudong New Area of Shanghai.

The offshore oil–gas fields in the South China Sea

The Liuhua 11-1 field is located in the 29/04 'cooperation area', about 220 km from Hong Kong. It primarily serves the southeastern part of China and also urban centers along the eastern sea coast. China Offshore Petroleum Nanhai East Company and Amoco Oil Company of the United States jointly operate and extract from South China Sea production centers. China's offshore cooperation area totals 317 square km, with the controlled reserves being 233 million tons. South China Sea is the largest known offshore oilfield in China. It started operating in March 1996, with an output of 1.68 million tons annually during peak period (see tables 2.5 and 2.6).

The Fanyu 4-2 oilfield was discovered in a basin in the east part of the South China Sea in May 1998. Fanyu was also developed by the American Santa Fe Energy Resources Co. Ltd. Fanyu's oil production capacity can reach 1102 tons per day. Fanyu's estimated recoverable reserves are around 20 million tons.

Table 2.5 Oil and gas in the South China Sea – comparison with other regions

	Proven oil reserves (billion barrels)	Proven gas reserves (trillion cubic feet)	Oil production (million b/d)	Gas production (trillion cubic feet/year)
Caspian Sea	17.2–32.8	232	1.6	4.5
North Sea	16.8	178.7	6.4	9.4
Persian Gulf	674.0	1,923.0	19.3	8.0
South China Sea	(est.) 7.0	(est.) 150.3	2.2	3.2

Notes: Proved reserves as of 1 January 2003. Oil production as of 2002. Oil supply includes crude oil, natural gas plant liquids, and other liquids. Natural gas production as of 2001.

Source: Energy Information Administration (various years).

Table 2.6 Oil and gas in the South China Sea

	Proven oil reserves (billion barrels)	Proven gas reserves (trillion cubic feet)	Oil production (b/d)	Gas production (billion cubic feet)
Brunei	1.4	13.8	189	366
Cambodia	0	0	0	0
China*	(est.) 1	(est.) 5.0	(est.) ~ 350,000	(est.) ~200
Indonesia*	(est.) 0.1	(est.) 32.0	(est.) ~ 323,000	(est.) ~50
Malaysia	3.0	75.0	751,973	1.895
Philippines	0.2	3.8	24,512	<1
Singapore	0	0	0	0
Taiwan	<0.01	2.7	3.3	26
Thailand	0.6	13.3	193,162	661
Vietnam	0.6	6.8	339,595	46
Total	(est.) 7.0	(est.) 150.3	2,174,542	3.244

Notes: *Only the regions around the South China Sea are included; there are no proved reserves for the Spratly and Paracel Islands. Proved oil and natural gas reserves are as of 1 January 2003. Oil production is a 2002 average. Oil supply includes crude oil, natural gas plant liquids, and other liquids. Natural gas production is the 2001 average. All figures cited for China and Indonesia are estimates of offshore reserves and production in the South China Sea only.

Source: Energy Information Administration (various years).

Yinggehai Yacheng 13-1 natural gas field surrounds the Yinggehai Basin and Qiongnan Basin, about 96 km to the south of Sanya city, Hainan Province. Yinggehai covers 53.58 square km and is the largest offshore natural gas field found in China. Yinggehai also has proven reserves of 96.8 billion cubic meters and a production capacity of 3.45 billion cubic meters per year. Yinggehai is estimated to be able to maintain stable production for twenty years. It is also jointly developed by Chinese and American corporations. Yinggehai Yacheng started operation in January 1996 and represents the main supplier of the Hong Kong Special Administrative Region with 2.9 billion cubic meters of natural gas per year. Yinggehai also supplies Hainan Province with 550 million cubic meters of natural gas per year. It is expected to have a minimum exploitation duration of 20 years.

The Dongfang 1-1 gas field is also located in the Yinggehai Basin, having a natural gas area of 229.6 square km. Dongfang was discovered in 1995 and is the second largest offshore gas field in China, with proven reserves of 80.1 billion cubic meters as of 2006.

CHAPTER 3

Natural Gas Availability and Utilization

According to the International Energy Administration World Energy outlook (2007), natural gas is China's primary energy resource. China's domestic reserves of natural gas amounted to 53.3 trillion cubic feet (Tcf) as of the beginning of 2005. Because China has been a net importer of oil since 1993 and it is estimated that oil exploration will reduce towards the 2030 limit, China has intensified the diversification into alternative products as well as the usage of cleaner products which are non-corrosive to the environment. Consumers can now use less polluting hydrogen vehicles.

Since China opened its borders in December 2001, there has been a lag in domestic infrastructures, policies and regulations. The Chinese energy system does not have an operational research program to optimize transportation costs between areas. Until the beginning of the twenty-first century, there was no real regulation of Chinese energy, and as far as is known China has never had a centralized energy ministry. Although according to the *People's Daily* online, of 3 December 2004, Chinese government departments have been restructured to replace the current Energy Bureau under the ministry-level National Development Reform Commission (NDRC). The Chinese Energy Bureau counts only a dozen members of staff and is not adequately resourced to manage the gargantuan energy industry with its total assets of more than 10 trillion yuan (US$1.2 trillion). The restructuring may continue until at least 2008. China created a Ministry of Energy in 1988 but it was dismantled five years later, in 1993. It had become overly bureaucratic and administrative and duplicated work from other departments, such as the State Development Planning Commission. Additionally, the energy sector faced increasing energy shortages, and insufficient infrastructures to contend with the growing manufacturing sector and increasing living standards.

The government implemented the Energy Bureau under the NDRC during reform of the administration in March 2003. Since then, more than two-thirds of China has suffered power blackouts. Coal mines are old fashioned and dangerous and are not well equipped to remedy to less polluting infrastructures. Finally, oil imports as well as other commodities such as steel, aluminum, and copper have, since 1993, sky rocketed. Oil and coal imports and exports, and oil market regulations are supervised by the Ministry of Commerce, while the Ministry of Land and Resources oversees resource exploration. As of 2004, according to the article, the Chinese Energy Bureau only counted some 20 personnel compared with more than 1000 employees in the Ministry of Energy in the United States. Natural gas prices are, therefore monitored and set arbitrarily by local authorities and local regulation. Most recently, however, the Chinese government has been drafting a new energy policy framework to legalize its internal regulations, policies, and taxation, and to manage the efficient distribution of energy to the general public. Such natural gas policies as existed have mostly been archaic and lacking in innovation. They were never efficiently ramified to neighboring countries, such as Russia, before the beginning of the twenty-first century. Since 2003, China and Russia have accelerated their cooperation programs to exploit Russian resources and support growing Chinese demand.

Until the 1990s, natural gas was primarily used for agricultural purposes, as a feedstock for fertilizer plants. However, it also started to be used as an alternative source of energy for electricity generation in urban areas. The International Energy Administration estimates that natural gas accounts for about 3 per cent of all Chinese energy consumption. (Consumption is expected almost to double by 2010 as the new middle class is formed.) Natural gas is recognized in China as a clean, abundant and comparatively cheap energy source. Carbon dioxide and Nox emissions from natural gas are only 50 per cent and 20 per cent respectively of emissions from coal pollution, and environmental pollution produced from natural gas burning is only 25 per cent of that from petroleum or 1/800 compared to coal. While Chinese energy consumption is rising, natural gas remains at 3 per cent of total consumption. In developed countries, natural gas represents, on average, 22 per cent of energy consumption.

In EIA (2007), it is stated that the rising domestic demand will be met by a combination of domestic production and imports via pipeline and in the form of liquefied natural gas (LNG). The largest reserves of natural gas in China surround western and northern central China. These reserves require greater investment in transportation and distribution and in the establishment of a pipeline infrastructure to feed eastern cities. The Chinese liquefied natural gas market is structured by regional zones. OilGas (www.oilgas.com.cn) issue a report that gives the main Chinese domestic prices by region (see chapter Appendix 3.1 below). EIA (2007) also reports that it is intended that imported liquefied natural gas (LNG) will primarily serve China's southeastern coastal region. A new intra-country market for LNG has developed and private investors now trade LNG indices. The main Chinese commodities indices are displayed by regions in Appendix 2.

Chinese indices have an upward trend over time and are correlated to those of their international counterparts. While both private and government entities in China are improving liquidity and market trading infrastructures, new projects are also under way. This is because the limited solutions at present available depend heavily on international pricing conditions. Examples of new infrastructures and projects to improve the Chinese gas market include the following. In January 2005, the China National Petroleum Corporation (CNPC) participated in the construction of the west-to-east pipeline, which transports natural gas to various urban centers along the southeastern coast of China but also to Shanghai in the western Xinjiang Province. The west-to-east pipeline also transports natural gas along its route from the Ordos Basin. This part of the pipeline was created around July 2002 and the eastern section of the Ordos Basin started operations in 2004. Most Chinese commodities firms have a stake in the project and own a minority interest share. Some foreign entities are also connected to the project. For instance, Shell participated in the development of Changbei natural gas deposit, which investment also partially contributes to the pipeline's infrastructure and architecture, with a seed investment of $600 million with cashflow intake until 2007.

Beside the major west-to-east pipeline project, in 2001 the Chinese discovered a gas field at Sulige in the Ordos Basin in the Inner Mongolia Autonomous Region near the Changqing oilfield. Estimated reserves of these resources are about 16–21 Tcf. Additionally, CNPC also invested in the building of substantial new infrastructural reserves, with the drilling of the Sulige fields in 2005. CNPC shares resources with other Chinese energy suppliers to participate in the proportional share allocations of those natural reserves. The Ordos Basin is intended to feed the west-to-east pipeline in order to distribute natural resources to populated areas and economic development regions.

In 1997, another ramified pipeline was completed surrounding the regions of Ordos Basin and Beijing. A further pipeline was constructed and ramified to this grid to feed the demand for natural gas in Beijing, Tianjin, and Hebei provinces. Beijing's main pipeline is also expected eventually to feed into the northeastern region. The west-to-east pipeline project is also planned to be ramified and branched with the Russian natural gas grid in Siberia extending to China and South Korea through a pipeline from the Kovykta gas fields near Irkutsk which hold reserves of more than 50 Tcf. The investment to connect the Chinese, South Korean and Russian lines is around $12 billion. The investment's forecast capacity is around 2.9 billion cubic feet per day (Bcf/d) with an estimated Chinese consumption of about 1.9 Bcf/d and South Korean demand of 1 Bcf/d. Kogas is the main South Korean company which joined the feasibility study in November 2000. Brother firms Kogas and CNPC agreed to participate in the project in November 2003. Western commodities companies also participated in the project, with a minority interest stake of 30 per cent in Russia Petroleum which is the licensing agent for the Kovykta gas field. However, geopolitical tensions surrounding the Korean peninsula mean that the pipeline transportation route to South Korea will bypass North Korea via an undersea pipeline from Dalian in China to the South Korean coast near Seoul. This new mapped circuit would also bypass Mongolia. Due to the portion of the pipeline through Russia, Gazprom also owns a minority interest stake in the 2004 investment agreement. Additionally, China and South Korea also have a stake in the pricing and terms for Russian natural gas imports.

The Chinese authorities have also influenced energy and commodities firms to accelerate the construction of other smaller natural gas deposits in order to respond to rising consumer demand. Thus, for example, a pipeline was completed in 2002 connecting the Sebei natural gas field in the Qaidam Basin with the city of Lanzhou. In addition, CNPC also completed a pipeline from natural gas resources in Sichuan region to feed the demand of southwestern urban areas in Hubei and Hunan regions and in central China. This project also cost around $600 million.

Offshore gas projects are also significantly expanding in order to build strategic cumulative reserves to add to the Chinese gas supply. For instance, the Yacheng 13-1 field was initiated and developed around 1995, and in 1996 was already producing energy for the main urban center of Hong Kong as well as Hainan Island. Chunxiao gas field, which is located in the East China Sea, is targeted to be one of the main energy contributors for southeastern urban areas from 2000 to at least 2010. Chunxiao gas field is near Xihu and is part of the Eastern China Sea. This area is very rich in natural resources and has already been the subject of geopolitical tensions between China and Japan as there are no well defined sea areas to enable allocation of economic goods disputed between China and Japan. As we shall see below, natural resources beneath seas and oceans are no longer the subject of governmental authorities but are shared privately-owned investment projects sponsored by various international corporations redistributing resources and cashflow to their respective 'national' entities.

The northern region of China is used as a contingency resource in case of Russian embargoes of energy feeds from natural resource centers. Guangdong Province implemented six regional energy generating resource centers with 320-megawatt (MW) gas-fired power plants that transform energy derivatives. From this, the center converts oil-fired plants with a capacity of 1.8 gigawatts (GW) to liquified natural gas. In March 2001, British Petroleum financed the construction of the first Chinese liquified natural gas plant to import resources and feed the urban area of Guangdong. In this project, British Petroleum owns an investing minority interest share of 30 per cent equity in the returning cashflow in partnership with China

National Offshore Oil Corporation (CNOOC), which owns 31 per cent. The remaining interest shares are distributed among smaller shareholder entities located in Guangdong and Hong Kong. As well as these various natural gas projects to sustain future rising demand, another supply agreement has been reached with the Australian North West Shelf liquified natural gas center. A liquefied natural gas line has been started in Zhangzhou in the Fujian region along the northern coast and is scheduled to finish later in 2007. British Petroleum also entered another contract to provide liquefied natural gas from Tangguh in Indonesia to China. Other residual, smaller projects also involve CNPC in Dalian, Hebei, and Jiangsu, and CNOOC in Zhejiang and Shanghai. There is also a Sinopec project in Shangdong. In addition to these projects, China owns a 3 million square kilometer offshore area with significant natural gas reserves. Sampled drilling shows that China owns seven gas basins of 14,000 billion cubic meters and 5000 billion cubic meters of economic value in offshore areas. Nansha Basin appears to be the richest with 8000 to 10,000 billion cubic meters of gas.

Liquefied petroleum is another alternative solution to insufficient global resources and to secure Chinese energy supplies. Liquefied petroleum gas is also utilized as a substitute or a complementary solution to alleviate energy prices. According to ChinaBidding (31 July 2006) and the Hong Kong Electrical and Mechanical Services Department, the price of liquefied petroleum gas (LPG) rose by about 14 cents per liter in August 2006. The prices at LPG stations were also revised, ranging from 2.89 to 3.26 Hong Kong dollars for a liter. Ceiling prices were adjusted to a specified pricing formula that has two elements – the LPG international price and the LPG operating price. The former is the LPG international price of the preceding month, while the latter is adjusted annually on the first day of February according to the movement in the Composite Consumer Price Index in the previous year.

The China National Offshore Oil Corporation (CNOOC) also aims to participate in more significant resource extraction projects in the future. Private Chinese energy corporations are now in line with global (as well as domestic) competition in terms of meeting domestic demand and intervening to affect global energy pricing and to create inflationary pressures on global middle classes. Natural gas demand is expected to reach 6.8 billion cubic meters (bcm) by 2010, accounting for 60 per cent of the entire domestic capacity. The Dalian LNG first-phase project was originally planned to open by 2011 to secure an annual supply of about 3 million tons of liquefied gas, or 3.9 bcm of un-liquefied gas. The capacity is expected to double in the second phase of development. China's main domestic oil corporations, Sinopec and China National Offshore Oil Corp (CNOOC), are also increasing their fuel supplies with planned LNG terminals along the east coast.

So while the country expands technology, capabilities and resource reserves, private Chinese firms are developing the following additional energy centers (according to the government of China).

Liaoning Province

JZ20-2 condensate gas field is the first gas field in offshore China to have a steady supply capacity of 400 million cubic meters of high-quality gas to Liaoning Province annually for 20 consecutive years through a 48-km sub-sea pipeline. Liaoning has utilized offshore gas to establish a plant, Jingzhou Gas Chemical Plant, producing 520,000 tons of urea and 300,000 tons of synthetic ammonia each year, partially for civil use.

Tianjin

Boxi gas field was set up and brought into production on 1 January 1998. About 100 million cubic meters of gas are transported to Tanggu, Tianjin through a 46-km sub-sea pipeline. Two hundred thousand local residents benefit from this natural gas pipeline. Local municipalities also built a gas-generated power plant with offshore natural gas. Tianjin Coastal Gas Power Plant officially started operating on 1 July 1999.

Shanghai

Pinghu gas field is situated about 265 kilometers offshore from Shanghai and started producing as of 11 November 1998. Pudong International Airport, Buick car manufacturing centers and the urban population of Shanghai get their energy resources directly from the East China Sea. Pinghu gas field has a generating capacity of 1.2 million cubic meters of gas per day for 15 years, and is able to maintain continuous supplies to one million consumers. The Pinghu gas field production center contributes to the reduction in consumption of 13.6 million tons of standard coal and emissions reductions of 43,400 tons of sulfur dioxide and 91,200 tons of carbon monoxide.

Hainan Province and Hong Kong

Yacheng 13-1 gas field is situated in the South China Sea a few kilometers offshore from Hainan Island. Hainan is the largest gas field ever discovered in China, with a natural gas capacity reserve of 100 billion cubic meters. CNOOC, ARCO and KUFPEK are the private energy companies contributing to an annual gas production of 3.4 billion cubic meters. They ramified to two sub-sea pipelines from the field to connect to onshore supplying centers near urban areas. The 800 kilometers of pipeline supply 2.9 billion cubic meters of gas to CAPCO in Hong Kong, while the 100-kilometer pipeline transports natural gas to Hainan Island for power generation. Yacheng 13-1 gas field also contributes to Sanya urban region.

Guangxi Zhuang Autonomous Region

Guangxi is situated in Beibu Bay. Wei Xinan gas field started producing as of 6 June 1999. Natural gas from this pipeline provides supplies to Weizhou Island, Beihai. After being cleaned and generated the natural gas is used for power generation and chemical projects, and is used for alternative energy gas projects.

Chinese government future projects

CNOOC, one of the major Chinese corporations, is building a domestic national transmission grid for natural gas to supply internal domestic urban centers in coastal regions and to distribute energy resources equally across the region. The energy grid aims to cover three economic zones along Bohai Bay, in the Yangtze River Delta and the Pearl River Delta. According to ChinaBidding, the Chinese authorities have developed an increasing reserve of gas, primarily exploiting the output in the East China Sea and the Puguang field in Sichuan Province and with the help of the west-to-east gas pipeline. The supply capacity will be between 5.5 billion and 6 billion cubic meters by 2010. In 2005, the natural gas supply was less than 2 billion cubic meters and was short of market demand. Nonetheless, 2005's 2 billion cubic meters was about seven times the level of 2002, meaning it could become the fastest growing energy sector in the city. Shanghai is also speeding up the construction of state and commercial energy reserves, especially for natural gas. Shanghai is scheduled to supply all households with natural gas instead of coal gas by 2010. There are many benefits to exploiting natural gas, which is widely also used to fuel power generation. Up to now, China has used natural gas primarily because it is a widespread natural resource, which is less expensive to exploit and extract than other energy sources, less polluting than coal, and cheaper than alternative energy resources. Additionally, it is easily transportable via underground pipelines and consequently less vulnerable to foreign geopolitical risks. ChinaBidding also reports that central government plans to boost the use of clean-burning fuels and reduce reliance on coal and oil. The Chinese Ministry of Land and Resources reported that a natural gas field discovered in the South China Sea in June 2006 is estimated to be China's largest offshore gas deposit.

APPENDIX 3.1: EXAMPLE OF DAILY MARKET RISK REPORT BY REGION

South China Market Description

On Tuesday, several parts of south China market shows up in a confused situation mixed with rise and decline.

Eastern Guangdong and Pearl River Delta market kept prices to firm and was equal to deals prices in the quiet trade today. South China could remain prices because the north materials could not swarm into south China and local materials consumed in local. But import LPG will be arrived in Pearl River Delta market, so there were increased pressures in stock.

Maoming petrochemical was balance between production and selling. Moreover output and inventory were without any change. So Maoming petrochemical and Zhenjiang Dongxing both kept prices unchanged. Guangxi area adjusted prices downwards by 50yuan/mt because Zhonghai Oil was in the cooling selling.

It is anticipated that although central China raised prices slightly today, prices were higher to sell to south China. So south China would keep prices unchanged these day.

Main Refineries		
Maoming Refinery	5080	
Guangzhou Refinery	5250	
Zhenjiang Dongxing refinery	5160	
Fujian Refinery	5100	
Zhenhai refinery	5260	T
Shanghai Petrochemical	5200	
Shanghai Gaoqiao	5150	
Nanjing Refinery	5110	
Yangtsi Refinery	5110	
Chief Terminals		
Sinobenny	5500	
Jovo BP	5500	
Nansha Huakai	5500	
Zhuhai Newocean	5520	
Shantou Caltex	5450	
Chaozhou Huafeng	5400	
Shantou Longpeng	5390	
Yangjiang CNOOC	5130	
Guangxi Zhonghai Oil	5050-50	
Fujian Huaxing	5500	P
	5200	M
Ningbo BP	5260-140	P
	5260-140	M
Wenzhou Huadian	4260	
Golden Conti	5400	
Shuzhou BP	5400	
Zhangjiagang Unocal	5400	

Fujian market

Fujian market kept prices unchanged on Tuesday. Refinery was in the low level in stockpile. Deals prices were in the range of 5100-5170yuan/mt.

Fujian refinery sold about 600mt and kept balance between production and selling in lower inventory. It was heard that reinforcements might arrived in terminals in 20th October. At present terminals was in the mid stock, so terminals must sell in average level for release pressure in inventory. Merchant terminal transacted as equal as that of yesterday. Fuzhou market, Fada and Guantou increased reinforcements in this month due to Wenzhou Huadian purchased more materials. Fuzhou BP still sold wholesale.

The market can be stable due to the low stock of refinery. But there were more materials in terminals, so the market might fall in weekend.

East China Market

East China extensively holds the price level unchanged. However import LPG cut prices one after another, it had existed pressure for nation LPG. So maintenance in Nanjing refinery was litter influence. Jiangsu market turned into slack supply.

Although Nanjing refinery was in maintenance and output was only 600mt, import LPG slashed prices so it led refinery to be in the cooling sentiment. Nanjing refinery sold only about 300mt yesterday. Up to now refinery was in the mid-low stock. Yangtze petrochemical was limited supply today due to maintenance. It was said that import LPG transacted at 5050yuan/mt in fact.

In Zhejiang market, leading refinery Zhenhai was turnaround bit by bit. The sales reached 1400mt by trucks yesterday. Moreover buying interests of users increased to pick materials up from Daxie terminal. More cargoes were arrived in Wenzhou area from Dalian. Shanghai market still was in the quiet trade. But the bullish sentiment in Jiangsu must impact sentiment in Shanghai market.

Materials were abundance in the market now, so east China will keep stable in short-term, but the perspective of the future is not prosperous for the long term.

Delivery schedule

Ref Terminals	10/6	7	8	9	10	11	12	13	14	15	16	17	18	19	20
Nansha Huakai			22												
Shenzhen Sinobenny	20					20					20				
Shantou Caltex															20
Chaozhou Huafeng		20											20		
Jovo BP															
Yangjiang CNOOC															
Suzhou BP															
Zhangjiagang Unocal															
Ningbo BP															
Golden Conti															
Ref Terminals	**21**	**22**	**23**	**24**	**25**	**26**	**27**	**28**	**29**	**30**	**31**	**11/**	**2**	**3**	**4**
Nansha Huakai															
Shenzhen Sinobenny															
Shantou Caltex			20												
Chaozhou Huafeng															
Jovo BP															
Yangjiang CNOOC															
Suzhou BP															
Zhangjiagang Unocal															
Ningbo BP															
Golden Conti															

Pre Terminals	10/6	7	8	9	10	11	12	13	14	15	16	17	18	19	20
Shenzhen Iwatani															
Shenzhen Guangju															
Huizhou Longpeng			1.8							1.8					
Zhuhai Gas															
Zhuhai Newocean							20								
Shantou Longpeng			0.8			0.8				0.8					
Shenzhen Merchant	0.8			0.8			0.8			0.8			0.8		

Shantou Chemical			0.8			0.8			0.8			0.8			0.8
Jieyang Luyuan		1.9				1.5				1.6				1.9	
Chimbusco	1.5				1.5			1.5			1.5				1.5
Shantou Gas		1									1				
Pre Terminals	21	22	23	24	25	26	27	28	29	30	31	11/	2	3	4
Shenzhen Iwatani															
Shenzhen Guanjju															
Huizhou Longpeng		1.8				15									
Zhuhai Gas															
Zhuhai Newocean					20										
Shantou Longpeng															
Shenzhen Merchant	0.8			0.8											
Shantou Chemical			0.8												
Jieyang Luyuan		1.5													
Chimbusco				1.5											
Shantou Gas					1										

units: kiloton

Name	Location	Peiord	Influence
Dagang Oil field	Tianjing province North China	From 2th Sept. for 40days	Totally cease the production
Yangtze Petrochemical	Nanjing provice East China	From 21st Sept. for 40days	Reduce 30-40%
Nanjing Refienry	Jiangsu Provice	From 9th Octo. For 50days	Output reduce half
Daqing Refinery	Helongjiang province Northeast China	From10th August	Totally cease the production
Hatu Oil field	Xingjiang province Northwest China	From 10th August to end of October	Reduce output to 250-300mt
Maoming Refinery	Guangdong province South China	From27th Sept. to 30th Oct.	Cut 80-100mt or so from LPG yield

Source: www.oilgas.com, 10 October 2006.

CHAPTER 4

Chinese Energy: Rising Demand, Consumption and Forecasts

According to the International Energy Administration, world oil consumption rose by about 1.2 million barrels per day in 2005, after an increase of 2.6 million barrels per day in 2004. The non-OECD countries accounted for 1.1 million barrels per day of the 2005 increase, and the OECD as a whole accounted for 0.1 million barrels per day. Conversely, in China in 2004 oil use rose by 0.9 million barrels per day and demand rose by only 0.4 million barrels per day 2005. The United States experienced a 0.4 per cent decline in oil demand in 2005. On a longer timescale though, despite the lower demand levels of a few random years such as 2005, United States' consumption continues to rise as a result of larger numbers of SUVs, disruptive weather conditions and geopolitical concerns. Nonetheless, 2005 saw the first decline in US demand since 2001.

According to the 2006 International Energy Administration report, growth in world oil demand will average 1.4 per cent per year from 2003 to 2030. The International Energy Administration also estimates that world oil prices in 2025 will be around 35 per cent higher than those projected in 2005. It is estimated that oil demand will rise more slowly to 111 million barrels per day in 2025, as compared to 2005 predictions of 119 million barrels daily. The 2006 International Energy Association report predicts that the total consumption of liquid petroleum will have increased by 115 to 120 million barrels daily by 2030. Consumption levels are expected to rise by at least a third of demand as of 2005 (Table 4.1 and Table 4.2).

Since the beginning of the twenty-first century, the increase in oil demand has encountered significant operational variation, especially between 2001

Table 4.1 World energy demand, 2004 and 2030 (million b/d oe and per cent)

	2004	(%)	2030	(%)	Increase
North America	55	25	69	20.6	25
Latin America	13	5.9	24	7.2	85
Europe	39	17.7	46	13.7	18
Russia and Caspian Region	20	9.1	28	8.4	40
China	26	11.8	52	15.5	100
Japan	11	5	12	3.6	9
Africa	12	5.5	19	5.7	58
Middle East	11	5	18	5.4	64
India	11	5	29	8.7	164
Rest of Asia Pacific	22	10	38	11.3	73
Total	220	100	335	100	52

Note: Figures are rounded.
Source: Bustelo (2005) and Exxon Mobil.

Table 4.2 Distribution of primary energy consumption by source, 1991 and 2004 (%)

	China 1991	China 2004	Japan 2004	Korea 2004	US 2004	World 2004
Oil	17.7	22.3	46.9	48.3	40.2	36.8
Natural gas	2.1	2.5	12.6	13.1	25	23.7
Coal	78.7	69	23.5	24.4	24.2	27.2
Nuclear	0	0.8	12.6	13.6	8.1	6.1
Hydroelectric and others	1.5	5.4	4.4	0.6	2.6	6.2
Total	100	100	100	100	100	100

Sources: British Petroleum (2005); Asia Pacific Energy Research Centre (2004); Bustelo (2005).

and 2004. These variations have been reflected in global macroeconomic statistics but not at national levels. In 2004, Chinese demand represented one-third of total global demand (see Table 4.3 and figures 4.1 and 4.2).

However, demand per capita in China remains significantly smaller and trivial in comparison to that of the United States, Russia or Japan. There are no substantial discrepancies when comparing the energy demand/consumption relative to the gross domestic product variable (Table 4.4 and figures 4.3 and 4.4).

Table 4.3 Increase in world and Chinese oil demand, 1994–2004 (thousand b/d compared with prior year, and in %)

	World	China	%
1994	1459	232	15.9
1995	1127	245	21.7
1996	1523	282	18.5
1997	1917	263	13.7
1998	446	112	25.1
1999	1616	370	22.9
2000	904	568	62.8
2001	501	46	9.1
2002	793	349	43.9
2003	1248	412	33.0
2004	2464	893	36.2

Sources: British Petroleum (2005); Bustelo (2005).

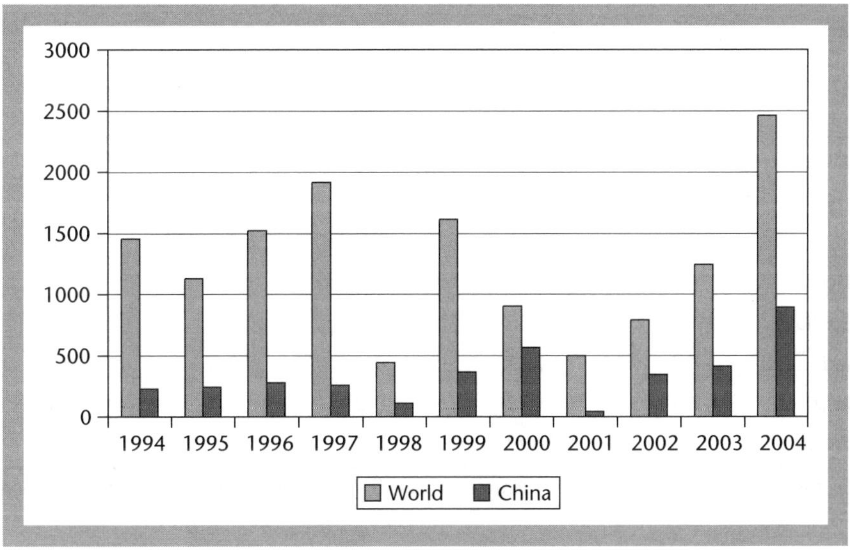

Figure 4.1 Comparisons between China and world oil demand, 1994–2004 (thousand barrels per day)
Sources: British Petroleum (2005); Bustelo (2005).

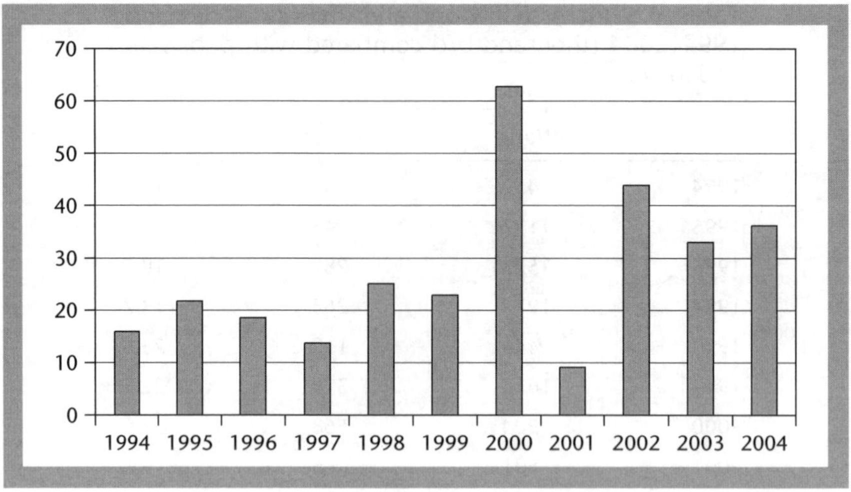

Figure 4.2 Capacity allocation of Chinese oil demand, 1994–2004 (thousand barrels per day, %)
Sources: British Petroleum (2005); Bustelo (2005).

Table 4.4 Energy polluting emission and energy economic data, by country

	GDP/energy (US$ in PPP)	Energy per capita (kg of oe)	CO_2 emissions/ GDP (kg per US$)	CO_2 emissions/ per capita (tons)	CO_2 emissions/ energy (tons per toe)
China	4.6	960	0.6	2.2	3.64
USA	4.4	7943	0.6	19.8	2.42
India	5.0	513	0.4	1.1	3.34
Japan	6.4	4058	0.4	9.3	2.3
Russia	1.9	4288	1.4	9.9	2.26
Indonesia	4.1	737	0.4	0.5	2.83
Brazil	6.8	1093	0.2	1.8	1.74
World	4.6	1699	0.5	2.1	2.53

Sources: World Bank (2005); British Petroleum (2005); International Energy Agency 2005; and Bustelo (2005).

Table 4.5 and Figure 4.5 show exponential oil consumption predictions. This demand has not been more statistically obvious since 2000. This significant operational rise coincides with the appearance of China in global statistical reports. It is interesting to note that statistics reported by private corporate entities, such as Goldman Sachs (see Figure 4.5), are significantly higher – or overestimated – than those of government agencies such as IEA or IEEJ.

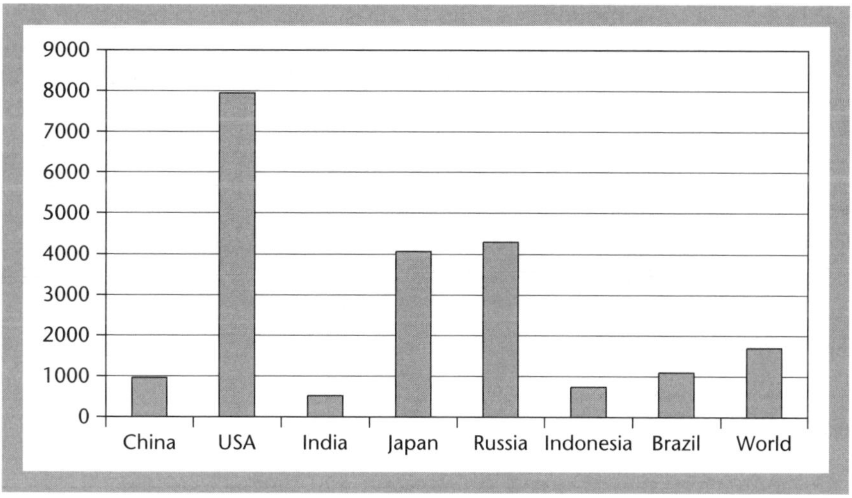

Figure 4.3 Energy per capita (kg of oe)
Sources: World Bank (2005), British Petroleum (2005), International Energy Agency 2005, and Bustelo (2005).

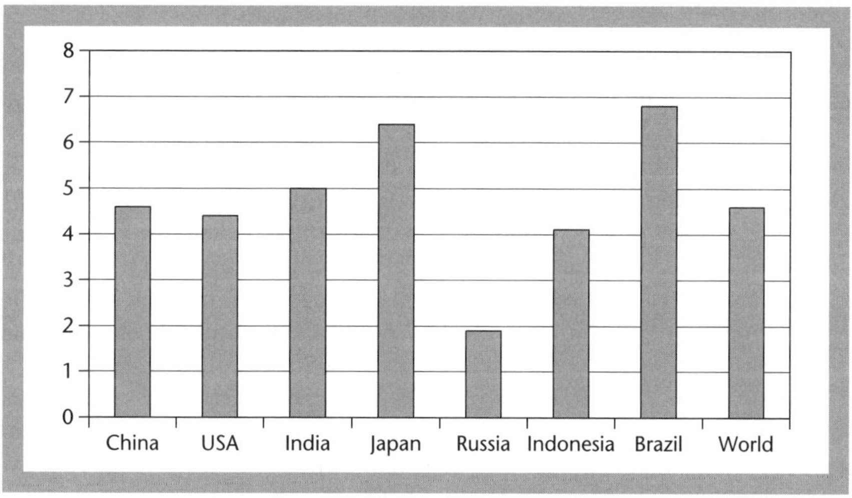

Figure 4.4 GDP/energy ($US in PPP)
Sources: World Bank (2005), British Petroleum (2005), International Energy Agency 2005, and Bustelo (2005).

Table 4.6 summarizes production and consumption balances for oil, natural gas, coal and electricity.

In Guizot (2006), the author considers the expansion of the Chinese economy under a situation of enclosed self-dependency during the 1990s and the capacity to mass produce manufactured goods at economically

Table 4.5 Oil consumption predictions, 2004–30 (millions of barrels per day)

	2004	2010	2015	2020	2025	2030
EIA, 2005	6.6	9.2	10.7	12.3	14.2	
IEA, 2004	6.6	7.5		10.1		12.8
EIA, 2004	6.6	7.6	9.2	11.0	12.8	
IEEJ, 2004	6.6	7.3		11.9		
GS, 2004	6.6	10.6	14.2	17.6	20.6	22.9
DRC, 2004	6.6	7.6		9.0–12.2		
DRC, 2003	6.6	6.4	7.2	8.8		
ERI-LBNL, 2003	6.6			10.4		
APERC, 2002	6.6			10.0		
Others (1)	6.6	7.1–9.6	8.3–11.4	8.4–9.5–12.8	10.6–15.2	

Sources: British Petroleum (2005) www.bp.com; International Energy Agency (2005) www.iea.org; Asia Pacific Energy Research Center (2005) www.ieej.or.jp/aperc; Energy Information Administration (2004 and 2005); ERI-LBNL (2005); Goldman Sachs Research (2004) www.gs.com; Institute of Energy Economic Japan www.ieej.or.jp/en/.

competitive prices and taking advantage of productivity efficiencies. The pegging of foreign exchange remained fixed for more than a decade and China was closed and thus prohibited to highly speculative instruments. This enabled China to finance significant capital projects at low cost and without significant international standards interventions. Thus, China created the biggest manufacturing hub in the world, supplying 50 per cent of goods to the global consuming classes. According to *The Economist* (19 August 2006), China's industrial production in 2006 rose by 19.5 per cent in the year to June and by 16.7 per cent in the year to July. The rise of industrial and manufacturing production centers consequently engendered a significant increase in energy consumption. Figures 4.6 and 4.7 model Chinese supply and demand of energy functions using the last ten years of actual historical data.

Clearly, consumption has been outrunning production at exponential rates since 1993. Despite the proliferation of international partnerships and advanced technological means to increase supplies, these plateaued or leveled off in relation to demand. *The Economist* online (19–25 August 2006) revealed that the tangible values of investment in fixed assets, such as factories and real estate, were 30.5 per cent greater during the first seven months of 2006 than in 2005. These economic developments have made China the world's second largest energy producer with a rising consumer middle class. As a response to rising oil prices worldwide as a result of geopolitical risks and intense international competition, the Chinese government has created a strategic oil reserve through international partnerships, cooperative drilling alliances and internal production. China reached a record high in domestic oil production and consumption in the first half of 2006.

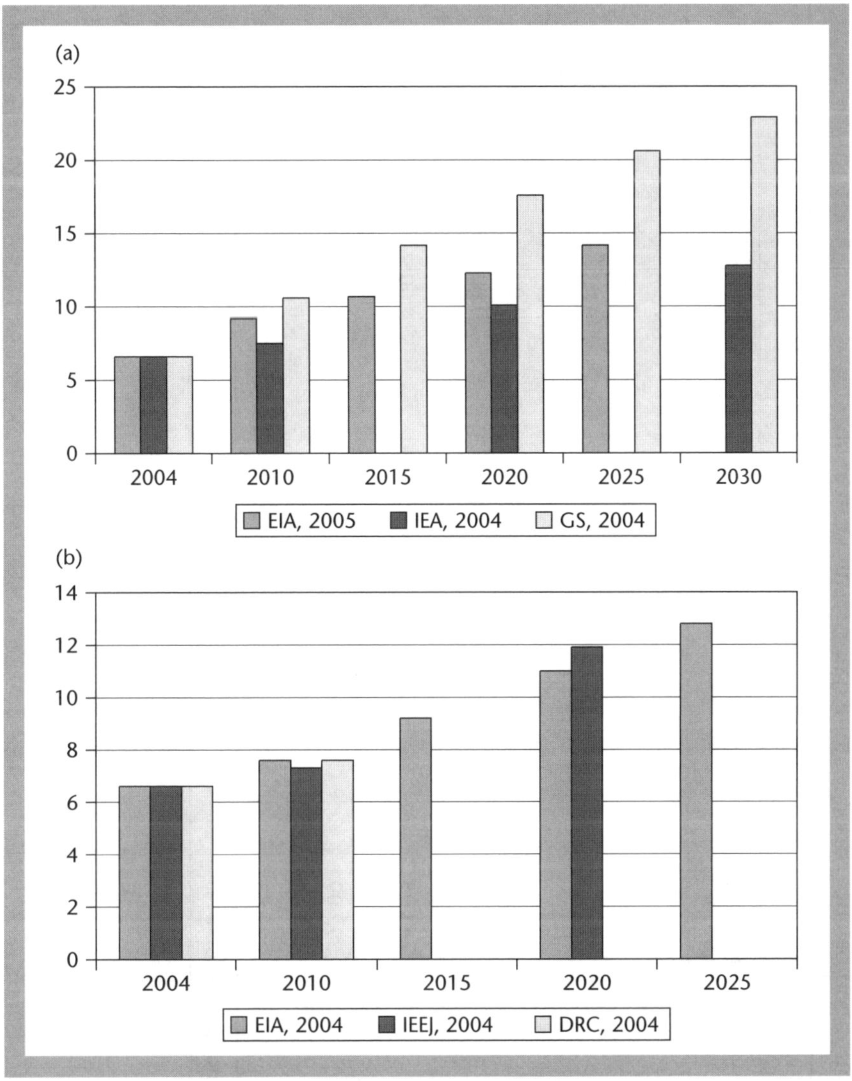

Figure 4.5 Oil consumption predictions, 2004–30 (million b/d)
Sources: British Petroleum (2005) www.bp.com; International Energy Agency (2005) www.iea.org; Asia Pacific Energy Research Center (2005) www.ieej.or.jp/aperc; Energy Information Administration (2004 and 2005); ERI-LBNL (2005); Goldman Sachs Research (2004) www.gs.com; Institute of Energy Economic Japan www.ieej.or.jp/en/.

China's apparently sudden recent economic rise has a long dormant history to which its exploding demand in energy goods is also correlated. The root of the rise primarily comes from China's deliberate, voluntary alienation and isolation from global affairs: a strategy utilized by many political

Table 4.6 Production and consumption balance for oil, natural gas, coal and electricity

(a) Oil (thousand barrels per day)

	Production	Refinery output	Imports	Exports	Stock build	Consumption
Crude oil	3408.87			1805.80	162.67	0.00
NGLs	0.00		0.00	0.00	0.00	0.00
Other oils	0.00		0.00	0.00	0.00	0.00
Refinery gain	150.13					
Gasoline		1119.62	0.00	176.26	0.00	944.59
Jet fuel		127.92	0.00	0.00	0.00	127.92
Kerosene		56.44	44.53	42.71	0.00	57.77
Distillate		1744.95	17.34	45.78	0.00	1716.97
Residual		430.85	437.09	13.89	0.00	854.05
LPGs		392.00	202.36	.76	0.00	587.25
Unspecified		1252.05	95.19	57.68	0.00	1289.56
Totals	3559.00	5123.83	2602.31	499.75	0.00	5578.11

(b) Natural gas (billion cubic feet and quadrillion Btu)

Gross Production (Bcf)	1211.30	Dry Imports (Bcf)	0.00
Vented and Flared (Bcf)	0.00	Dry Exports (Bcf)	67.88
Reinjected (Bcf)	0.00		
Marketed Production (Bcf)	1211.30		
Dry Production (Bcf)	1211.30	Dry Production (Quads)	1,4075
Dry Consumption (Bcf)	1143.43	Dry Consumption (Quads)	1,3287

(c) Coal (thousand short tons and quadrillion Btu)

Production (1000 Tons)	Imports (Quads)	Exports (1000 Tons)	Stock build (Quads)	Stock build (1000 Tons)	Stock build (Quads)	Stock build (1000 Tons)	Stock build (Quads)
Hard Coal			12233	.2312	103649	2.5465	−5566
–Anthracite	335314	6.8592					
–Bituminous	1427726	29.2051					
Lignite	74513	.5633	0	0.0000	0	0.0000	0
			2	.0001	16227	.3925	−4111
Coke	1837553	36.6276	12235	.2313	119876	2.9390	−9677
Total Coal Consumption: (1000 Tons)	1720236	(Quads)	33.7066				

(d) Electricity (million kilowatts, billion kilowatt hours, and quadrillion Btu)

	Capacity (million kW)	Generation			(Billion kwh)	(Quads)
		(Billion kWh)	(Quads)			
Hydroelectric	86.075	278.517	2.8523	Total Imports	2.975	.0102
Nuclear	4.468	41.661	.4210	Total Exports	10.339	.0353
Geothermal and Other	.468	2.348	.0240	Losses	126.473	
Thermal	265.547	1484.232				
Totals	356.558	1806.758		Consumption	1672.91	

Source: International Energy Administration (2003).

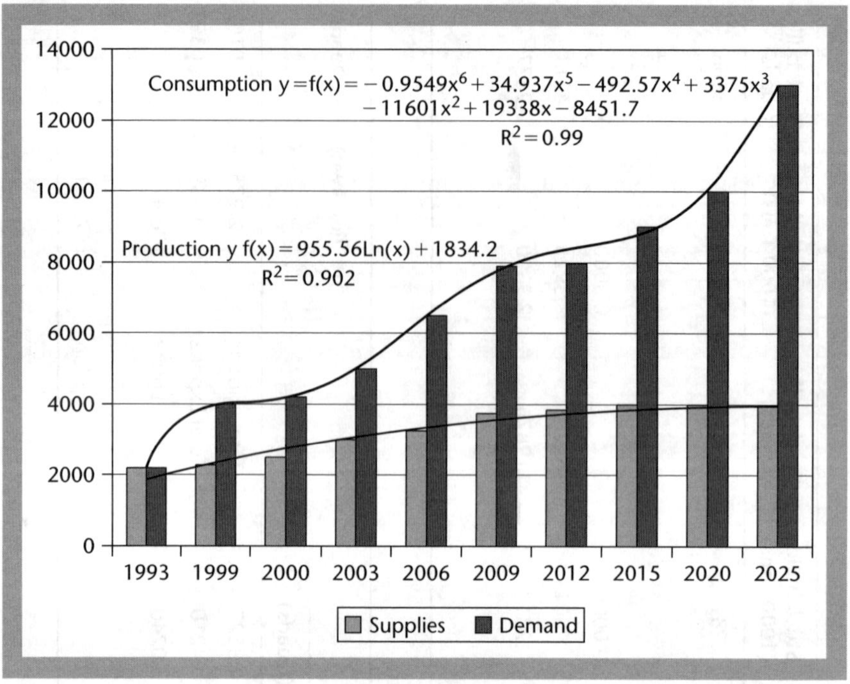

Figure 4.6 Oil consumption and production in China, 1990–2025 (estimations in thousand b/d)
Sources: British Petroleum (2005); Information Energy Administration (2005); and author's estimates.

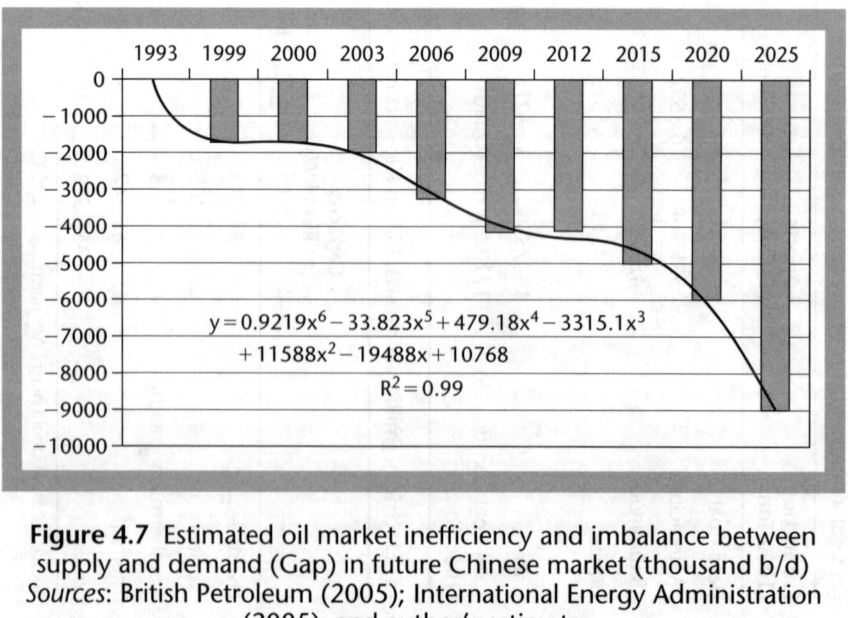

Figure 4.7 Estimated oil market inefficiency and imbalance between supply and demand (Gap) in future Chinese market (thousand b/d)
Sources: British Petroleum (2005); International Energy Administration (2005); and author's estimates.

global players who aim to remain leaders for as long as they can despite the vanishing economic credibility of their systems. This isolation meant that the calibration of financial indices did not take place among various financial markets in 2001, leaving, as a result, significant operational pricing discrepancies between market players. In addition, although China entered the World Trade Organization in winter 2001, foreign exchange data and statistics were excluded from global financial statistics reported by the Bank of International Settlements (www.bis.org) from 2001 to 2004. Foreign exchange information started to appear in international reports as of the introduction of the yuan in July 2005 as a connected variable to other currencies.

Why did China wait to join the WTO until December 2001? Why not join the WTO in 1990 while it had the capacity to be at the same economic and productivity levels as some of its democratic counterparts? China could then have been included in parallel with peer market players; instead it was excluded and financial markets exhibited a consequential operational lag in infrastructures, reforms and efficiencies in the first few years of the twenty-first century. Inevitably, it is impossible to know how demand would have been in reality had Chinese foreign exchange data been included in global financial reports in the 1990s. The calibration of market indices and valuations is not included in the exponential rise of demand. Demand depends on the financial ability to pay for the goods. Effectively, at this point in time, there are no known market efficiencies primarily because there is no enforcing single regulator. Globalization has been so far synonymous with deregulation at higher future prices. China's energy efficiency was 25.9 per cent in 1980 and 29 per cent in 1992 – an increase of 3.1 per cent in 12 years. In 1995, after reforms in finance, taxation, prices, investment and financing, China's energy efficiency reached 34.1 per cent, increasing by 5.1 per cent in three years. China has maintained an energy saving rate of 3.6 per cent from 1981 to 1990.

While the Chinese government witnessed a subtle economic rise, it also initiated changes in internal domestic policy planning through various forecasting estimations. First, in the 1990s China participated in energy price adjustments and reformulated an energy market mechanism that was in line with supply and demand in order to be able to keep global energy prices in balance. The Chinese government's annual average energy saving rate managed to reach 5.6 per cent during the Eighth Five-Year Plan and continued to rise to 7.1 per cent during the Ninth Five-Year Plan. Mainly due to the country's domestic productive and production capacities, the Chinese middle class contributed to a gargantuan rise in oil consumption, which doubled between 1984 and 1995, from 1.7. million to 3.4 million barrels per day. Energy goods – mainly oil demand – doubled from 1995 to 2005. The author estimates that it is currently around 6.8 million b/d. Chinese oil consumption reached 6.6 million b/d in 2004 which represents 33 per cent of the US consumption with 20.5 million b/d, but as the Chinese population is seven times that of the US the scaling balance is expected to reverse. Chinese demand represents a 16 per cent increase since 2003. Commercial and manufacturing demand represent the largest concentrations in the rise of Chinese energy consumption (see Table 4.7 and Figure 4.8 for figures on rural energy demand).

The International Energy Agency has revised its January 2007 oil market report by raising China's 2006 and 2007 demand forecasts from 6.98 to 7.35 million barrels per day to 7.11 to 7.54 million b/d. These changes are due to

Table 4.7 Chinese rural energy consumption, 1991–2000

	1991	1995	2000	Increase rate in 1991–95 (%)	Increase rate in 1995–2000 (%)	Increase rate in 1991–2000 (%)
Commercial energy, Mtce	90.5	109.7	160.0	4.93	7.84	6.53
Non-commercial energy, Mtce	285.7	266.6	219.1	−1.71	−3.85	−2.90
Total, Mtce	376.2	376.3	379.1	0.01	0.15	0.09

Note: Non-commercial energy consumption includes a little of producing energy use in the rural areas.
Source: Chinese Energy Ministry, Energy and Technology Symposium (2003).

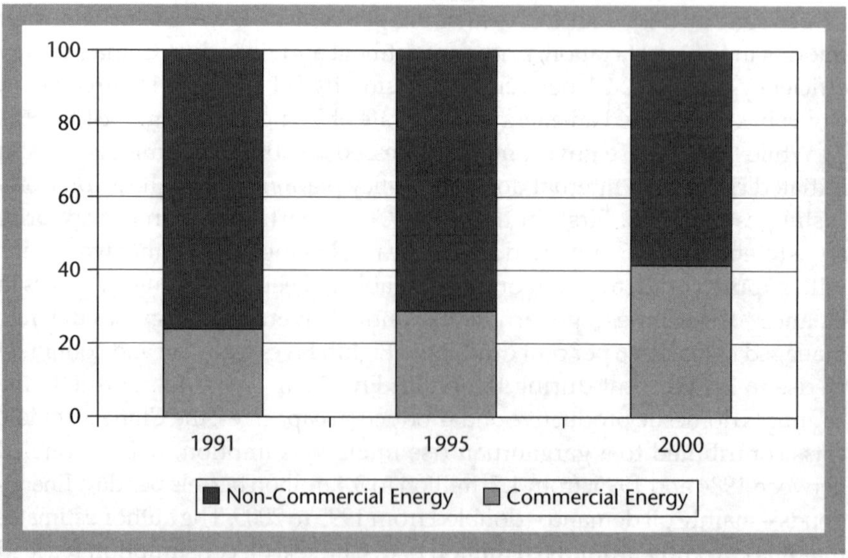

Figure 4.8 Change in rural energy consumption, 1991–2000 (%)
Source: Chinese Energy Ministry, Chinese Energy and Technology Symposium (2003).

preliminary understatements of evaluated Chinese oil demand data highlighted in the January 2007 oil market report. Newly submitted 2005 data and more detailed refinery throughput data led to the revisions. IEA also revised upwards China's first and second quarter 2007 demand projections by 0.16 and 0.37 million b/d to 7.27 and 7.76 million b/d. The IEA reports that state-owned refinery output grew rapidly in the second half of 2006 primarily due to increased capacity and a more favorable pricing environment for the refineries. State-owned refineries are defined by the agency as relatively large refineries owned by local governments or as joint ventures with international oil companies. The IEA defines refineries as very small unsophisticated units usually located close to supply sources. The agency added that under current forecasts, China is expected to account for nearly half of all Asian demand by 2011, when the agency projects China's requirements at 9.39 million b/d.

On a global scale the lack of significant changes from 2003 consumption levels seems to have a minimal impact on global statistical forecasts, and other government agencies report less substantial changes than the Chinese Energy Ministry and US Department of Energy for the next decades with respect to Asian energy demand (Table 4.8 and Figure 4.9).

However, the research did not consider consumption per capita, which will soon become of greater relevance in comparing countries' oil consumption levels than comparisons of consumption at either country or macro levels. At the individual level, the ratios are reversed, so that individual per capita consumption in China is at least seven times less than that of the US. This individual consumption ratio ought to be a way of measuring countries' consumption in terms of global citizenship and morals, taking into account territorial variations or geographical idiosyncracies. The Chinese population is expected to adopt similar living standards to those of American citizens; consequently, the Chinese population is expected to need an average of 150 million barrels per day by 2010 to 2014. This projection represents 22–23 times

Table 4.8 World oil consumption by sector, 2003–30 (million b/d)

	Residential	Commercial	Industrial	Transportation	Electricity
2003	5.22045	2.48543	26.06731	41.19233	5.04239
2010	6.08503	2.59073	30.17086	46.56570	6.22229
2015	6.33334	2.72143	33.16457	49.50239	6.59973
2020	6.44925	2.81493	35.96807	51.90088	6.94660
2025	6.60108	2.90822	38.41879	55.52224	7.27805
2030	6.74788	3.00253	40.73825	59.91817	7.55301

Source: International Energy Administration, *International Energy Outlook* (2007).

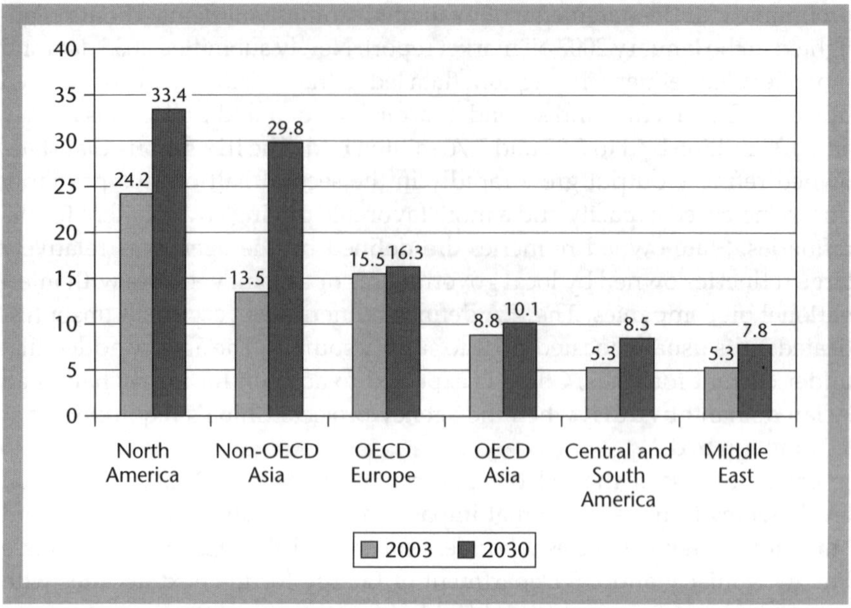

Figure 4.9 World oil consumption by region and country group, 2003 and 2030 (million b/d)
Source: International Energy Administration, *International Energy Outlook* (2007).

the consumption levels of 2006. This estimation is supported by the fact that energy consumption in China has been rising by at least 10 per cent per year since its entry into the World Trade Organization and its floating of currencies on international financial markets. While consumption levels will rise exponentially in China, they are estimated to plateau in the United States, with a mature population that will maintain its current living standards or at least undergo no drastic changes. China has been a net importer of oil since 1993 and China's gross oil imports have increased to 3.4 million b/d; this represents more than 50 per cent of the total country's consumption. By 2000, Chinese imports accounted for 38 per cent of the total consumption with a daily input of 1.9 million barrels. China has been pressured to raise its internal production through new deals and production centers in order to secure its own reserves for futuristic needs. As shown in figures 4.10 and 4.11 and tables 4.9 and 4.10, China's reserves remain minimal in comparison to global competitors.

In her first book *China's Economy: Financial and Banking System, Foreign Exchange and Interest Rate Risks*, the author vastly expands on the dimensions of Chinese economic development principally due to its enclosure and self-dependence during the 1990s. The pegging of China's foreign exchange was

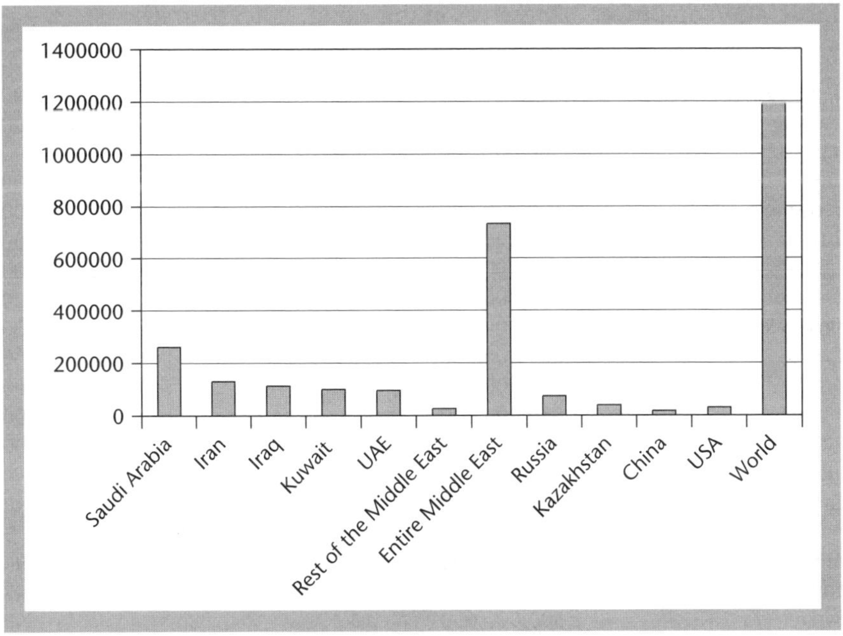

Figure 4.10 Oil reserves, 2004 (million of barrels)
Source: International Energy Administration (2004).

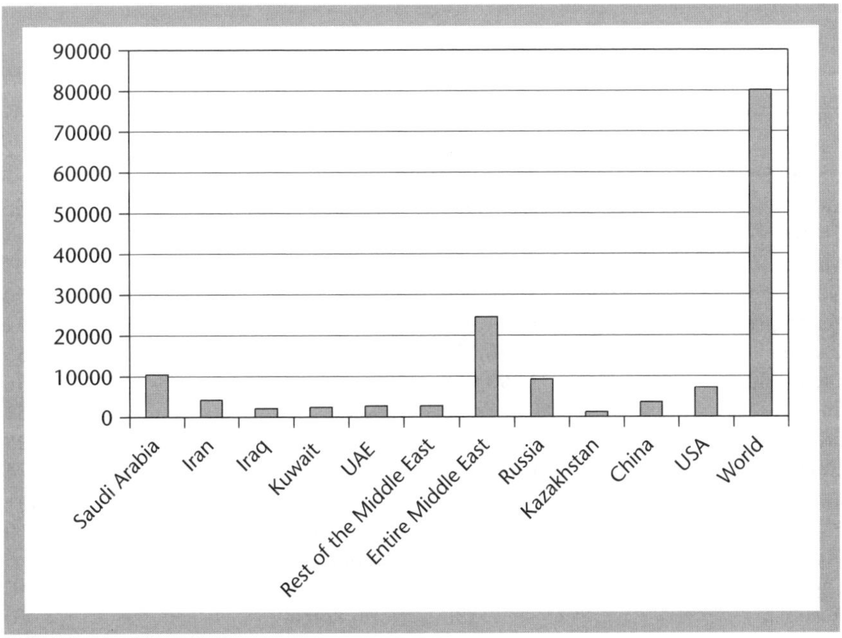

Figure 4.11 Oil production, 2004 (thousands b/d)
Source: International Energy Administration (2004).

Table 4.9 Top proven oil reserves in the world, 2005 (barrels)

Saudi Arabia	262,700,000,000	Thailand	583,000,000
Canada	178,900,000,000	Germany	395,800,000
Iran	133,300,000,000	Ukraine	395,000,000
Iraq	112,500,000,000	Pakistan	341,800,000
Venezuela	75,590,000,000	Turkey	288,400,000
Russia	69,000,000,000	Philippines	152,000,000
Libya	40,000,000,000	France	144,300,000
Mexico	33,310,000,000	Poland	142,400,000
United States	22,450,000,000	Japan	29,290,000
China	18,260,000,000	Spain	10,500,000
Brazil	15,120,000,000	Greece	4,500,000
Norway	9,859,000,000	Taiwan	2,900,000
India	5,700,000,000	Israel	1,920,000
Indonesia	4,600,000,000	Lebanon	NR
Great Britain	4,500,000,000	Sweden	NR
Australia	3,664,000,000	Afghanistan	0
Argentina	2,950,000,000	South Korea	NR
Egypt	2,700,000,000	North Korea	NR
Syria	2,500,000,000	Nepal	NR
Italy	586,600,000		

Note: NR = not reported.

Source: www.globalfirepower.com.

Table 4.10 Top oil producers in the world, to December 2005 (b/d)

Saudi Arabia	9,475,000	Germany	158.7
Russia	9,150,000	Italy	136.2
United States	7,610,000	Japan	120.7
Iran	3,979,000	Ukraine	85.66
China	3,504,000	France	76.3
Mexico	3,420,000	Pakistan	63
Norway	3,220,000	Turkey	50
Venezuela	3,081,000	Spain	24.54
Canada	3,070,000	Poland	24.53
Great Britain	2,393,000	Philippines	14.36
Iraq	2,093,000	Taiwan	8.354
Brazil	2,010,000	Greece	5.805
Libya	1,643,000	Israel	2.74
Indonesia	1,094,000	Sweden	2.441
India	785	Afghanistan	0
Argentina	745	Lebanon	0
Egypt	700	South Korea	0
Australia	530	North Korea	0
Syria	403.8	Nepal	0
Thailand	230		

Source: www.globalfirepower.com.

Table 4.11 Main oil consumers in the world, to December 2005 (b/d)

United States	20,030,000	Turkey	715.1
China	6,391,000	Egypt	566
Japan	5,578,000	Venezuela	530
Russia	2,800,000	Ukraine	491.7
Germany	2,677,000	Poland	476.2
India	2,320,000	Argentina	450
Canada	2,300,000	Greece	435.7
South Korea	2,168,000	Pakistan	365
Brazil	2,100,000	Iraq	351.5
France	2,060,000	Sweden	346.1
Italy	1,874,000	Philippines	335
Saudi Arabia	1,775,000	Israel	270.1
Mexico	1,752,000	Norway	257.2
Great Britain	1,722,000	Syria	240
Spain	1,544,000	Libya	236
Iran	1,425,000	Lebanon	102
Indonesia	1,155,000	North Korea	25
Taiwan	915	Nepal	15.4
Australia	875.6	Afghanistan	5
Thailand	851		

Source: www.globalfirepower.com.

fixed for more than a decade and the country remained closed to the introduction of highly speculative instruments. Thus China slowly created the biggest manufacturing hub of the world supplying 50 per cent of goods to the global market. As shown in Table 4.11, China has 1.5 billion citizens and currently consumes three times less oil than the United States (300 million consumers) or only three times more than France (35 million consumers). In retrospect, China's consumption remains very low because it has not yet reached average globalized living standards.

According to ChinaBidding news of 28 July 2006, the Ministry of Commerce reported that China's dependence on foreign oil reached 47.3 per cent in the first quarter of 2006, up 4.4 per cent from the first quarter of 2005. Net imports of crude oil amounted to 70.33 million tons as of June 2006, which represents an increase of 17.6 per cent over the same period in 2005. Refined oil products totaled 12 million tons, 48.3 per cent up on 2005. The Ministry of Commerce declared that China's first 2006 quarter produced 91.66 million tons of crude oil, up by 2.1 per cent. According to *China Studies* China depends on imported oil, and its rising middle class will correlate with a rising dependence by at least 50 per cent in 2010.

In order to meet its rising demand, China's major national oil companies and international contracts continue to exploit reserves. Since 2003, China has also gone out of its way to partner with Russian neighbors to increase the oil inflows for its rising population, and China is even now subcontracting and offshoring deals in Africa and South America to leverage its labor markets in science, technology, and engineering.

The China Petroleum and Chemical Industry Association reports that, during the first quarter of 2006, China's domestic production of crude oil amounted to 92 million tons, a 2.1 per cent rise on the first quarter of 2005, and that domestic oil production over the same period totaled 85 million tons, up 5.6 per cent. Various researches provide forecasts of oil consumption. The prediction model in Figure 4.12 is issued by a private corporation (British Petroleum) and shows that consumption rose by a coefficient of 8 from 1965 to 2005.

Oil consumption in China could reach 12 million b/d by 2020 and 16 million b/d in 2030 while imports rise at an increasing rate, reaching 7 million b/d in 2020 and 11 million b/d in 2030. Given the developing pace of the new Chinese middle class, it is foreseeable that China will continue to increase its consumption of crude oil by about 4.5 per cent per year for at least twenty years into the future, without taking into account any additional inflationary premium, financial engineering overvaluations from speculations, real intergovernmental financing rates and the rise in daily living costs. The added premiums could accentuate the exponential shape of the demand function. By 2020, consumption per capita will be of equal standing with those of Western societies. Part of this growth will come from increasing car utilization, from 20 million vehicles in 2004 to at least 130 million in 2020. NBS (2005) calculates that the number of individual vehicles – cars, trucks, and motorcycles – increased from 6.3 million in 1990 to 12.9 million

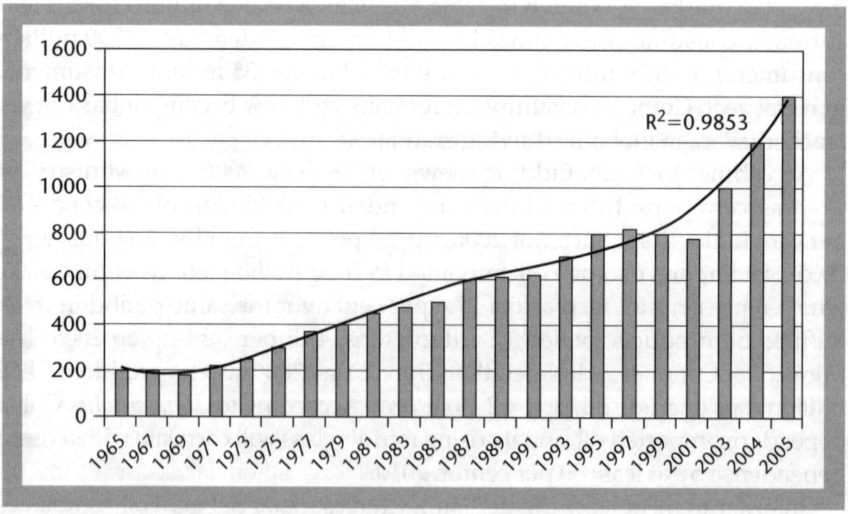

Figure 4.12 Approximate energy consumption in China, 1965–2005 (million toe)
Source: BP (2005) and author's calculations.

in 1995, from 22.3 million in 2000 to 36 million in 2003. This substantial rise in the number of registered transportation vehicles is primarily due to the rise of individual per capita income. Some estimates reckon that the 20 million vehicles registered in 2004 will rise to 60 million in 2010, 80–100 million in 2015 and 130–165 million in 2020. In the short term, Crompton and Wu (2005) predict that energy consumption will rise to 1521.4 Mtoe in 2010 with an annual growth rate of 3.8 per cent. By 2004, transportation accounted for 33 per cent of oil consumption, and this could rise to 57 per cent by 2020 according to Dorian (2005). In 2003 China's consumption in global terms represented 27 per cent of global steel consumption, 20 per cent of the copper share, 19 per cent of the aluminum, and 8 per cent of oil. According to Bustelo (2005) Chinese oil imports increased from 4 million b/d in 2005 to 7 million b/d in 2020, to 8 million b/d in 2025 and to 11 million b/d in 2030. The security of oil reserves held by large global leaders and the sharing of natural resources are topics found in a number of national strategic military plans. Bustelo (2005) also provides information about Chinese energy consumption, noting that it rose from 766 Mtoe to 1386 Mtoe, rising from 8.4 per cent to 13.6 per cent of world consumption in 2004. Annual growth in energy consumption averaged 16 per cent from 2000 to 2004 a period during which China caused a 54.2 per cent increase in world energy consumption.

The rise in consumption is a result of the substantial rise in manufacturing and industrial energy demand in the 1990s when foreign capital flows were heavily injected into China for production under cheaper labor. Until 2005, the depreciated local foreign exchange favored foreign direct investment in China and contributed to prosperous production centers. Since 2005, the appreciation of the yuan necessary to rebalance its value with competitive counterparts (US and EUR) has made oil imports more expensive and international partnerships more costly. Chinese commodities companies are currently scrambling to partner and make deals in foreign markets while the foreign exchange rate remains at favorably low levels as has been the situation from its floating in mid-2005 until beginning 2007. China will encounter more expensive dependencies in the future, due to both rising demand and to rising prices from the appreciation of foreign exchange valuations. This double rise will put internal stability at stake. For this reason China relies heavily on public transportation infrastructures and alternative transportation systems. This is also why, according to Logan (2005), the government decided in 2004 to start building a progressive energy reserve with a daily stockpiling of 35 days' imports or 100 million barrels by 2008, 50 days' imports or 300 million barrels by 2015 and 90 days' imports or 600 million barrels by 2020.

Let us now review Chinese consumption levels in comparison to specific geopolitical counterparts. The International Energy Agency (2004) estimates that China will consume 2100 Mtoe in 2020 – more than the estimate for Western Europe – and 2500 Mtoe in 2030, close to the current consumption of the US and Canada combined. Per capita energy consumption in 2004 was 1.1 toe in China, 4.0 toe in Japan and Germany, and 4.3 toe in the US. Thus in 2004, on a per capita basis China consumed four times less than Japan, Germany or France and seven times less than the United States. Various research centers, including the Center for Energy Economics and Development and Beijing Energy Efficiency Center, suggest that between 2002 and 2025 oil consumption will increase at an average annual rate of 4.5 per cent in China versus 3.5 per cent in India, 1.4 per cent in the United States, 0 per cent in Japan, 1.3 per cent in South Korea and 1.9 per cent in the world as a whole. The fact that China is raising its current production to very high levels compared with past trends is not alarming; it aims to reach average peer levels in global per capita terms. China remains a below average consumer. In 2004, oil consumption was 236 kg in China, 452 kg in Brazil, 473 kg in Argentina, 1500 kg in Germany, 1551 kg in France, 1876 kg in Japan and 3171 kg in the United States. Part of the reasons underlying diversification in China and the intense rise in productivity is to alleviate its dependence on coal and other types of energy, with some of them perceived as more polluting and less energy efficient. The Institute of Energy Economics (2006) also forecasts that the proportion of oil used as a primary source of energy will increase from 22 per cent in 2004 to 29 per cent in 2020, while the proportion of coal will diminish from 69 per cent to 56 per cent.

The International Energy Agency (2004) also estimates that Chinese GDP will grow between 6 per cent and 7 per cent over the next twenty years based on a number of forecasts. In 2004, the IEA forecast 8.4 per cent growth in 1971–2002, 6.4 per cent in 2002–10, 4.9 per cent in 2010–20 and 4 per cent in 2020–30. The US Energy Information Administration (2005) estimated a rise of 6.2 per cent in 2001–25 in the basic scenario with a possible high of 7 per cent and lows around 5.3 per cent. The Institute of Energy Economics of Japan (2004) predicts 7.2 per cent in 2000–2020, versus 2.7 per cent for the entire world. Exxon Mobile (2004) estimates that between 2004 and 2030 Chinese energy demand will double. In the meantime, world demand is also estimated to double and European demand estimated to grow by 18 per cent. Exxon Mobile also reports that Indian demand will outgrow all counterparts, with an estimated demand of 164 per cent, but, by 2030, it would still be lower than China, with 29 million barrels a day of oil versus 52 million.

The Energy International Agency (2004 and 2005) also noted that China will no longer be self-sufficient in natural gas production since consumption will rise to 2.6 trillion cubic feet (Tcf) in 2010, 3.4 Tcf in 2015, 4.2 Tcf in 2020 and 6.5 Tcf in 2025 while production will grow much less: 1.6 Tcf in 2010, 1.9 Tcf in 2015, 2.3 Tcf in 2020 and 3.1 Tcf in 2025. It is also estimated that China's imports will reach more than half of its natural gas production capacity. China is expected to become a net importer of natural gas by 2010.

British Petroleum also provided some forecasts in 2005: its production level forecasts have a logarithmic shaped curve leveled off and plateaued at less than a 4 billion b/d limit line in 1993, since when consumption accelerated at an exponential pace to reach 3 billion b/d in 1993, 4 billion b/d in 1999, 5 billion b/d in 2000, and 7 billion b/d in 2004. Effectively, the mathematical curve of the supply versus demand for oil demonstrates that the market is no longer efficient as was the case in 1993 when the curves crossed each other. Since the intersection, consumption has an exponential shape while the production side has flattened, with an apparent limit of 4000 b/d. With this assertion, the adequacy of energy resources comes into question once Chinese consumption per capita rise to Western levels and Chinese citizens adopt Westernized standards of living and middle-class consumer life styles.

British Petroleum also evaluated the function of coal production and consumption in China. This appears in line or correlated with others and reflects market efficiency, as of 2004, between coal supply and demand. Andrews-Speed (2005) reports that in the short to medium term, global energy policy-makers believe that markets are manageable and sustainable despite drastic shifts between supply and demand since 1993 which have created globally unbalanced and inefficient markets, both at the tangible and the intangible level. However, most policy-makers are also looking to find alternative solutions to cope with the longer term taking into account current global standards of living, assuming no further progress in response to the growing demand and in line with global emissions policies. However, coal remains a major energy source in China. The International Energy Agency (2005) estimates that Chinese carbon dioxide emissions will reach 8.13 billion tons in 2025 or 21 per cent of world emissions, compared to 3.32 billion tons in 2002, or 14 per cent of world emissions. Of these 8.1 billion tons, about 5.8 billion comes from coal consumption. According to British Petroleum and the National Bureau of Statistics (Bustelo, 2005), Chinese energy intensity is converging only slowly and is stabilizing; its lognormal decreasing at an accelerating rate, was, in 2003, the same as that of mature Western economies. Despite its increased consumption, China also contributed to the reduction of carbon dioxide emissions, up to 770 million tons, and 19 million tons of sulfur dioxide from 1981 to 2002, due to energy saving and reduction in energy use.

China is also making progress in advanced technologies, including hydro systems, since the 1980s. It has also established energy saving technologies in rural areas to reduce pollution levels. Energy-saving buildings have been erected in urban areas. New energy sources, such as solar, wind and geothermal energy have been under development and utilization. The benefits brought by these measures in the first decade of this century are energy savings of 30 Mtce conventional commodity energy, equal to a carbon dioxide emission reduction of 20 million tons each year. Thus some progress has been made with respect to the sustainable development path.

According to BTM (2005), in 2003, China needed 832 toe to produce $1 million of GDP at given foreign exchange rates, which is four times more than the United States with 209 toe, or six times more than Germany (138 toe) and seven times more than Japan (118 toe). The International Energy Agency (2005) also reported that by 2025, per capita carbon dioxide emissions are estimated to be lower than US emissions, at 5.6 and 22.7 tons, respectively. Total US emissions will reach 7.98 billion tons, but China will be producing the world's largest carbon dioxide emissions between 2020 and 2025. The amount of energy consumption has only doubled while GDP has quadrupled. Consumption of primary energy amounted to 1480 Mtce in 2002, ranking second in the world. While China's GDP grew by 9.7–10 per cent between 1980 and 2000, its corresponding average annual energy consumption rose by only 4.6 per cent. Thus, energy consumption remained below the economic growth rate, especially during the 1990s. The elasticity coefficient of energy consumption was only up to 0.47. About 1260 Mtce was saved from 1981 to 2002, meeting the energy supply goal for economic growth to be achieved half by exploitation and half by saving. Such a low elasticity coefficient of energy consumption is rare both in developing countries and developed countries.

Energy usage has become more efficient and individual citizens have become more conscious and aware of energy waste. Energy consumption per unit of GDP has continuously decreased. Energy consumption (by 10,000 RMB GDP), calculated on the basis of constant price of the year 2000, fell from 4.28 tce in 1980 to 1.45 tce in 2000. GDP produced per ton of coal equivalent rose from 2335 RMB in 1980 to 6880 RMB in 2000. The energy consumption per unit output was reduced by 64 per cent with an average annual conservation rate of up to 4.6 per cent during the 20 years from 1980 until 2000. The world average energy consumption per unit output fell by 19 per cent during the same period. The Organization for Economic Cooperation and Development countries experienced an average drop of about 20 per cent. Energy consumption per unit product also fell in high energy consuming industries and sectors, for example, metallurgy, the chemical industry, building materials, petrochemicals, and power. The unit consumption indicator of general energy consumption per ton of steel, copper metallurgy, and compound ammonia and oil consumption of internal combustion engines fell by more than 30 per cent, narrowing the energy consumption gap of major energy consuming products with international advanced levels. For example, coal consumption by thermal power generation decreased from 32.5 per cent in 1980 to around 21 per cent in 2006, and the comparable energy consumption per ton of steel has also dropped from 70.4 per cent in 1980 to approximately 20 per cent in 2006.

China's Social Sciences Academic Press forecast that China's dependence on imported oil will reach 50 per cent in 2010. These predictions are in line with car production and rising daily living standards of the new Chinese middle class. Bustelo (2005) reveals that China became the second biggest world consumer of oil in 2003 and actually surpassed Japan. According to his research, only 30 per cent of the consumption is used for transportation.

China has still not reached its full potential with respect to consumption targeted for transportation as the world average is 48 per cent. The Chinese government also reports that China's net crude oil imports reached 70 million tons, up 17.6 per cent, and China's net import of processed oil reached 12 million tons, up 48 per cent. The Commerce Ministry revealed that Chinese imports totaled about 47 per cent of total oil consumption in the first quarter of 2006. The NDRC said that China relies primarily on domestic energy supplies and that its oil production will remain approximately 180 and 200 million tons a year for at least a decade.

China is currently producing a new middle class with high expectations of improved standards of living. In 2005, China ranked second to the US, surpassing Japan, in global oil consumption, with demand of around 314 to 318 million tons of oil per year, and used approximately 6.69 million b/d (Bustelo, 2005). Car sales rose by almost 50 per cent in the first quarter of 2006 which contributed to a 20 per cent rise in demand for gasoline, to 1.2 million b/d, in April 2006. More cars are needed to service a more mobile middle class migrating from rural areas to expanding city centers where job opportunities are rising and individual wealth is created. American and German car markers have exported production centers to China to facilitate rising demand and to cut costs. Goldman Sachs (2004) estimates that car production in China will rise from 19.2 million in 2005 to 42.5 million in 2010 to 131.6 million in 2020 and to 198.8 million in 2030, overtaking the number of automobiles in the US between 2020 and 2025. Although China (2004) will surpass the US in absolute numbers of automobiles, individual ownership will remain far below US levels, with around 188 per thousand individuals in China versus 538 in 2030, compared to 32 versus 513 in 2010.

The Global Association of Risk Professionals reported that China incurred an increased demand for 318 million tons of oil in 2005, largely due to growing car production and improvements in living standards. An estimated 123 million tons of this was connected to life style changes. The fact that the yuan remains slow to appreciate in relation to other currencies is also due to this sudden rise in living standards of a billion and a half individuals. To avoid internal social distortions and revolts, the rate of appreciation is increasing at a decelerating pace. According to the International Energy Agency (IEA), the Chinese demand for oil is estimated to increase by 20 per cent until 2020. The IEA also forecasts that China's oil consumption will rise from 2 million barrels in 2006 to near 10 million barrels in 2030, almost 80 per cent of which represents domestic oil demand. *China Studies* forecast that dependence on imported oil in China will rise 50 per cent from current levels by 2010. Consumption climbed by nearly 10.8 per cent in April 2006, which is the strongest rise since 2004. This acceleration in energy consumption correlates to a sudden burst in manufacturing production enterprises. The Chinese National Bureau of Statistics reports that energy consumption as of December 2005 totaled about 2.22 billion tons of standard coal equivalent, up

9.5 per cent since 2004. Of this total, coal comprised 2.14 billion tons, up 10.6 per cent, consumption of crude oil was 300 million tons, up 2.1 per cent, that of natural gas 50 billion cubic meters, up 20.6 per cent, that of hydroelectric power 401 billion kilowatt hours, up 13.4 per cent, and that of nuclear power 52.3 billion kilowatt hours, up 3.7 per cent. Energy consumption for producing 10,000 yuan of GDP was 1.43 tons of standard coal equivalent, the same level as in 2004.

China's refineries also shifted production capacity to diesel and cheaper gasoline grades. China's National Development and Reform Commission raised gasoline and diesel prices to record highs of 300 and 250 yuan per ton, respectively. China's domestic oil prices are adjusted with reference to those of the international markets and in correlation with foreign exchange valuations and domestic debts and demands. The International Monetary Fund had foreseen a scattered regional and geopolitical fall in the global oil supply in the middle of the first decade of the twenty-first century until infrastructural investments with Russia and in the neighboring drilling seas would be re-established and enable distribution of greater oil supplies. Reported oil supply levels are also a function of transparency of political pricing structure formulas, unregulated and unenforced growing global financial speculative trading activities, infrastructural extraction capacities, transportation costs, geopolitical and natural events, pricing unit, alternative energy usage, rising living standards of the mass consumption, individual economic purchasing power and geopolitical partnerships' drilling activities. The global pricing of oil from 2001 to 2006 has been more strongly influenced by speculative trading activities by financial institutions and hedge funds dealing in commodities and energy derivatives instruments than by OPEC. The price of oil tripled from mid-2001 to summer 2006 and doubled its value from 2001 to 2007.

Since 2002, Russia, China and other eastern countries rich in oil have been cooperating to implement adequate infrastructures for oil flows in responses to geopolitical requirements for distribution and resource sharing. On 31 July 2006, PetroChina announced that a first batch of crude oil was floated in pipes from Kazakhstan to its reserves in northwestern Xinjiang Uygur Autonomous Region through the China-Kazakhstan grid. This was the first crude oil import operation between the countries. The transnational grid pipe is 962 km long and carries 10 million tons of crude oil per year. It starts at Atasu in west Kazakhstan and enters China at Alashankou port on the Sino-Kazakhstan border before reaching its destination at PetroChina's Dushanzi Petrochemical Company. This project is one of PetroChina's first and is in partnership with China National Petroleum Corporation, China's largest oil producer and the parent company of PetroChina. This is aimed to strengthen China's oil supply security by offering an alternative option in case of future shortages.

Aside from creating infrastructures to diversify oil production in response to rising demand, deals are also being made with Shell, Esso, British

Petroleum, Total and others to take minority stakes and shares in new emerging markets such as Africa or South America. Another factor in the growing demand for oil is the increase of national strategic reserves, similar to that of the US, which is used to influence monetary policies in case of abnormal market pricing conditions. The Chinese national strategic reserve was initiated in 2005 in Zhejiang, Shandong and Liaoning provinces. Reserves will rise from 100 million barrels in 2008 to more than 600 million barrels by 2020. Reserves are also created as part of the Three Gorges Dam Project in relation to its intended reduction in oil energy consumption, from 5.4 per cent in 2004 to 2.7 per cent in 2020.

Rising oil prices are exacerbated by shortages in supplies that are unable to meet growing demands despite improved technological progress, international partnerships and energy saving efforts. The inability of supplies to keep up with demand is also due to geopolitical factors, which are sometimes already discounted in the prices and financial trading speculations of trading institutions with tangible and intangible commodities instruments. The proportion of actual demand versus supply, financial speculations and geopolitical factors affecting the price per barrel is unknown. Higher energy prices are directly correlated to the rising intangible fictitious trading activities of financial entities indulging in high speculation during periods of global restructuring during which time arbitrage opportunities by hedge funds are proliferating unlawfully and illegally. The National Bureau of Statistics reported that in the first half of 2005 China's gross production of crude oil was 89.797 million tons; this represents a rise of about 5 per cent since 2004. Chinese imports of crude oil also rose by about 4.1 per cent to 63.552 million tons; this represents a growth rate of about 35.2 per cent since 2004. This is possible because China has multiplied its foreign partnerships with South America, Africa, and Russia since 2003. These relationships and alliances, coupled with lower foreign currency rates, have enabled China to build up a significant strategic reserve while there are no global regulations in force or limits as to what a country is able to import per year. According to *TIME Asia*, China's oil imports have doubled since 2001 and grew by 40 per cent in the first half of 2004. The Global Association of Financial Engineers also reported that competitive international financial oil trading and international oil organizations purposely reduced and slowed down oil production centers to push prices up so that the buying of energy, notably oil, for domestic reserves is also reduced. OPEC uses a non-transparent mechanism of altering daily oil production to manipulate oil prices and regulate global economies. Yet, China has been shrewd in diversifying partnerships globally and geopolitically in not relying exclusively on a single region, rather diversifying, for instance, in Africa, where Chinese manufacturing companies are also now subcontracting labor and creating new wealth production centers. Chinese oil companies are multiplying deals and partnerships to build reliable reserves for a rainy day and to ensure pricing security.

China is also trading with emerging trading partners such as Nigeria and other African countries, not only to locate cheaper manufacturing centers but also to create oil production drilling centers to sustain competitiveness and to keep its regional prices steady.

In 2001, China created an open market, but domestically it is not sufficiently mature to fully integrate its oil prices with international markets. The National Development and Reform Commission reported, on 9 July 2005, that China intended to launch reforms on the price forming mechanism of refined oil products to transform the price from delayed to real time to enhance trading liquidity across China. The Chinese authorities are also forecasting an exponential demand to meet the overdevelopment in car manufacturing since 2002 and have formed alliances with Russian oil production centers in northeastern Russia to transport petroleum to China and facilitate the supply of reserves. China is also boosting its domestic oil infrastructural reserves and creating oil production centers. The state press reported on 19 June 2006 that construction of the first facility for strategic oil reserves had been completed. The *China Daily* reported that the plant has the capacity to hold about 5.2 million cubic meters or 32.7 million barrels of crude oil, and is located in the coastal city of Zhenhai in China's eastern province of Zhejiang. The National Development and Reform Commission reported that another plant was under construction in Zhejiang and operations should be begun in early 2007. There are already plants operating in the eastern provinces of Liaoning and Shandong. Chinese authorities expect to use those reserves to maintain half a year's supply of oil within 10 years. China is now investing in production centers in the Chinese Sea, the Straits of Malacca and near the Philippines.

According to Gary Vasey in 'Oil and China' of 15 August 2005, riskcenter.com, China is also a leader in forming economic alliances. China took a $150 million share via CNOOC in the Alberta Oil sands, and PetroChina signed a commercial partnership with Enbridge for half of the supply from a $2 billion project to transport oil from Alberta to Prince Rupert in British Columbia. New emerging democracies such as China or India are taking the lead in colonizing less transparent and infrastructurally weaker and poorer cultures to import natural resources and construct strategic reserves. While Western democracies use sophisticated transportation means, modern technologies, cleaner alternatives and hydrogen cars, and diminish dependencies on foreign partners, emerging countries are adopting their old ways of securing their strategic reserves. China is also getting involved in Africa and South America. Major Chinese oil and gas companies are making enormous investments abroad, most particularly in Latin America, Africa and Australia to ensure sufficient reserves in advance for its demand. Other energy companies are also in pursuit of investments on which returns are anticipated on capital in ten or twenty years. China's energy demand fell in late 2005 due to the start of foreign exchange currency adjustments, reappreciations, inflationary pressure on the industry sectors in fixed assets and manufacturing sectors, and an overheated economy from 2002 to 2005. The creation of new manufacturing and industrial production centers engenders rising energy demand. Electricity consumption is directly correlated to the rise in industrial and manufacturing production and manufacturers accounted for 74 per cent of 2005's total electricity production.

China's foreign exchange adjustment occurred in July 2005 and consequently inflows of capital have decreased, creating slowdowns in industrial and manufacturing production output in December 2005 and continuously since then. The Chinese economy overheated from 2002 to 2006, and the March 2007 Asia market downturn was foreseeable in preliminary indicative signs such as plateaus in the industrial and manufacturing sectors with electricity consumption reduced by 14.4 per cent in December 2005. This growth rate was further eroded in March 2006 with electricity consumption down by 19.4 per cent. As a consequence of further reforms and tightening of regulation, the industrial growth rate has fallen steadily since late 2005.

Weather conditions also impact energy prices and demand, and domestic electricity demand reached 2.43 trillion kilowatt hours, a rise of 12 per cent, since 2005. It should be noted, however, that in 2005 oil demand decreased at least in part as a result of the initial adjustment and appreciation in the Chinese currency. It is expected that if weather remains normal without catastrophic conditions, the industrial sector's electricity demand will average an increase of about 14 per cent per year in the next few years. The government has installed electricity generation infrastructures capable of producing about 70 million kilowatts. So demand and supply rose in parallel with shortages of supplies in the summer and winter due to typical peak seasonal demands. Regional markets within China are more affected than others; with, for instance, provinces in eastern China such as Jiangsu and Zhejiang having less regular supply availability. Electricity prices also rose due to shortages of coal and due to shortages of the substitute; this accounts for a premium of 4 per cent alone. Coal consumption in the industrial sector accounts for more than 90 per cent of the total energy. And the major industries, among them electricity, metallurgy, building materials and oils consumed about 75 per cent of the total coal availability. Like other substitutes, demand for coal rose with the overheated economy for about three years. Consequently, the electricity industry increased its coal consumption by about 120 million tons in 2005 and the coal industry increased its output by about 150 million tons the same year. Total coal output in 2005 reached about 2 billion tons, about 9 per cent increase from 2004. Pollution, global warming, safety and infrastructures are all of major concern within the coal mining sectors of China where numerous explosions have occurred since 2002. Coal prices in China were kept artificially high due to unsustainable overheating and economic variables. The metallurgical sector increased consumption by about 30 million tons.

Maintaining a stable oil supply and obtaining the highest level of profits are in accordance with the interests of major oil exporters. The increasing interdependence of oil exporting countries and main oil consuming countries means that it is beneficial for all sides to maintain and to strengthen geopolitical and economic cooperation. This global cooperation is part of a consensus in sharing global oil supplies and contributing to finding energy derivatives. It is also less costly.

CHAPTER 5

Market Forecasts and Research Predictions

The largest regional difference between 2006 and 2007 – and the only instance in which IEA regional growth projections fall outside the range defined by the *International Energy Outlook 2007* low and high macro-economic growth cases – is for China. The IEA anticipates that China's energy demand growth will slow to 2 per cent per year for the final 15 years of the outlook, whereas the *IEO 2007* reference case expects that China will maintain a 2.7 per cent annual growth rate in energy demand through the end of the projection period. The IEA growth projection for energy use in China from 2015 to 2030 is lower than the *IEO 2007* low macroeconomic growth case projection. The IEA asserts that world oil demand will grow from 80 million barrels per day in 2003 to 98 million barrels per day in 2015 and 118 million barrels per day in 2030. With rising living standards in China, world demand will reach 150 million barrels per day by 2015. China has represented at least 30 per cent of the global energy demand from the end of 2001. If this consumption trend continues at a rate greater than 20 per cent per year, the 150 million barrels per day estimate becomes a conservative understatement. Demand will continue to increase strongly despite world oil prices that are expected to be 35 per cent higher in 2025 than in 2005's *Outlook*. Much of the demand is expected to occur in parallel in the nations of non-OECD Asia. Basically all markets enjoying an economic surplus will profit from future strong continuing exponential economic conditions, and global oil demand will rise at greater pace than ever before. Non-OECD Asia, including China and India, are also expected to contribute to 43 per cent of the total increase in world oil consumption until at least 2025. The International Energy Administration also estimated that to maintain 2006 energy levels, the total petroleum supply in 2030 will need to increase by 38 million barrels per day to 118 million barrels per day in 2030.

These are conservative estimates considering that despite the rise of new technologies and advanced alternative renewables, rampant consumerism and the growth of the market for higher-fuel more expensive, consuming cars (such as SUVs) outrun the global resourcing abilities in the foreseeable future. Some countries have become more dependent on cars because of new routing infrastructures. Other nations rely more on public transport systems as a result of insufficient infrastructures and expensive energy prices. OPEC producers are expected to provide 14.6 million barrels per day of the increase. Higher oil prices cause a substantial increase in non-OPEC oil production, as much as 23.7 million barrels per day; this represents 62 per cent of the increase in total world oil supplies to 2015. The International Energy Administration also estimates that production increases reflect actual proved reserves. The International Energy Agency evaluated global oil demand at around 83.7 million barrels per day in 2005 or an increase of 1.7 per cent from 2004. Figure 5.1 shows exponential net consumption functions outpacing the supply function. Figure 5.2 illustrates predictions by Chinese energy experts that the domestic consumption function will increase by more than 20 per cent from 2005 to 2020.

The growth rate of oil demand effectively started to slow down, by 3.2 per cent, in 2004. However, China's finished oil demand has grown rapidly since 1993, as the new Chinese middle class bought more consumer goods, domestic machinery and cars. Consequently, the use of diesel for

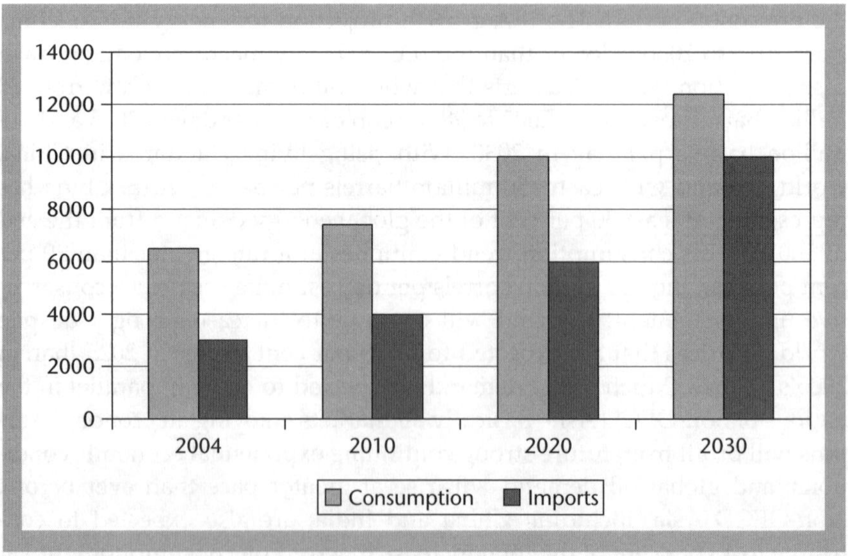

Figure 5.1 Chinese oil consumption and imports to China, 2004–30 (thousands barrel/day)
Source: International Energy Administration (2004).

generating electricity also increased demand. The Chinese authorities also developed new public transportation infrastructures to facilitate communication between regions and from rural to urban areas. This also created a higher demand for finished oil products. Finished oil consumption is forecast to rise by at least 10 per cent per year on average. These energy predictions are correlated and embedded in the pricing of financial instruments.

Various studies have participated in the independent evaluation of predictions on oil consumption and the expected national availability of oil in given global regions as part of a global strategic security safeguard. These studies and policy variations are readjusted annually according to new deals and changes in global companies' progress and international partnerships. The International Energy Administration carried out surveys in 2004 and

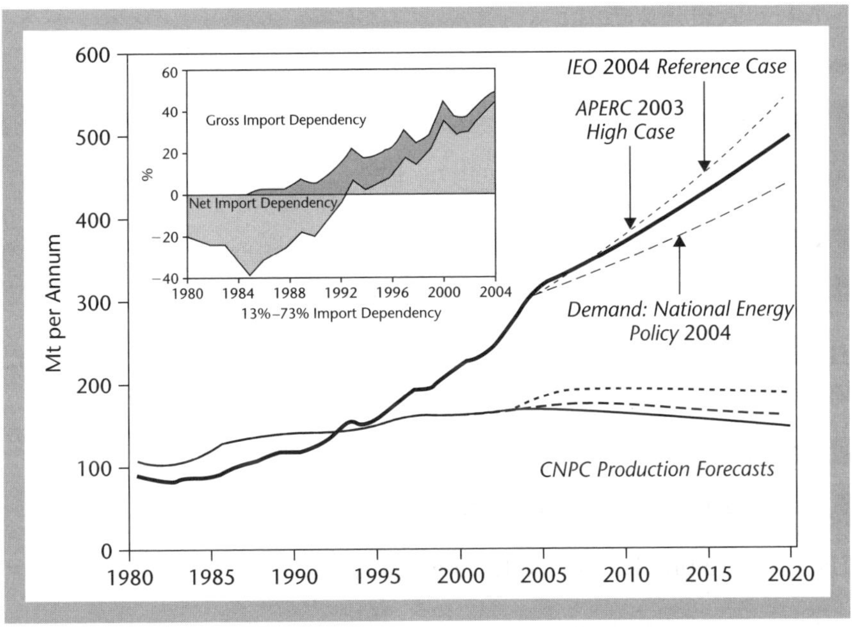

Figure 5.2 China's consumption and production functions, 1980–2020
Note: China's oil use is expected to rise by 50% or more over current levels by 2020, while production is expected to stagnate, leading to ever-greater import dependence.
Sources: (Domestic Production) China Statistical Yearbook; (Crude Imports) MOFERT Almanac of Foreign Trade; China's Latest Economic Statistics; China's Custom Statistics; (Product Imports) 1949–64: Soviet and Chinese trade statistics. Figures for 1980–84: calculated as (consumption) − (total crude) + (product exports); Figures for 1985–present: from *China Economic News*, monthly figures; China's Customs Statistics; (Crude and Product Exports) MOFERT Almanac of Foreign Trade; China's Customs Statistics.

2005 and with IEEJ in 2004 to evaluate future projections of oil demand. Most 2004 estimates predict that oil demand in China will reach 7.5 million b/d in 2010, or 11 million b/d in 2020 and 13 million b/d in 2030. EIA (2005) projects energy demand in China at:

- 9 million b/d in 2010
- 12 million b/d in 2020
- in each case applying a growth rate of 3.9 per cent in 2025–30 same as in 2020 to 2025
- 16.4 million b/d in 2030

This equates to a projected rise of:

- 6.4 per cent in 2002–10
- 4.9 per cent in 2010–20
- 4 per cent in 2020–30

And forecasts of:

- 6.1 per cent in 2001–25
- with alternative energies of 5.1 per cent and 7.1 per cent

The EIA also reissued the following re-evaluations, despite major variations from the earlier year's predictions:

- 6.2 per cent in 2002–25
- with alternative energies of 7 per cent and 5.3 per cent

The IEEJ also provided the following data:

- 7.2 per cent in 2000–20
- with alternative energies of 5.6 per cent and 8.3 per cent

EIA (2005) predicts in imports that:

- China would consume 14.2 million b/d that is about 12 per cent of world consumption
- China could import 10.7 million b/d of this consumption level
- Oil imports could reach 11 million b/d in 2030; this represents 85 per cent of domestic consumption

- Imports could rise to 8.6 million b/d in 2025; of this, 5.7 million b/d or 69 per cent could come from the Persian Gulf and 1.7 million b/d or 20 per cent could come from the Soviet countries.

The APEC (2005) predicts that:

- The Middle East's share of world production could rise to 46 per cent in 2030

Wu (2004) forecasts that:

- The Middle East's share could increase to 70–80 per cent by 2015

Crompton and Wu (2005) predict output of 7.3 million b/d in 2010. Frisch (1982) provided estimates by commodities types (Table 5.1).

The consumption function is directly correlated to imports. Figure 5.3 shows countries in need and by capacity of imports relative to others to 2030.

Figure 5.4 provides another similar function, highlighting the average annual increases in industrial natural gas consumption until 2030.

Looking at the research, it does appear that two outliers stand out: Goldman Sachs with significantly higher than normal predictions and the

Table 5.1 Estimates for the energy exploitation time limit in the world

Sort	Proved reservation (PR) and assumption reservation (AR)	Used up day
Coal	900(PR)	About 2200
	2700 (AR)	
Oil	100 (PR)	Before 2020
	36 (AR)	
Natural gas	74 (PR)	About 2040
	60 (AR)	
Uranium	According to the thermal reactor calculation 60 (PR + AR)	According to the thermal reactor calculation 2073
	According to the breeder reactor calculation 1300 (PR)	According to the breeder reactor calculation 2110–2120
	1600 (AR)	
Total non-renewable	1100 (PR)	About 2200
Energy	300 (AR)	

Source: Frisch (1982).

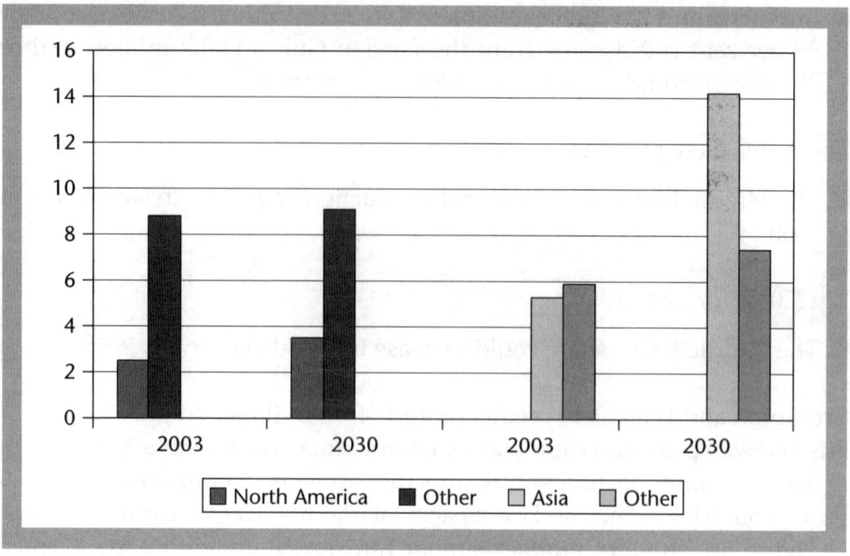

Figure 5.3 Imports of Persian Gulf oil by importing regions, 2003 and 2030 (million b/d)
Sources: 2003: Energy Information Administration (EIA), International Energy Annual (2003), May–June 2005, www.eia.doe.gov/iea; 2030: EIA System for the Analysis of Global Energy Markets 2006.

Development Research Center of the Chinese State Council (DRC) whose 2003 predictions were overtaken by events. It predicted that consumption in 2010 would be around 6.6 million b/d, levels actually reached in 2004.

According to the International Energy Agency (2004), Chinese consumption in 2030 will reach about 12.8 million b/d of a world total of around 121 million b/d. China's proportion of oil consumption will increase from 8 per cent in 2004 to 11 per cent in 2030. And according to the International Energy Agency (2005), China's proportion will rise to 12 per cent in 2025 assuming that Chinese consumption in 2025 is around 14.2 million b/d and world consumption is 119.2 million b/d.

As of 2004, China was in the top five world oil producers including international offshore operations. China's production counts 3.5 million b/d, and according to the Energy Information Administration (July 2006), in 2004 China surpassed Canada's total of 2.4 million b/d and Norway's total of 2.9 million b/d. India produced 0.8 million b/d in 2004 and imported 70 per cent of its supplies, that is, 2.5 million b/d. India reserves a total of 5.6 billion barrels as of 2004. As of mid-2006, China's production totaled 3.7 million b/d versus 2.5 million b/d in Norway and 2.3 million b/d in Canada. Only the North Sea counts 4.5 million b/d. Yet, China remains well behind the largest global producers: Saudi Arabia, Russia and

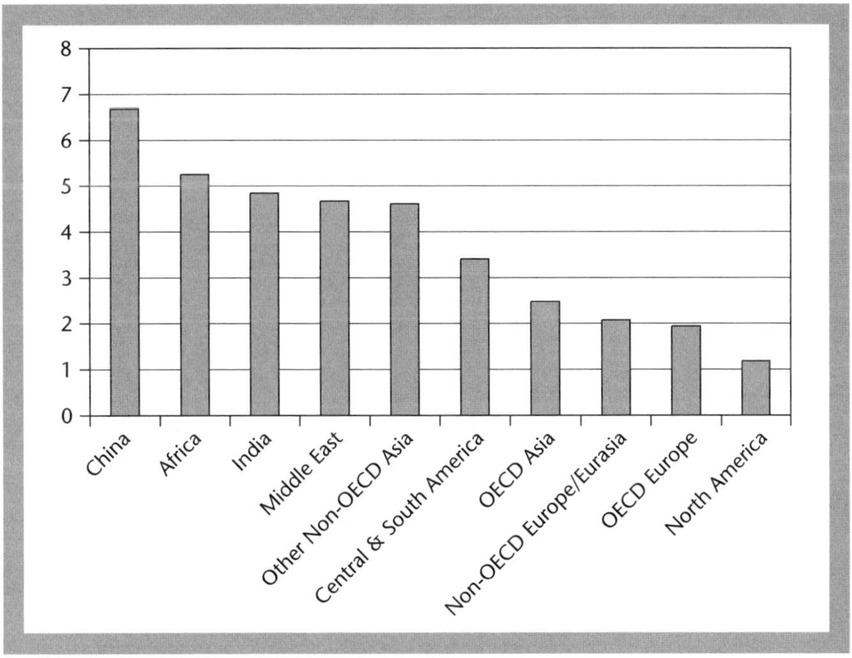

Figure 5.4 Average annual increases in industrial natural gas consumption by region and country 2003–30 (million metrics mmcf/day)
Sources: 2003: Energy Information Administration (EIA), International Energy Annual 2003; 2030: EIA, System for the Analysis of Global Energy Markets (2006).

the US. According to British Petroleum (2005), Chinese reserves total 17.1 billion barrels; this represents a proportion of 1.4 per cent of the global reserves (see Figure 5.5).

In line with other curves reflecting the supply side of global markets, while demand has grown exponentially, the supplies leveled off and flattened around 2003. Production rose from 3.3 million b/d to 3.5 million b/d from 2000 to 2004. The average production growth rate of 1.8 per cent per year is outstripped by the average exponential consumption rate of 7.6 per cent.

According to the International Energy Agency (2005), petroleum firms predicted that production would reach a maximum of 3.7 million b/d in 2010 and asserted that it would fall to 3.5 million b/d in 2025. Chinese oil companies are aware of possible future production saturation given rising demand function and thus are already creating adequate infrastructures to sustain the consumption function. China partnered with Russia and Kazhakstan to import Russian oil into various parts of western China. However, China is keen on developing and maximizing domestic resources

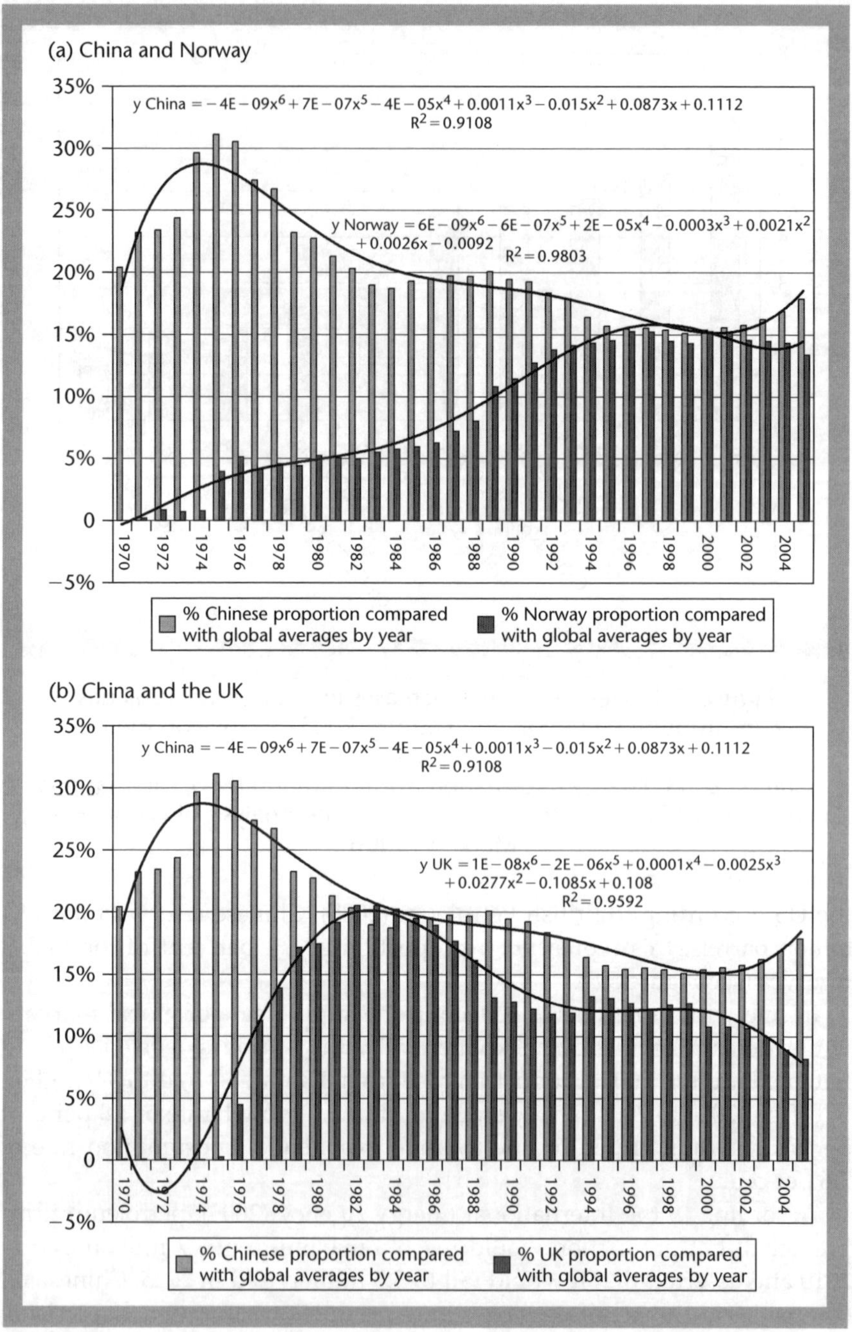

Figure 5.5 Individual countries' rates of production of oil over time as a percentage of global rate
Note: Equations represent the oil production rate functions.
Source: EIA, *International Petroleum Monthly*, www.eia.doe.gov/emen/ipsr/source4.html.

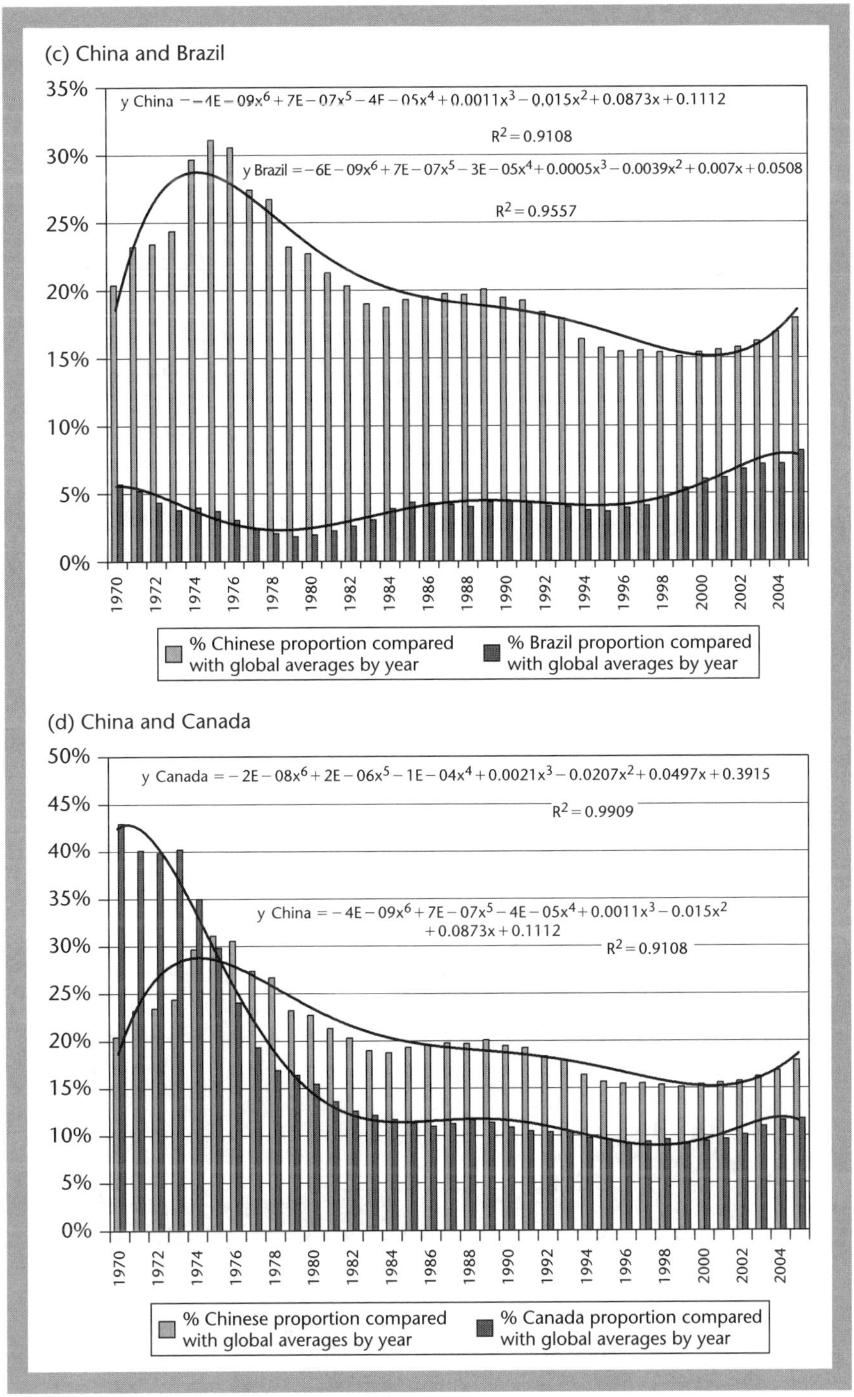

Figure 5.5 (Continued)

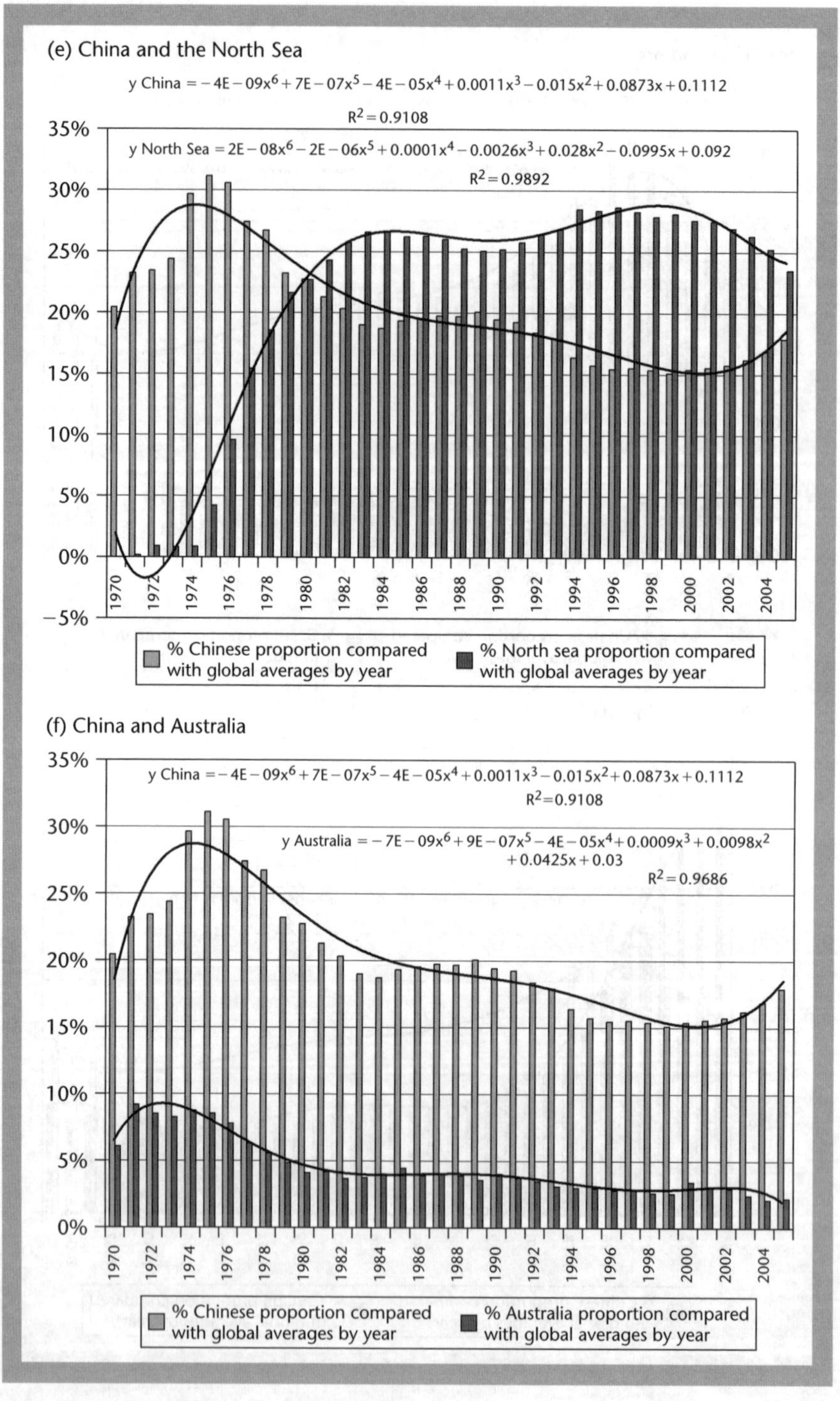

Figure 5.5 (Continued)

while also multiplying international partnerships to secure internal demands in case of future international embargoes. The east coast relies on actual Chinese production coming from the eastern sea side. China's oil resources are currently primarily in Daqing, Shengli and Diaohe in the northeast. Yet, research (Kenny, 2004) also shows that new infrastructures and production centers in the East China Sea, the Gulf of Bohai and the South China Sea are small in response to the exponential demand. KPMG (2005) also asserted that western production centers such as Xinjiang are scheduled to produce 1 million b/d by 2008. A study conducted by British Petroleum and the Energy Information Administration (2005) shows that there exists a growing gap between the production (flat at less than 3.5 million b/d) and the consumption levels (rising at 6.5 million b/d) as of mid-2006. The new production centers, despite their proliferation, are not sustaining actual levels of consumption or demand. Assuming all conditions remain the same between 2006 and 2025 in infrastructures and partnerships within Chinese energy companies, it is assumed that the gap between production and consumption will widen as follows:

- 2011: 3.7 million b/d production, versus 7.8 million b/d consumption
- 2020: 4 million b/d production, versus 10 million b/d consumption
- 2025: 4 to 6 (conservative) million b/d production, versus 13 million b/d consumption

New peer democracies are poorer than China in both natural resources and in productivity levels, thus devouring some of the international demand for oil. For instance, while India and neighboring Asian countries are growing in population and improving average standards of living, their oil production remains the same and their imports are expected to rise, creating higher competing prices.

Also while those larger emerging countries provide statistical data, new Eastern European countries do not yet figure in the reporting agencies' statistical tables. They are rich in natural resources but are as yet unexploited infrastructurally and economically.

The Energy Information Administration (2004) estimates that China's imports of crude oil will rise from 3.4 million b/d in 2004 possibly to 5 million b/d in 2010 (sustaining half of demand from internal production) to 7 million b/d in 2020 and 11 million b/d in 2030. As a net oil importer since 1993, current levels do not as yet reflect the actual change and impact of rising living standards and foreign investment capital injections of the 1990s. The Energy Information Administration (2005) conservatively reveals that the imported part of oil consumption will rise from 51 per cent in 2004 to 66 per cent in 2010 and to 85 per cent in 2030.

On the same lines, the Institute of Energy Economics in Japan revealed in 2004 that net imports of 3.5 million b/d would represent 52 per cent of Chinese consumption in 2010 and 9 million b/d would represent 75 per cent in 2020. The Energy Information Agency also predicted in 2004 that gross imports of 8.6 million b/d in 2025 with consumption of 12.8 million b/d would represent an import share of 67 per cent. The Energy Information Agency predicted in 2005 that imports will reach 10.7 million b/d in 2025 to equal about 75 per cent of consumption. The future of oil consumption and the global sharing of resources will depend much on large corporations' ability to gain global market shares and partnerships via temporary permission to drill on national soils and through financial stakes in corporations. Yet, it does appear that oil is a matter of concern in national reports with respect to military plans. The fact that alternative forms of transportation such as hydrogen cars or oil substitutes have not been pushed further is also of concern. Research shows that geopolitical resources are shared on a regional basis with, for instance, a focus on Russia, Eastern European countries and China, while Europe focuses on hegemony with North African and Scandinavian countries and the United States' influence is connected with Venezuela and South and Central American geopolitical entities.

CHAPTER 6

Geopolitical Balance of Powers and Private–Public Equity Deals

Ever since its entrance to the World Trade Organization in December 2001, China has intensified its global partnerships and proliferated its international infrastructural developments towards the creation of a more balanced and diversified portfolio of geopolitical and economical partners. The competitive race for strategic reserves and natural resources' cumulative dependencies is coupled with a significant rise in domestic and international private equity deals between global commodities corporations. Deals are multiplying now more than ever, particularly since the beginning of this century, deregulation favors corporations, foreign exchange currencies give China a significant advantage and the Chinese need a strategic reserve to secure the future. The new Chinese middle class exponentially consumes energy in various forms in correlation with the rise of their living standards.

This chapter aims to provide more transparency about these intra- and supra-national partnerships and calls for more corporate governance and implicit risk management in relation to these deals, which are occurring now at a pace never seen before. It also questions the ongoing risk-monitoring of these private equity deals and their ultimate economic efficiencies to reward us with a safer future. Figure 6.1 illustrates the evolution of Chinese oil imports from various foreign entities.

6.1 INTERNATIONAL PARTNERSHIPS: INTENSIFICATION AND PROLIFERATION

Dannreuther (2003) provides more information about the Qingdao initiative on energy cooperation that was approved in June 2004 by foreign ministers

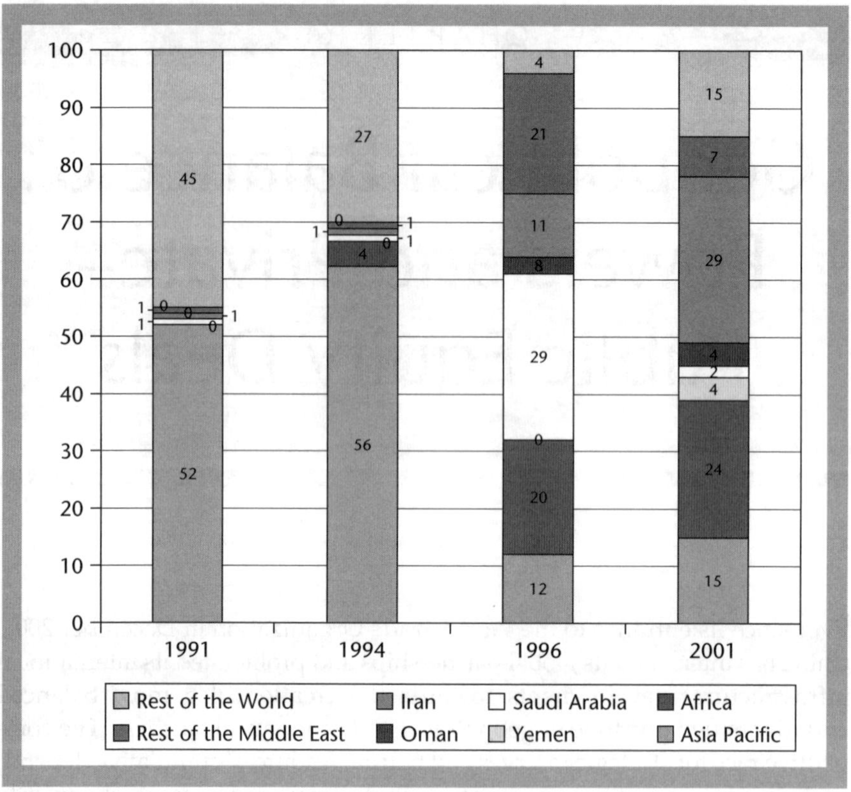

Figure 6.1 Approximate geographic distribution of Chinese oil imports, 1991–2001 (%)
Sources: APERC (2005); Bustelo (2005).

of some 22 countries of the Asian Cooperation Dialogue. The ACD includes China, Japan, India, and South Korea. The international partnerships consist of greater coordination and cooperation towards the further development of energy exploration and extraction, conservation, efficiency, creation of renewables, the construction of new pipelines, strategic reserves accumulation, and transportation security.

Another similar example is the 2002 Code of Conduct agreement between Beijing and Hanoi on the sharing of energy resources in the Southern Chinese Sea. In 2005, there was a tripartite accord to facilitate the sharing of the Southern China Sea, and the Spratly and Paracel Islands energy resources.

China is also pressing India to allow a Chinese naval presence in the Indian Ocean and attempting to reach closer ties with Myanmar. In 'China and India: a Rage for Oil' (2005), it is shown how both expanding Asian economies are fighting to establish connections with Angola, Russia and Kazakhstan. Prior to making major investments and constructing infrastructures between national and regional states to facilitate the transportation of various energy alternatives to relieve China's growing dependence on foreign oil, the Chinese government and firms developed a variety of partnership schemes and international cooperative arrangements. During the 1990s, China extended international cooperation and partnerships with Russia, Kazakhstan, Canada, Australia, Sudan, Tunisia, Iran, Azerbaijan, Peru, Brazil, Argentina, and Venezuela through trade agreements and acquisition of minority stakes in corporations.

Sinopec has made a significant gas discovery in the northern part of the South China Sea with an estimated reserve of more than 100 billion cubic metres of natural gas. The National Offshore Oil Company and Canada's Husky Energy Corporation confirmed the discovery. Husky is involved in China's oil and gas exploration in partnership with China National Offshore Oil. Together, they announced a major gas discovery in June 2006 in the Liwan 3-1-1 field in the Zhujiangkou Basin, 250 kilometers (155 miles) south of Hong Kong.

The Information Energy Administration (2000) distinguished three main stages in the process of international acquisition and energy policy-making in China with the three main Chinese corporations:

- China National Petroleum Corporation (CNPC)
- China National Offshore Oil Corporation (CNOOC)
- China Petroleum and Chemical Corporation (Sinopec)

The main policy periods are:

- Prior to the 1990s, China was interested in securing its domestic demand and was simply exploring whatever new energy resources its domestic soil could offer. It is not until the beginning of the 1990s that China began to become involved in international resources, not least because during that decade, China attracted many foreign corporations to invest in its massive growth in manufacturing. With this major economic change, China foresaw that it would also eventually depend on international counterparts principally to buy resources that it does not have that are natural goods.
- From 1992–95, Chinese firms entered experimental arrangements in Thailand, Canada, Indonesia, Peru, and Sudan.
- From 1997–2002, Chinese companies entered arrangements in Sudan, Venezuela, Iraq, Kazakhstan, Algeria, Azerbaijan, Chad, Ecuador, Indonesia, Iran, Myanmar, Niger, Oman, Uzbekistan, Russia, Syria, Tunisia, and Turkmenistan.
- As of 2003, since the United States' involvement in Iraq, China has entered new deals with Algeria, Gabon, Egypt, Brazil, Argentina, Iran, Venezuela, and Canada.

The fact that China has quietly developed strong relationships and acquired minority interests in all regions of the globe increases competition between all partners in the new era for resources. And China has shrewdly diversified its portfolio of importers in all regions of the globe, including South America and Africa, with minority stakes in all partnerships and deals. This ensures that China secures a global geopolitical presence that is also one of well balanced partnerships (see Figure 6.1 above).

Other agreements are intended to gain control of oil transportation grid routes. The Shanghai Cooperation Organization (SCO) is made up with the participation of China, Russia, Kazakhstan, Kyrgyzstan, Tajikistan, and Uzbekistan to secure energy and security controls. The 2005 meeting of SCO in Astana, Kazakhstan, suggested that China also wants to propose a new pipeline project from Iran to India passing through Pakistan and leading to China. In 2005, China also signed a contract agreement with Uzbekistan.

The acquisition of stakes in various parts of the world shows that China aims to develop its control of the transport and the servicing of energy in the world in the next decades. This control will also allow China to influence economic and geopolitical policy in peer global corporations. However, it is also possible that the cost of transportation will rise as a result of geopolitical tensions and intensifying international competition. Figure 6.2 shows Japanese involvement in the competitive race for natural resources. Its strategic location near China, size and maturity forced Japan to maintain global alliances with the Middle East to secure its own strategic reserves.

This dilution of energy dependencies is aimed to provide a more balanced model of geopolitical influence in order to enable negotiations between economic, industrial and geopolitic interests in all regions of the

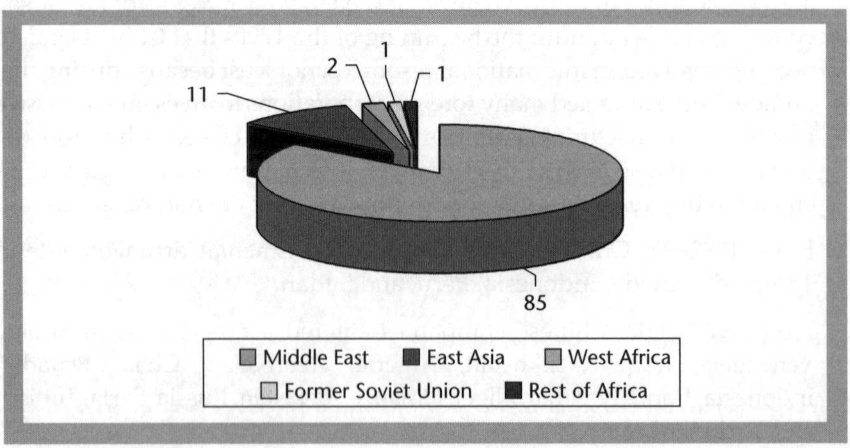

Figure 6.2 Geographic distribution of Japanese oil imports, 2006 (%)
Note: +/−2% approximations with Japanese daily imports of 5.2 million b/d
Source: International Energy Administration.

globe. This is a new form of shared influence, different to a long-term focus on a concentrated area as the US maintained towards the Middle East. The Middle Eastern region controls global oil markets through the OPEC pricing mechanism which influences financial speculation on commodities as the OPEC countries are the main investors in those corporations. For instance, OPEC countries are vested in large financial institutions and hedge funds and they alter the pricing structure of energy markets by creating pricing volatility. Financial entities (corporate banks and hedge funds) trade their portfolios in coordination with such market interferences. This is why so many financial entities became involved in the business of commodities trading.

As of 2004, about 60 per cent of global oil reserves represent 37 per cent of Chinese imports. Chinese imports are projected to represent between 20 to 30 per cent of global demand. This rise will create geopolitical dependencies between the Middle Eastern region countries and China. While OPEC controls the pricing of oil, the added Chinese demand will exacerbate volatility pressures on the oil pricing structure. Unlike China, other countries, such as the US, are not expected to change their economic needs significantly by 2010. And if there is no transparency in the economic mechanism of global oil pricing by then, further geopolitical tensions will develop. The pricing of oil between regions could exhibit much wider valuation spreads, providing ample opportunities to profit from more inefficient market arbitrage, thus creating greater insecurity in the world. APEC (2005) estimates that the Middle East could have increased production levels by 46 per cent in 2030, depending on geopolitical availability and opportunities. There are also many undiscovered markets yet to be exploited to enable a more balanced sharing of wealth and resources in Africa. Yemen, Nigeria, and neighboring regions have oil capacities underground but have not yet been infrastructurally exploited and invested. Since 2002 and the start of the global war against terrorrism, oil reserves and transportation are exacerbating geopolitical tensions – the full extent of this may not be apparent as of yet but it will certainly be real in the future. The US hegemony in Iraq created a significant change, suggesting that control of geopolitical regions through force no longer brings an economic advantage; the result has been highly disadvantageous in macroeconomic affairs, causing the price of oil to become more volatile, the US deficit to widen, and the sharing of resources with new geopolitical competitive partners more difficult to achieve. However, new partners are invited to join the combat against terrorism and to prepare for their own future needs and in this situation, developing partnerships throughout the world, as China did, provides a route to exercising control in those regions on other matters as well as oil. Partnerships are valuable means of negotiation and bringing pressure to bear in other deals and other industries, especially when it comes to merging and acquiring large corporations across various industries. This is primarily why the development of transparency in the Middle East and the pricing mechanism of oil for adequate global distribution is crucial. In addition, assuming that the United States does not change its current high levels of consumption and that new emerging countries, such as China or India or Eastern Europe, increase by a factor of 22 times their 2006 levels within the next few decades, it is also foreseeable that an exchange of currency pegging in the pricing of oil will occur in the future. Figures 6.3 and 6.4 from *National Geographic* (2004) provide more evidence about the implications of oil in the future and the partitioning of resources between various foreign entities.

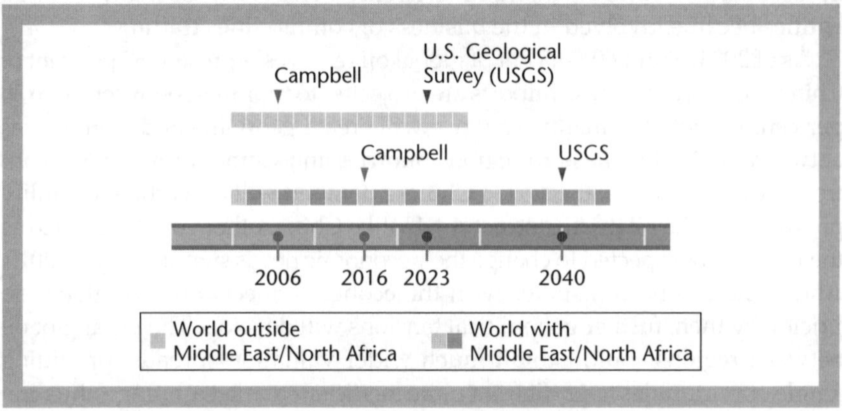

Figure 6.3 Two views of the coming peak

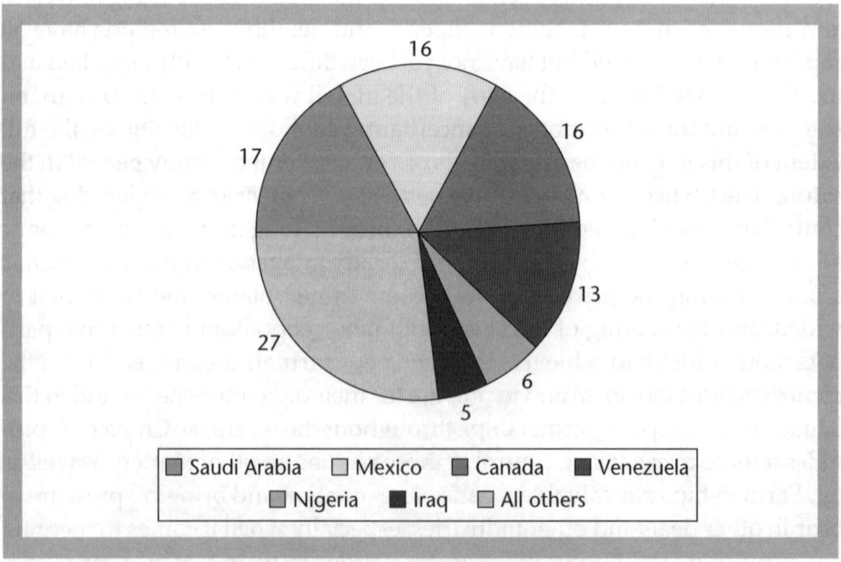

Figure 6.4 Crude oil imports, 2002 (%)
Source: 'The End of Cheap oil', *National Geographic*, June 2004.

Mature markets peg their currencies against developing economies such as China in order to maintain and protect their domestic economic and social conditions. They are less affected by the rising middle classes' potential for eroding the purchasing powers of their own middle classes. A potential reverse of foreign currency pegging would fix the dollar and immunize its system against a floating yuan that would result for Chinese society in the fluctuating inflationary costs exacerbated by the increasing consumption of its own new middle class. And to diminish the disparities between concentrated regional demands, pricing spreads weaken with the diversity of energy resource distribution in South America, Canada, the North Sea, Africa, the seas of Asia and the Middle East.

The International Energy Administration (2006) expects that Iraq will be of particular interest in the next decade, once it has developed its own multisecular governing structure. Iraq expanded its production capacity to 2.8 million barrels per day in 1999 to reach $5.2 billion in oil exports. According to the International Energy Administration (2006), Iraq planned to maintain its current oil production capacity of about 2.5 million barrels per day in 2006 despite global oil pricing volatility. It also aims to expand its production capacity more aggressively once its infrastructures are reconstructed and private corporations are allowed to negotiate partnerships and to produce more than 6 million barrels per day. The oil pricing structure is not expected to be subject to higher volatilities and would also be required to provide more democratic transparency vis-à-vis rising global competing countries. However, OPEC continues to use its political influence to alter volatility; this was especially the case in January 2007 with oil prices falling back to just below $50 a barrel or near to $78 in July 2007. The transparency of the oil pricing mechanism will occur in line with the restructuring and re-engineering of those OPEC countries currently under arbitrary autocratic dictatorships. Outside investors and private international corporations are also contributing to the economic changes in the region and in the distribution of future resources. Preliminary discussions of exploration projects have already been held with a number of potential outside investors. Dubai is an example of such economic development. In addition, significant increases in Iraqi oil exports would ease market tightness and would foster domestic economic developments. Iraq's oil production reaches 5.5 million barrels per day in 2030 in the reference case. Implementing oil infrastructures, drilling in foreign territories or controlling the construction of a 4000-kilometer oil pipeline throughout the whole of the Middle Eastern region is part of a strategic economic plan to globally balance resources and establish future leadership.

The Asian region will increase its dependency on those Middle Eastern countries that remain the heaviest oil producers. In 2003, the Asia Pacific Energy Research Center (APERC) asserted that this dependency will become increasingly significant as costs of drilling in Siberia and importing Siberian oil increase. It is also expected that a major premium in the oil pricing structure will arise as a result of more costly transportation through the regions of the Middle East, Pakistan, or India. China has not yet gained as much influence as its Western counterparts in the Middle East, although it is increasing its involvement in building relationships and creating partnerships with Saudi Arabia, Iran, and Iraq. This is increasingly important as global competition for oil rises, especially among the US, Japan, India, and the EU, according to Izraelewicz (2005).

Security of oil transportation and long routes to conduct energy supplies are all subject to higher costs as the war on terrorrism intensifies. As 75 per cent of Chinese oil imports go through the Strait of Malacca, this concentrated dependence on one route makes Chinese imports strategically vulnerable. According to Kellner (2005), the Strait of Malacca opens strategic alternatives in energy transportation routes, enabling:

- Implementation of a 250-kilometer oil pipeline along the Isthmus of Kra in Thailand from the Andaman Sea to the Gulf of Siam.
- Utilization of the Sunda Strait between the Sumatra and Java.
- Leverage of the Lombok Strait in Eastern Java through the Makassar Strait. Transportation costs could rise to more than $2 a barrel in addition to the raw barrel oil price as the transportation route is longer and subject to greater risks.
- Implementation of an oil pipeline from Myanmar to China from the port of Sittwe in the Bengal Gulf to Kunming in Yunnan Province.

According to the *China Daily* (2004), the recent intense diversification of energy development, especially within the Asian region, creates tensions between the national interests of Asian countries, such as Japan, Australia, Taiwan, and Hong Kong and the mainland policies of Thailand, Vietnam, and the Indonesian region.

Wealthier countries are also perceived as growing competitors, in particular in forming alliances with Russia, Iran, Kazakhstan, and Sudan. For instance, the Economist Intelligence Unit reports on 1 September 2006 in 'China and Japan Compete for Gas and Oil', that a visit by Japanese prime minister Koizumi to Kazakhstan and Uzbekistan underlined the significance of the economic and strategic influence of Japan in ensuring energy security and maintaining an international Japanese diplomatic profile. Japan relies on imported supplies of fuel (80 per cent) and it aims at diversifying away from the Middle East. Kazakhstan's oil reserves totaled 40 bn barrels at the end of 2004, 3.3 per cent of the world's total and second only to Russia outside the Middle East. It also had 3 trillion cubic meters of gas reserves. Uzbekistan has 100 m barrels of oil and 1.86 trillion cubic meters of gas. Both Kazakhstan, and Uzbekistan also have rich reserves of uranium, and both countries' domestic mining capabilities are backed by Japanese investments.

Sumitomo, Japan's third-largest trading firm, and Kansai Electric Power Co. (KEPCO) have signed an agreement to develop a uranium deposit with KazAtomProm, the state-owned Kazakhstan uranium monopoly. Sumitomo gets a 25 per cent stake and KEPCO 10 per cent, with the combined initial investment expected to be around US$100 million. Itochu, another Japanese trading firm, has also signed a deal with KazAtomProm. Kazakhstan could provide as much as 25 per cent of Japanese demand for uranium in a few years. In 2005, China outbid Japan to develop uranium mines in Kazakhstan. In December 2005, a US$800 million oil pipeline was opened linking Atasu in central Kazakhstan with Alashankou in the northwestern Chinese province of Xinjiang. The pipeline has a capacity of 10 million tons a year (China's total consumption last year was 300 million tons). This is expected to double by 2010. In October 2005 the state-run China National Petroleum Corporation (CNPC) bought the Canadian-owned firm PetroKazakhstan, which operates the Kumkol oilfield. Joint ventures were developed in Uzbekistan with the state-owned oil and gas monopoly, Uzbekneftegaz. China National Oil Development Corporation (CNODC) also agreed to provide geological exploration work at five hydrocarbons blocks. CNODC plans to invest US$209 million. Figure 6.5 shows cumulative strategic oil reserves. The proportional production by participating countries indicates (Figure 6.6) that there may also be more production centers outside the Middle East. However, prior to expanding internationally with neighboring markets, Chinese corporations are also attempting to maintain internal national demand and to secure domestic demand (Figure 6.7).

According to ChinaBidding of 31 July 2006, Chinese energy corporations, such as the China National Petroleum Corporation (CNPC), intend to speed up oil exploration in the Tarim basin and upgrade China's prospecting

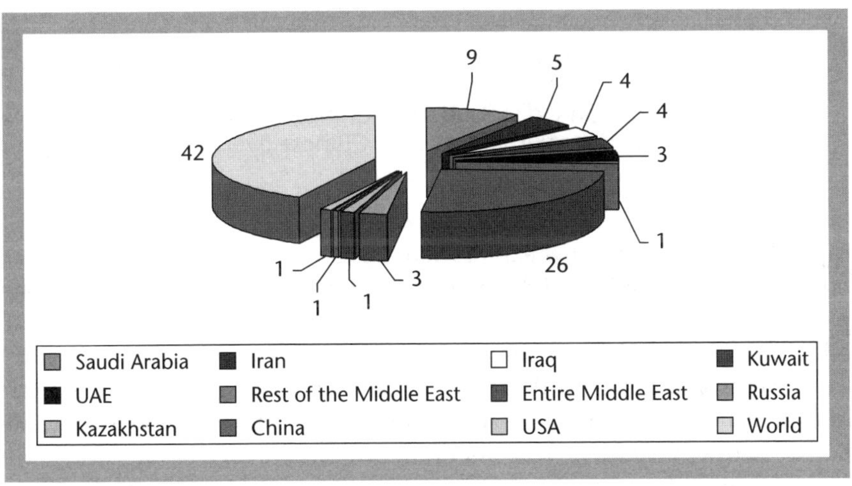

Figure 6.5 Proportion of oil reserves by countries of interest (%)

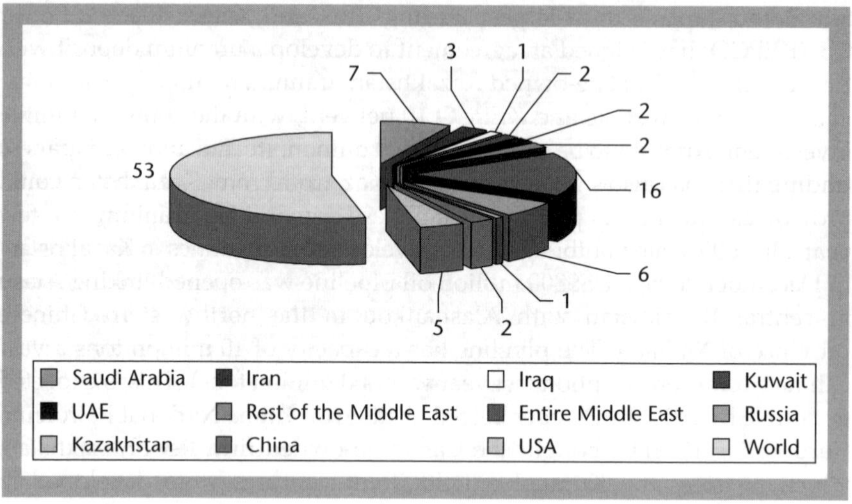

Figure 6.6 Global oil production proportion (%)

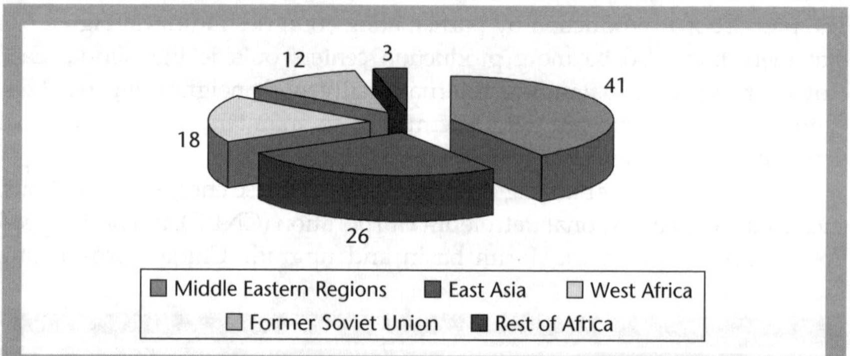

Figure 6.7 Geographic distribution of Chinese oil imports, 2006 (%)
Note: +/−2% approximations with Chinese average import of 3, 4 million b/d.

capability. About nine new exploration zones will be open to international oil firms. CNOOC initiated the first LNG terminal in Southern Guangdong province in 2005. It also plans to invest in four projects in Fujian, Shanghai, Ningbo, and Zhuhai. The plan is to base them all along the east and south-eastern coasts by about 2010. CNOOC intends to implement a second phase of the Guangdong terminal to increase productivity and capacity by 6.2 million tonnes per year (tpy). The other four projects are each intended to reach 3 million tpy from around 2010. CNOOC will need to buy about 18 million tpy of LNG to ensure the successful implementation of those infrastructures and to secure imports to South Korea, which is the world's second largest buyer of LPG. China intends to complement its energy production of oil with gas, and is planning to double its gas production share by 2010. CNOOC also faces intra-market competition with Sinopec Corporation and PetroChina. They both also intend to establish about six terminals along the northern coast. In addition to competition between the main Chinese energy companies, new private entities are also forming, with private investment, to increase LNG supplies until at least 2012, pre-sold to Japanese or Western buyers. Private Chinese entities also finance new projects for export to Australia and Qatar. Due to the increasing intramarket and international competition for supplies, China is also looking to secure renewable short-term contracts – 5–10 years rather than the usual 25–30 years – with less established suppliers with short-term availability. China sees potential in Indonesia, which hedged its domestic energy supplies with Japanese counterparties in deals that will expire in 2010, and Iran, which also aims to enter the LNG market. The Chinese National Energy Commission also intends to approve another 25-year agreement for one of China's big three state-owned energy companies to buy liquefied natural gas (LNG) from Petronas, the Malaysian state-owned energy corporation. According to the *Xinhua China Securities Journal* and the China National Offshore Oil Corporation (CNOOC), Petronas will supply the Shanghai LNG terminal at a contract price of US$5–6 per million British thermal units (MBtu) from the opening of the terminal in 2008. This particular project is designed to receive six million tons of LNG annually at what will be China's third LNG terminal in Shanghai, built in two phases. The first phase is targeted for completion by 2008 to enable the processing of three million tons of LNG and to supply four billion cubic meters of natural gas annually to the Shanghai municipality. Part of the project is financed by the Shenergy Group, a state-owned energy company based in Shanghai, which holds a 55 per cent stake, and part by CNOOC, China's largest offshore oil company, owning a minority interest with 45 per cent. Insurance and freight costs are additional price inflationary pressures. For instance, the world's largest LNG importer, Japan, increased its costs of LNG primarily as a result of increased insurance and freight, from $3.05 per MBtu in 1998 to $6.05 in 2005. International LNG prices increased primarily due to increased consumption, most particularly outside city centers, but were also affected by trading speculations, financial engineering, and insufficient guidelines with respect to risk management infrastructures in the trading of global commodities. For instance, the Shanghai supply price is expected to be higher than for CNOOC's LNG projects in Southeastern Fujian and Shenzhen city, in Guangdong province. Shenzhen prices were around $3 per MBtu as of June 2006. Fujian Province prices are negotiated with Indonesian suppliers.

Geopolitical partnerships have also been proliferating. For instance, according to ChinaBidding of 27 July 2006, China's biggest oil producer, PetroChina, will submit a feasibility study to the government to establish a liquefied natural gas (LNG) terminal in northeast China to maintain fuel supplies from Australia, Iran, and Qatar. Another ChinaBidding report of July 2006 says that PetroChina was in negotiations with more than 10 international suppliers to source fuel for three terminals, including Dalian. PetroChina has received governmental approval to contract a feasibility study to build LNG facilities in Dalian, Tangshan in Hebei Province and Rudong in Jiangsu. Apart from its Petronas deal discussed above China National Offshore also received governmental approval for two LNG terminals, one each in Guangdong and Fujian. The Guangdong venture has already received its first shipment of 60,000 tons of LNG from Karratha in Australia. BP Gas China also agreed with China National Offshore to supply 2.6 million tons of LNG annually for 25 years to its Fujian terminal, and they were waiting for final approval from the Indonesian government. Partnerships are also internationalizing, with a proliferation of commercial contracts. For instance, according to a Dow Jones report of July 2006, Shell submitted a development plan to PetroChina to help secure production in the largest gas field in southwest China's Sichuan Province with 58 billion cubic meters of proven reserves. According to ChinaBidding of 17 July 2006, Royal Dutch Shell, Europe's second-biggest oil company, and Shenhua Ningxia Coal Industry Group agreed to study the feasibility of spending as much as US$6 billion on a Chinese plant to convert coal into alternative fuels and chemicals. Shell and Shenhua Ningxia, a unit of Shenhua Group Corporation, China's biggest coal producer, will study the technical and commercial viability of building a 70,000 barrel-a-day plant in northern China's Ningxia province.

For another example, ChinaBidding reported on 6 July 2006 that China and Russia aim to build a pipeline in Altay in Xinjiang Uygur Autonomous Region to transport natural gas from Russia to the coastal areas of east China. According to the report, Altay region Communist Party secretary Li Xianglin said that China and Russia are accelerating their partnerships in the energy field. The Russian Natural Gas Industry Corporation is the country's largest firm in natural gas drilling and is allying with China for pipeline construction. The final version of the negotiation agreement had been handed over to the Xingjiang Transportation Bureau and local government for approval. The Russian Natural Gas Industry Company aims to construct two pipelines to China which would transport 68 billion cubic meters of gas from Russia to China each year. The western part of the pipeline will go through west Siberia in Russia to Altay, and then connect to the west-to-east natural gas transmission pipeline, which stretches along the coastal areas of China. This project aims to support the economic region of Xinjiang. The western part of the pipeline represents the main transmission line, as the natural gas fields in west Siberia were close to the existing natural gas infrastructure. Most newly constructed pipelines are branched from these main 'arteries' to feed the domestic regions and rural areas whose infrastructures are yet to be improved. It is expected that the western pipeline will transport 30 billion cubic meters of gas every year. ChinaBidding also reports that Chinese policy-makers have mapped out China's natural gas pipeline network, which is to be implemented in the next 20 years. Natural gas, as cheaper than oil and the most easily available alternative source of energy, is seen as the best solution to alleviate petroleum supplies and inflation on global prices. The Chinese natural gas framework began construction in the 1990s with the Shaanxi-Beijing pipeline and west–east gas pipeline that runs from Lunnan Oilfield in the Tarim Basin in Northwest China's Xinjiang Uygur Autonomous Region via eight provinces and regions to Shanghai. The Zhongwu line transports gas out of Sichuan Province with a second Shaanxi-Beijing pipeline.

According to research at the University of Petroleum in China, China is now partnering with Russia and Kazakhstan to build a basic framework of gas transportation pipelines. In 2006, China's natural gas pipelines measured a total of 24,000 kilometers. According to research at the Guangdong Oil and Gas Association, this will increase to 36,000 kilometers by the end of 2010. The Petroleum Reserve and Transportation Committee of the China Petroleum Society also reports that China's natural gas pipeline industry is undergoing major investment and re-engineering to compensate for rising oil demand and energy needs anticipated until at least 2020, and in order to alleviate potential inflationary pressures on global oil prices which could reduce global competitors' average purchasing powers and contribute to more expensive credit costs. The new Chinese middle class market is expected to mature around 2025, at which time they would probably share the same life style as their Western counterparts. Natural gas and oil are both in demand and China is expected to experience some natural gas shortage by 2020, perhaps by around 90 billion cubic meters. China expects to compensate for these shortages with rising imports of liquefied natural gas (LNG) and natural gas via land pipelines according to research at China National Petroleum Corporation (CNPC)'s Natural Gas and Pipeline Branch Company. As a result of this and projected demands from China's new middle classes, China has already started several LNG import projects in association with Australia, one of the main regional exporters. In mid-2006, 11 coastal provinces, municipalities and autonomous regions had participated in the construction of large-scale LNG import projects to feed the region of Guangdong. In addition, about 12 new LNG projects have been proposed to the Chinese government to expand resource reserves.

Oilfields such as Daqing, which accounts for 33 per cent of China's oil production will become depleted, and the government is pinning oil hopes on vast virgin fields in Xinjiang. Xinjiang sits on 30 per cent of the nation's total oil reserves and 34 per cent of its natural gas resources, and the government's energy plan is to further expedite oil exploration in areas such as Karamay, Turpan Basin and Tarim Basin. *China Daily* reported that the oil and gas industry accounted for 60 per cent of the incremental industrial value of Xinjiang as of 2005. For the previous two years Xinjiang had maintained a gross domestic product growth of more than 10 per cent annually, representing a level beyond the national average.

CNOOC is increasing its partnerships primarily because of rising global energy prices but it seems unaware that a major source of inflationary pressure is new forms of financial engineering speculation and disguised trading gambling. According to the EIA's historical pricing time series, LNG contract prices approximately doubled between the end of 2001 and the end of 2006. Natural gas and oil are the two primary natural resources that are exploited by financial engineering speculators and hedge funds and whose rising prices are the result of speculative trading by commodities trading advisors and commodities pool operators. The underlying prices of commodities are volatile and can be exploited to magnify the profits of financial institutions. The equivalent of those speculative financial instruments are now active in China and are starting to trade with their Western counterparts, mostly since November 2005.

China cooperates with Russia and Central Asia to remediate and balance its rising demand. In October 2006, Russia also entered deals with European counterparts to hedge demand and distribute natural resources equitably. In March 2006, China and Russia signed a commercial contract to transport natural gas between Russia and China. From 2011, Russia was then expected to be the first and largest global natural gas exporter to supply more than 60 to 80 billion cubic meters annually to China. Exports will be transported in two ways: the west pipeline will enter China's Xinjiang province through Siberia and Altai and will connect with the west–east gas pipeline to supply natural gas to China's coastal areas. The east pipeline is also expected to transport natural gas from Kovykta or Sakhalin and Chayandinskoye gasfields located in Yakut. Russia and two of China's major oil companies, CNPC and Sinopec, are also scheduled to build the 3000-kilometer west pipeline. This is expected to transport gas within five years and to supply 30 to 40 billion cubic meters of natural gas to China per year. The projects are significant and intended to ramify the east and west markets. Turkmenistan will supply China with 30 billion cubic meters of natural gas per year, starting in 2009, under an agreement signed in April 2006 between China and Turkmenistan during President Saparmurat Niyazov's visit to China. Under the 30-year contract, imports from Turkmenistan will begin two years earlier than from Russia under the March 2006 contract. The contract indicates that the China-Turkmenistan natural gas pipeline will be completed by 2009. Because China and Turkmenistan do not share any borders, financial and logistical arrangements had to be agreed to transport the natural gas through Uzbekistan and several Kazakhstani territories before finally connecting with the west–east gas pipeline. Uzbekistan and Kazakhstan also have rich natural resources, especially in natural gas. Both aim to export to China to alleviate the rising oil inflationary trends affecting global economic conditions and their peers' average purchasing powers. Once the China-Turkmenistan pipeline is constructed, all three Central Asian countries will be able to supply natural gas to China simultaneously.

According to the *China Daily* of 11 November 2002, the Lanzhou-Chengdu-Chongqing oil product transportation pipeline is the country's longest oil pipeline, with the largest diameter and highest pressure. The PetroChina branch in southwest China's Sichuan Province announced its completion at the end of 2002. This pipeline is part of 12 major projects that form a strategic plan to 'go West' to connect with more western Asian neighboring countries and on to Eastern Europe. This project aimed to forestall population migrations from western areas to eastern urban centers that are already saturated as a result of rising economic growth, thereby increasing economic distortion inside China. The balanced distribution of economic and energy resources to the western regions aims to promote further developments in those 'desertified' areas. Oil consumption in Sichuan and Chongqing is 4.4 million tons a year and more than 90 per cent of the oil is still transported through old infrastructures that are inefficient, insecure, and inadequate to cope with increasing demand. The transportation grids in China are currently being overhauled and modernized to adjust to new rising consumption needs. In addition to ensuring a steady oil supply, PetroChina also reports that the main 'go West' pipeline could save up to 150 million yuan or $18 million per year in transportation fees.

In 2006, PetroChina invested almost 4 billion yuan or $483 million in the 'go West' pipeline project. The annual transportation capacity is more than 5 million tons and the 1250-kilometer pipeline starts in Xigu District in Lanzhou, passes through 40 counties, cities and districts in Gansu, Shaanxi and Sichuan provinces and the Chongqing Municipality and ends in the Dadukou District in Chongqing. It traverses diverse topography, from Loess Plateau, Mount Qinling, the Daba Mountains to the Chengdu Plain. The pipeline is equipped with ultrasonic waves to differentiate various energy goods and is the most sophisticated advanced technology in China.

In 2004, Russia initiated in cooperation with China the Tayshet-Skovorodino Nakhodka oil pipeline that runs from Eastern Siberia to Perevoznaya Bay in the Pacific, including a stretch of over 2300 kilometers along the Angarsk-Daqing route in China. The route runs from the south of Lake Baikal to the main Chinese distribution center in the northeastern province of Heilongjiang. According to Kenny (2004), China National Petroleum and Chemical Corporation (CNPC) and the Russian firm Yukos entered an agreement in 2003 against competition from a Japanese firm. Finally, the Russian firm Transneft won the contract to construct the Tayshet Nakhodka pipeline to connect to the Chinese domestic grid. To compensate China for the loss of their stake in this, Russia agreed in July 2005 to offer Beijing priority participation in the establishment of railway transportation to deliver oil from Skovorodino to the Pacific, scheduled to finish by 2010, and a pipeline from Skovorodino to Daqing, due to be effective by 2020.

However, pipelines through insecure eastern regions are subject to geopolitical risks. For instance, since 2002, American forces reallocated to Kyrgyzstan and Uzbekistan have been considered a potential risk to the Atasu-Alanshankou pipeline between Kazakhstan and Xinjiang. This pipeline also connects the eastern Chinese coast to the Caspian Sea through pipes from Charjou in Turkmenistan. Additional geopolitical risks and archaic infrastructures make development and construction more expensive in these regions. Further, many geopolitical players are increasing their efforts to secure new resources as transportation costs and prices are rising.

According to Kellner (2005), in 2006, Russian and Kazakh pipelines were conducting about 30 million tons of natural gas each year. This strategic infrastructure is aimed to alleviate petroleum dependence on the Middle East, which provides support half of the planet's demand.

At the same time, China is investing heavily in significant drilling projects in the neighboring seas and is developing deep water ports in Myanmar to access the Andaman Sea, on the southwestern coast of Pakistan in Gwadar, and on the Sea of Oman near the Strait of Hormuz to control crude oil imports from the Middle East and Africa.

With regard to international projects, some of them aim only to transport energy while others aim to produce and drill to obtain energy in response to growing demand. According to APEC (2005) and KPMG (2005), the following projects primarily aim to improve energy distribution:

- Gas pipeline from Irkutsk to Korea and China with Russia also distributing oil from Irkutsk as of 2010.
- Natural gas pipelines aimed to distribute energy to the Chinese province of Shenyang from Sakhalin.
- Liquefied natural gas (LNG) import terminals servicing the southern provinces of Guangdong and Fujian to receive LNG imports from Australia and Indonesia from 2007.
- Oil pipeline from Kazakhstan to the Xinjiang region.
- Oil pipeline transporting oil from Skovorodino to Daqing.
- West–east power transmission grid to transport electric energy produced from coal and hydroelectric power, effective as of 2020.
- West–eastern pipeline transporting natural gas from Sichuan, Xinjiang and Shaanxi to Beijing and eastern and northeastern cities.
- West–eastern oil pipelines transporting oils through Shanshan, Lanzhou and Gansu.

Other international partnerships include that of Sinopec Corporation in Teheran in November 2004 to explore and develop the Yadavaran Oil field in Iran. The Chinese government also became involved in Caracas, Venezuela, in December 2004, in a partnership to operate in fifteen oil drillings in the Eastern Venezuela region. This alliance was reinforced with more partnerships in 2005 and 2006. According to ChinaBidding on 19 July 2006, Venezuela plans to enter into a joint venture with China to manufacture oil drills instead of buying them from American companies. ChinaBidding also disclosed that President Chavez and the Chinese president planned to diversify geopolitical partnerships to reduce concentrations in areas of geopolitical risk. Since the beginning of this century, China has worked at multiplying alliances and at securing oil reserves for its growing domestic demand. The Chinese government thus also helps in the financing of the construction of manufactories, and factories to build oil drills. Venezuela is the world's fifth-largest oil exporter and is used primarily to support the United States. However, since 2003, states have made an effort at diversifying risk concentration and at reducing concentrated geopolitical economic dependencies. This is also to avoid specific states taking oil for granted and to force global citizens to be more efficient in their oil consumption. China offers a proactive example. The Chinese invented cheap low fuel-consumption cars, which have been available globally since 2006. At less than $5000 they are accessible to the general public. They are much smaller than other cars and run on hydrogen. Chavez also aims to alleviate one-way oil dependency with the United States, as relations between the countries have deteriorated since the beginning of this century. As the result of new global and geopolitical alliances with China, Venezuela agreed to sell 160,000 barrels a day of crude and diesel to China, with the aim of doubling that by the end of 2006.

Furthermore, in August 2005, according to the *International Herald Tribune* of 23 August 2005, in an article entitled 'Chinese Beat India for Kazakh Oil Fields', the CNPC took over the Canadian oil company PetroKazakhstan for $4.18 billion, superseding the Indian offer by ONGC Mittal Group. The *Straits Times* of 7 January 2005 also reported that China and other Asian countries were attempting to gain hegemony in Indonesian and Malaysian oil and gas and with Singapore refining capacity. Berger (2004) notes that Chinese firms relied on foreign companies in sharing resources in border areas, such as Japan in Diaoyu and Senkaku islands, and Vietnam in Nansha, Spratly, Xisha and Paracel islands.

Major energy corporations are also partnering to share intelligence, means and resources. According to ChinaBidding on 18 July 2006, China's biggest oil firm PetroChina will invite major foreign oil companies to help the state-owned company explore for oil and gas in northwest China's Xinjiang Uygur Autonomous Region. The new projects are mainly distributed throughout the southwestern, central and eastern parts of the Tarim Basin, which has total proven reserves of 6 billion tons of oil and 8 trillion cubic metres of natural gas.

This private capitalistic initiative marks the most significant cooperation with foreign giants in oil and gas exploration in the past 12 years. PetroChina reveals that more gas discoveries in the northwestern region will increase energy supplies to the 3800-kilometer natural gas pipeline which runs from Xinjiang in the west to coastal cities in the east. PetroChina intends to partner with several major international oil companies through product sharing contracts to share drilling technologies, alternative energy intelligence processes, and natural resources. The foreign listed companies also reported that they would enlist state-of-the-art technology worldwide to tap the huge energy potential in the northwestern region. According to ChinaBidding, the foreign oil companies dealing with the main Chinese entities in those geopolitical plans are Exxon Mobil, Total, Chevron, and Royal Dutch Shell. The intelligence report also reveals that over the past 20 years, PetroChina and its parent firm CNPC teamed up with as many as 46 overseas oil companies, including Total, ConocoPhillips and Shell, to work on 16 oil and gas projects in China and to strategically alleviate resource shortage. These joint efforts produced 22 million tons of oil and 2.5 billion cubic metres of natural gas as of 2006.

6.2 THE NEW WORLD ORDER DOES NOT COME FREE OF POLITICAL TENSION

The creation of international partnerships and the prosperity of these partnerships will partially depend on the economic stability of global oil pricing and the main variables impacting this balance. The pricing of oil depends on supply, demand, technology, alternatives and geopolical tensions. But pricing disparities and various spreads will also depend on transportation costs and security issues that could substantially change the pricing structure of future oil curves due to geopolitical concerns. Since 2003 and the entry of China to the World Trade Organization, the new world order has also provoked new tensions and suspicions among players. For instance, some American analysts have suggested that the proliferation of energy partnerships may also be used as weapons of exchange for military and arms deals or used as political levers and to provide possible veto influences in international negotiations far into the future.

The creation of these new axes of economic dependencies also increases the turmoil within established international institutions, such as the United Nations, which are too centralized and bureaucratic to monitor the proliferation and continuity of partnerships. Its role has changed in a period of private globalization evolution, and diminished in the face of rising private corporate controls. More independent research centers are being created to monitor more objectively the corporate partnerships between national players in various strategic corners of the world, substituting for the role of the centralized, old-fashioned and mature United Nations.

Herberg (2005) considers the negative connotations of China's rising power in the face of its neighbors, and the National Bureau of Asian Research in Seattle also looks at the rising tensions resulting from China's increasing control of energy in the Asia Pacific region. Articles such as 'China Gorging and Japan-China Resource and Energy Conflicts' (Shimbum, 2005) and 'Caspian Oil Heading East' (Sukhanov, 2005), discuss the ways in which China's rising power and correlated energy needs contribute to added tensions between Russia, China and Japan. China once suspended exports from Daqing to Japan and Russia is open to preferential usage of the Taishet-Nakhodka oil pipeline. Despite the fact that regional zones are currently mostly experiencing cooperative partnerships and alliances in relation to resources, it is nonetheless anticipated that the demands of the new Chinese middle class will exacerbate tensions among neighboring countries which are experiencing diminishing resources. In the eastern Chinese Sea, resources are already starting to become a focus of disputes. For instance, the Chun Xiao, Dunquiao and Tianwaitan offshore gas fields are operated by China. Yet, Japan claimed partial ownership due to its geographical proximity to the median line. Japan considers the median line as a border to an exclusive economic zone while China claims the entire ownership of soil and oil drilling operations.

Harrison (2005) also notes that the Senkaku or Diaoyu islands, which generate 95 billion barrels of oil per year are the subject of dispute between Japan and China. The sharing of oil production centers in the region goes as follows: the South China Sea and the Gulf of Tonkin are shared by Vietnam and other East Asian countries and territories (although in 2003, China and the Philippines reached a partnership agreement for joint explorations in the South China Sea, which created tensions with Vietnam); China, Vietnam, and Taiwan claim shared interest stakes in the Paracel Islands and the Spratly Islands (Table 6.1). The Philippines, Brunei, Indonesia and Malaysia also own partial minority stakes in these production centers.

PetroChina is involved in the drilling of the South China Sea on behalf of the Chinese government. The Energy Information Administration provides more data about the region's geopolitical tensions due to the sharing of resources (tables 6.2 and 6.3).

As we have seen, China has come a long way in diversifying its strategic resources to be able to respond to its future growing energy demand and to alleviate geopolitical tensions. It has also come a long way in the creation of partnerships with international players to integrate the global mechanism of oil pricing. However, these partnerships have mainly been created by private corporations, principally the three main Chinese energy corporations, CNOOC, Sinopec and PetroChina. The main challenge remains how to integrate Chinese government policies with the corporations. The Chinese government is still highly centralized and aims to keep the structure of corporations vertical, with all reporting eventually to the central government of China. Yet, with the latest multiplication of international partnerships, despite their minority stakes, it is foreseeable that China will experience a breakdown of authority or its centralized leadership due to the infiltration of foreign entities into the fabric of the Chinese system. Thus, the policy framework that we are about to consider will be subject to overhaul and re-engineering in the decades to come depending on the levels of involvements of foreign entities in Chinese private corporations and in the influencing of the oil pricing structure.

Table 6.1 Territorial claims in the Spratly and Paracel islands

Country	Claim
Brunei	Does not occupy any of the islands, but claims part of the South China Sea nearest to it as part of its continental shelf and exclusive economic zone (EEZ). The boundary lines are drawn perpendicularly from two outermost points on the Brunei coastline. In 1984, Brunei declared an EEZ that includes Louisa Reef.
China	Refers to the Spratly Islands as the Nansha Islands, and claims all of the islands and most of the South China Sea for historical reasons. These claims are not marked by coordinates or otherwise clearly defined. China also claims the Paracel Islands (referred to as the Xisha Islands), and includes them as part of its Hainan Island province. Chinese claims are based on a number of historical events, including the naval expeditions to the Spratly Islands by the Han Dynasty in AD 110 and the Ming Dynasty from 1403–33. Chinese fishermen and merchants have worked the region over time, and China is using archaeological evidence to bolster its claims of sovereignty. In the 19th and early 20th centuries, China asserted claims to the Spratly and Paracel islands. During World War II, the islands were claimed by the Japanese. In 1947, China produced a map with nine undefined dotted lines, and claimed all of the islands within those lines. A 1992 Chinese law restated its claims in the region. China has occupied eight islands to enforce its claims. In 1974, China seized the Paracel Islands from Vietnam.
Indonesia	Not a claimant to any of the Spratly Islands. However, Chinese and Taiwanese claims in the South China Sea may extend into Indonesia's EEZ and continental shelf, including Indonesia's Natuna gas field.
Malaysia	Its Spratly claims are based upon the continental shelf principle, and have clearly defined coordinates. Malaysia has occupied three islands that it considers to be within its continental shelf. Malaysia has tried to build up one atoll by bringing soil from the mainland and has built a hotel.
Philippines	Its Spratly claims have clearly defined coordinates, based both upon the proximity principle as well as on the explorations of a Philippine explorer in 1956. In 1971, the Philippines officially claimed eight islands that it

(Continued)

Table 6.1 (Continued)

Country	Claim
	refers to as the Kalayaan, partly on the basis of this exploration, arguing that the islands: (1) were not part of the Spratly Islands; and (2) had not belonged to anyone and were open to being claimed. In 1972, they were designated as part of Palawan Province, and have been occupied.
Taiwan	Taiwan's claims are similar to those of China, and are based upon the same principles. As with China, Taiwan's claims are also not clearly defined. Occupies Pratas Island in the Spratlys.
Vietnam	Vietnamese claims are based on history and the continental shelf principle. Vietnam claims the entire Spratly Islands (Truong Sa in Vietnamese) as an offshore district of the province of Khanh Hoa. Vietnamese claims also cover an extensive area of the South China Sea, although they are not clearly defined. In addition, Vietnam claims the Paracel Islands (the Hoang Sa in Vietnamese), although they were seized by the Chinese in 1974. The Vietnamese have followed the Chinese example of using archaeological evidence to bolster sovereignty claims. In the 1930s, France claimed the Spratly and Paracel Islands on behalf of its then-colony Vietnam. Vietnam has since occupied 20 of the Spratly Islands to enforce its claims.

Note: The South China Sea is defined by the International Hydrographic Bureau as the body of water stretching in a southwest to northeast direction, whose southern border is 3° south latitude between South Sumatra and Kalimantan (Karimata Straits), and whose northern border is the Strait of Taiwan from the northern tip of Taiwan to the Fujian coast of China.

Source: Energy Information Administration (2003).

Table 6.2 Disputes over drilling and exploration in the South China Sea

Date	Countries	Disputes
1992	China, Vietnam	In May, China signed a contract with US firm Crestone to explore for oil near the Spratly Islands in an area that Vietnam says is located on its continental shelf, over 600 miles south of China's Hainan Island. In September, Vietnam accused China of drilling for oil in Vietnamese waters in the Gulf of Tonkin.
1993	China, Vietnam	In May, Vietnam accused a Chinese seismic survey ship of interfering with British Petroleum's exploration work in Vietnamese waters. The Chinese ship left Vietnamese block 06 following the appearance of two Vietnamese naval ships.

(*Continued*)

Table 6.2 (Continued)

Date	Countries	Disputes
1993	China, Vietnam	In December, Vietnam demanded that Crestone cancel offshore oil development in nearby waters.
1994	China, Vietnam	Crestone joined with a Chinese partner to explore China's Wan' Bei-21 (WAB-21) block. Vietnam protested that the exploration was in Vietnamese waters in their blocks 133, 134, and 135. China offered to split Wan' Bei production with Vietnam, as long as China retained all sovereignty.
1994	China, Vietnam	In August, Vietnamese gunboats forced a a Chinese exploration ship to leave an oilfield in a region claimed by the Vietnamese.
1996	China, Vietnam	In April, Vietnam leased exploration blocks to US firm Conoco, and ruled out cooperation with US oil firms that signed Chinese exploration contracts in disputed waters. Vietnamese blocks 133 and 134 cover half the zone leased to Crestone by China. China protested, and reaffirmed a national law claiming the South China Sea as its own in May.
1997	China, Vietnam	In March, Vietnamese issued a protest after drilling by the Chinese Kantan-3 oil rig near Spratly Islands in March. The drilling occurred offshore Da Nang, in an area Vietnam calls Block 113. The block is located 64 nautical miles off Chan May Cape in Vietnam, and 71 nautical miles off China's Hainan Island. The diplomatic protests were followed by the departure of the Chinese rig.
1997	China, Vietnam	In December, Vietnamese protested after Exploration Ship No. 8 and two supply ships entered the Wan' Bei exploration block. All three vessels were escorted away by the Vietnamese navy.
1998	China, Vietnam	In September, Vietnamese protested after a Chinese report stated that Crestone and China were continuing their survey of the Spratly Islands and the Tu Chinh region (Wan' Bei in Chinese). (The dispute over this area was resolved by an agreement between China and Vietnam concluded in December 2000.)
2003	Malaysia, Brunei	In May 2003, a patrol boat from Brunei acted to prevent TotalFinaElf from undertaking exploration activities in an area offshore from Northern Borneo disputed by the two countries.

Source: Energy Information Administration (2003).

Table 6.3 Military clashes in the South China Sea, 1974–99

Date	Countries	Military action
1974	China, Vietnam	Chinese seized the Paracel Islands from Vietnam, with 18 of its troops killed in clashes on one of the islands.
1988	China, Vietnam	Chinese and Vietnamese navies clashed at Johnson Reef in the Spratly Islands. Several Vietnamese boats were sunk and over 70 sailors killed.
1992	China, Vietnam	Vietnam accused China of landing troops on Da Luc Reef. China seized almost 20 Vietnamese cargo ships transporting goods from Hong Kong from June–September.
1994	China, Vietnam	China and Vietnam had naval confrontations within Vietnam's internationally recognized territorial waters over Vietnam's Tu Chinh oil exploration blocks 133, 134, and 135. Chinese claim the area as part of their Wan' Bei-21 (WAB-21) block.
1995	China, Philippines	China occupied Philippine-claimed Mischief Reef. Philippine military evicted the Chinese in March and destroyed Chinese markers.
1995	Taiwan, Vietnam	Taiwanese artillery fired on a Vietnamese supply ship.
1996	China, Philippines	In January, Chinese vessels engaged in a 90-minute gun battle with a Philippine navy gunboat near Capones Island.
1997	China, Philippines	The Philippine navy ordered a Chinese speedboat and two fishing boats to leave Scarborough Shoal in April; the Philippine navy later removed Chinese markers and raised its flag. China sent three warships to survey Philippine-occupied Panata and Kota Islands
1998	Philippines, Vietnam	In January, Vietnamese soldiers fired on a Philippine fishing boat near Tennent (Pigeon) Reef.
1999	China, Philippines	In May, a Chinese fishing boat was sunk in a collision with a Philippine warship. In July, another Chinese fishing boat was sunk in a collision with a Philippine warship.
1999	China, Philippines	In May, Chinese warships were accused of harassing a Philippine navy vessel after it ran aground near the Spratly Islands.
1999	Philippines, Vietnam	In October, Vietnamese troops fired upon a Philippine air force plane on reconnaissance in the Spratly Islands.
1999	Malaysia, Philippines	In October, Philippine defense sources reported that two Malaysian fighter planes and two Philippine air force surveillance planes nearly engaged over a Malaysian-occupied reef in the Spratly Islands. The Malaysian Defense Ministry stated that it was not a stand-off.

Source: Energy Information Administration (2003).

CHAPTER 7

Government Policies and Developments in Energy and Commodities

7.1 HISTORY AND CHRONOLOGY OF GOVERNMENT ENERGY POLICIES IN CHINA

According to Yao et al. (2005), China's energy policies have three levels or may be classified into three various categories. The first level of policies provides general guidance and directions; the second level deals with objectives and develops plans and strategic structures; and the third consists in producing specific rules and guidelines, for instance non-compliance penalties. Most Chinese policy plans are defined by time-scales, every five or ten years, depending on the strategies and budgetary conditions set by the Chinese government. Yao et al. report that general energy policies center on a five-year forecasting plan. According to the NREL (2004) and Yao et al. (2005), the Chinese government began to be concerned about the country's environmental pollution emissions around 1978. Beginning in the 1980s, the sixth, seventh, eighth, ninth, tenth and eleventh five-year plans demonstrated a growing attention and awareness to Chinese national energy policy. The sixth five-year program, from 1981 to 1985, required that 10 per cent of energy supply profits are reinvested in reserves for energy conservation projects and future growing need. This policy continued through the seventh five-year plan, though the percentage was decreased to 8 per cent. In the ninth five-year plan, from 1996 to 2000, China summarized its Agenda 21 response, encouraging the development of renewable energy production, savings, and efficiency as fundamental state policy. According to Yao et al. (2005), this ninth five-year program enabled the National Environmental Protection Agency (NEPA) to set up a long term 'Green Project Plan'. The tenth five-year policy program, from 2001 to 2005, also emphasized renewables and introduced a plan for sustainable development, consolidated with the enactment of the February 2005 law. This policy program also developed the tenth five-year plan for renewable energy commercialization development. NREL (2004) asserts that this energy regulation program promotes renewable and alternative substitutes. Energy using coal, solar, wind, geothermal, and hydro sources should be enhanced to meet rising demand and offset the inflationary pressure of oil prices. Zhao Jiarong (1998) provided ample information on renewable energies in 'Economic Incentives Policy Study for China Renewable Energy Development'. More information about Chinese policy development is available in 'China National Energy Strategy and Policy 2020', Zhenmin et al. (2005), who are based at the Development Research Center at the State Council of the People's Republic of China. Andrews-Speed (2004) also considers similar policies in *Energy Policy and Regulation in the People's Republic of China*.

Yao et al. (2005) provide a chronology of the evolution of energy policies in China. First level policies include:

- 1983: Reinforcing the development of rural energy.
- 1992: China's Agenda 21 response and the release of the ten strategies for China's environment and development.
- 1995: State Science and Technology Commission (SSTC) Blue paper No. 4 'China Energy Technology Policy'; Outline on New and Renewable Energy Development in China, State Planning Commission (SPC), SSTC, State Economic and Trade Commission (SETC); Electric Power Law.
- 1996: Guidelines for the ninth five-year plan and 2010 long-term objectives on economic and social development of China; state energy technology policy.
- 1997: Energy Saving Law.
- 2003: Renewable Energy Promotion Law.

The second level of policies includes:

- 1994: Wind program produced by SPC.
- 1995: New and ongoing energy development projects in mainland China from 1996–2010 sponsored by the State Science and Technology Commission (SSTC), State Power Corporation, and the State Economic and Trade Commission (SETC).

- 1996: Ninth five-year plan and 2010 plan for energy conservation and new energy development by the state power corporation; ninth five-year plan for industrialization of new and renewable energy by SETC.
- 1998: Incentive policies for renewable energy technology by State Development and Planning Commission SDPC and Ministry of Science and Technology (MOST).
- 2001: Tenth five-year plan for new and renewable energy commercialization development by SETC.
- 2003: Rural energy development plan to 2020 for western areas.

The third level of policies includes:

- 1997: Circular from the communication and energy department of SPC on the provisional regulations on the management of new energy capital construction projects.
- 1999: Circular from MOST and SDPC on further supporting the development of renewable energy.
- 2001: Adjustment of value added tax for some resource-intensive products by Ministry of Finance (MOF) and State Tax Administration; electricity facility construction in non-electrified townships in the Western Provinces of China with a coordination program by SDPC and MOF.

With respect to trading and contractual agreements, the Chinese government now lets private trading entities generate their own legal bidding with independent compliance for independent firms. The legal aspect of the Chinese energy markets has become increasingly open to foreign partners and trading firms since 2005 and most legal contracts are performed by international counterparty law firms to ensure independence and to monitor objectively the terms of the contractual agreements. Yet, in the case of rising international competition and enforceability of rights, it is becoming more and more difficult for independent law firms to rule objectively. In 2003, China National Aviation Corporation's Singapore (CNAC) subsidiary lost around $500 in oil trading. This raises issues regarding control in Chinese oil trading, and the subsequent disowning of the contracts by the parent company raises issue regarding the enforceability of legal contracts. These issues follow on from a case involving Lehman's losses in metal trading with a Chinese entity in the late 1990s. Chinese energy markets are becoming increasingly open and the interaction with international independent law firms is also gaining importance in commercial contractual agreements and rulings. This is all part of the open market strategy that the Chinese market is adopting.

On 17 April 2005, Doug Ogden, director of the China Sustainable Energy Program in Beijing and San Francisco and a policy activist warning about global warming, released the 'China Sustainable Energy Program' policy document. China's National People's Congress adopted a Renewable Energy Law on 28 February 2005 to improve and accelerate domestic energy infrastructures and policies. The new policies require utilities to purchase 100 per cent of new substitute renewable energy available – using hydro (under 25 Megawatts), wind, solar, geothermal, and biomass. This compliance created a national fund to promote renewable energy developments with discounted lending and tax preferences for renewable energies. The usage of renewables is also encouraged through tariff reductions, special discounts and tax deductions. In 'Renewable Energy Policy Options for China', Wiser et al. (2002) report on the implications of tariffs, taxes, and hidden costs in the development of renewables. In addition to such advantages, some investments promoting the developments of alternative energy are interest-free, and lending discounts favor advanced technologies towards renewables. The National Development and Reform Commission (NDRC) developed renewable energy programs at discounted tax rates or marginal costs to spread the national grid and balance energy consumption in rural and urban areas. NDRC is in charge of developing new solutions and regulations to ensure the secure and safe development of energy provision for the rising middle class. NDRC developed a renewable energy plan that includes a national electricity target of 10 per cent renewables by 2020, up from 3 per cent in 2005. NDRC approved the third batch of wind concession projects totaling 450 megawatts, a $400 million investment in new wind energy technologies. The eleventh five-year plan, from 2006 to 2010, supports the implementation of the Renewable Energy Law starting 1 January 2006. This policy overhaul is part of the wider overhaul of energy policy to renew and complement the usage of oil and natural gas, and is a major step in the history of Chinese energy regulation. The Chinese Center for Renewable Energy Development (CRED) aims to meet short-term energy needs with long-term clean energy objectives through the development of a widespread national grid and via purchase obligations according to the IFC (2006). The Renewable Energy Law announced a goal of producing 60 GW or 10 per cent of power supply with renewables by 2020 according to Jing (2005). The *China Daily* (2005) reported that it would raise by 15 per cent the goal set in March 2006.

NDRC represents the main regulatory body for national energy development. It also fosters growth and development through tax programs and incentives including a tax break of 50 per cent for solar, wind, and biomass energy investments, according to Jing (2005). The Renewable Energy Law drastically changed traditional policy-making, providing a new trajectory for future energy policy towards a more diversified program of energy substitution and diversification. The NDRC is the equivalent of a ministry of energy, remaining inaccessible to the general public and lacking transparency. Most of the information about energy policy comes from private research centers and institutes.

7.2 IMPLICATIONS FOR GOVERNMENT IN ENERGY POLICIES

Energy policy-making in China is not as centralized as peer governmental entities. It lacks a ministry and is instead governed by the NDRC relying mainly on research from a variety of private research centers. Although there have been calls for China to establish an energy ministry, as yet what centralization of policy that exists comes from research centers and cooperative studies. Meanwhile, a variety of energy strategy plans have been formulated.

In 2004 the Development Research Center of China's State Council concluded the National Energy and Strategy Report (NESP). The China Energy Group's China National Energy Strategy and Policy 2020 (NESP 2020) recommended new ways in which China's government might link high-level policy goals. The report includes energy security, economic growth, equity, and specific ideas for changes to improve energy investment, supply, and efficiency. The report prioritizes energy policy goals, reviews the role of the state in achieving them, and provides specific recommendations for key energy subsectors. Energy efficiency at the beginning of this new century is central to ensuring pricing stability and to meeting demands, as well as providing all these benefits within environmental pollution limits. To achieve these efficiency goals within a more-sustainable energy supply mix, the Chinese government and energy agencies focus on objectives that finance energy policies via low-cost loans, tariffs, taxes, educational advocacy through private and governmental programs, and regulation of market activities. NESP's goals also aim to integrate institutional and corporate restructuring to establish a new potential ministry of energy (MOE).

This entity would aggregate and monitor all policies and set future energy pricing to balance supply and demand and in response to international partners. The entity would be intended to set market efficiencies at the Chinese commodities levels, thus also connecting Chinese futures and commodities exchanges to ensure the development of trading in coordination with international commodities exchanges. In fact, it is even more important to have a governmental agency overseeing this particular segment of the market because commodities are, in effect, the trading product par excellence, and oversight is needed so as not to reflect its 'tangible' value via added speculations on undeliverable products. The suggested ministry (currently unknown to the general public and internet research engines) would act as an independent entity to monitor large state-owned energy supply companies, resolve monopolistic situations, and manage disputes between international corporations about resource sharing, sponsorships and partnerships, and strategic global energy resources reserves in order to negotiate oil pricing with foreign entities. This would include the future negotiating with American or OPEC counterparts in the price fixing of oil at various geographical points of the global grid. The ministry would also ensure prices remained acceptable and affordable by the Chinese middle class to ensure that citizens are not subject to too many inflationary pressures. Thus, China's transition from policy to applications has been characterized by gradual liberalization and decentralization of state power toward a more liberal institutional style.

The shift or transition in energy policies has been marked by a transfer of power and resources from centralized agencies to state-owned energy companies such as the China National Petroleum Corporation (CNPC), PetroChina and Sinopec. Along the same lines, the 2004 National Energy Strategy Report proposes multiple strategies to run alongside the sustainable energy development strategy.

On the production side, the report suggests improving energy security through various means, such as expanding energy diplomacy and partnerships, creating strategic petroleum reserves and limit benchmarks, reducing energy imports through additional alternatives, and raising energy efficiency to encourage Chinese citizens to consume less and save more. Another part of the policy supply side includes increasing natural gas reserves and developments, promoting hydro and nuclear power resources, promoting renewables to provincial legislations, and improving domestic oil drilling and infrastructures. The policy also suggests managing risk instead of regulating and monitoring oil pricing to implement and to ensure efficient pricing of oil throughout the domestic areas of the grid. Another part of the supply side is to promote investments to favor production and reserves.

The demand side of the policy involves changing energy types and using cheaper resources, such as coal and fossil fuels, promoting energy efficiency, prioritizing energy savings, raising public awareness, implementing economic programs to save energy and monitor power pricing, and implementing energy efficiency with advanced technology and new infrastructures and equipment.

Other government policy plans have developed over 5–10-year policy strategy guidelines on a rolling basis. However, the nature of these programs seems more bureaucratic than substantially innovating and overhauling. The various plans aim at enhancing energy conservation, reducing consumption, and promoting economic growth. The Chinese State Energy Council produced various research and made recommendations to formulate policies to promote market efficiencies. For instance, CSEP developed policies to force industrial enterprises to cut energy use in line with global international standards, and when the Chinese government entered the World Trade Organization in December 2001 it also promised to bring internal infrastructures in line with those of international standards. CSEP started some pilot programs in Shandong Province with two large steel enterprises.

The Chinese State Council implemented the Energy Conservation Law and the Clean Production Law with the aim of improving energy efficiency at the same rate as over the past 20 years. The Chinese State Council had brought in 22 administrative measures, 7 standards, 8 plans and 14 policies to enhance energy-saving technologies between 1980 and 2000. Part of the market efficiency program is to reduce energy demand as well as increasing energy supply. The fact that emerging markets such as China are 'late' in entering the WTO and introducing upscale infrastructures to create energy market efficiencies creates grave market inefficiencies in their first few years of entry to the global markets. A mature, over-regulated energy policy is also vulnerable to global market efficiencies (for example, Californian power defaults in the late 1990s versus the deregulation of European energy markets). China also experienced some challenges in smoothing its market, with significant power shortages in 24 of 31 provinces in 2004. China also suffered an energy crisis by expanding power plant construction, and by adding over 40 GW of capacity per year. Increasing capacity proved to be too costly without energy efficiency policies incorporated within the new energy frameworks. Investments also started to dry up since the introduction of the yuan to global forex markets meant that it became more costly to investors to inject capital into Chinese energy markets as of 2005. During the 1980s, energy conservation investment schemes represented 10 to 13 per cent of supply investments. The 1990s saw energy conservation investments decrease by 7 per cent. China invested RMB 424 billion in energy supplies and only RMB 23 billion in energy conservation in 2004. This century shows falling investments in the conservation of energy. Policy priorities shifted to investment in infrastructures and energy creation rather than in energy saving as alternative ways to meet the exponential increasing demand (see Figure 7.1).

Rising demand for energy creates a problem at both industrial and manufacturing levels. National surveys show that industry consumes 70 per cent of energy while industrial energy consumption rises by 4.2 per cent per year.

The National People's Congress passed the Energy Conservation Law (ECL) to renovate the outdated legislation from the 1980s, which was no longer applicable to the new era's technologies and living standards. The Shanghai Energy Conservation Supervision Center created the new framework to help regenerate outdated technology.

Another part of the efficiency framework is to introduce industrial equipment standards for electric motors, compressors, fans and other machinery. Electric engines consume the majority of China's electricity, especially those in old manufacturing infrastructures. CSEP is supporting new engine standards that could save 240 million tons of carbon by 2020 and cut the need for 37 500-megawatt coal-fired power plants. In addition, as part of the energy efficiency saving plan, the Chinese Energy State Council produced a list of building codes to reduce energy waste. Codes require buildings to use modern materials, insulation, and advanced windows to reduce energy leakage. Those new rules also catalyze economic development in new industries set up to produce advanced, energy efficient products that meet the

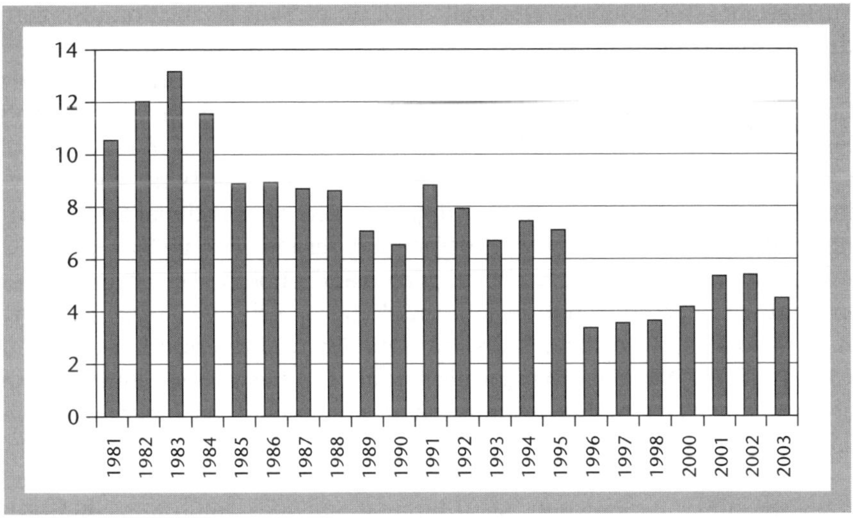

Figure 7.1 Percentage of energy investments for conservation purposes
Sources: National Bureau of Statistics of China (2004); *China Statistical Yearbook* (2004); Lawrence Berkeley National Laboratory estimates.

code requirements. Energy efficient building codes have been developed in both Central and South China which should decrease carbon emissions by 250 million tons by 2020 and thus displace the need for 50 500-megawatt coal-fired power plants. The Ministry of Commerce has required that Chinese national commercial buildings apply the energy efficiency rules since 2005. The new rules applied to all industrial and manufacturing buildings will save 50 million tons of carbon by 2020. In additional to new legislation on energy saving, China has implemented rules applicable to individual consumer appliances such as refrigerators, air conditioners, light bulbs, and various engines throughout the country. The savings on air conditioners in China in five years is equivalent to the entire electricity output of the massive Three Gorges Dam. China also requires emission reductions from coal-fired electricity plants. Advanced technologies are less corrosive to the environment and much more efficient because of the use of new materials. Energy investments have improved efficiencies in the manufacturing of refrigerators, air conditioners, televisions, fluorescent lamps, clothes washers, and gas appliances which displaces the need for 51 500-megawatt coal-fired power plants by 2020.

From 1 March 2005, it has been a requirement of the Chinese Energy State Council that all home appliances have regulation labels indicating energy consumption and informing users about energy efficiency awareness. For instance, home refrigerators and room air conditioners manufactured, sold or imported into China must carry an energy efficiency label.

Furthermore, the Chinese National Institute of Standardization also conducted due diligence audits to evaluate the practicality of new energy restrictions starting on 5 March 2005. According to the China Sustainable Energy Program, the State Bureau of Quality, Technical Supervision and NDRC's Energy Conservation Division recommended the integration of such restrictions into the Chinese population's daily practices. Energy saving regulations on labels are also supplemented by additional pricing tariffs and reforms with regard to electricity usage. For instance, NDRC reforms at the end of March 2005 regulated and smoothed electricity prices in terms of interconnection fees to the domestic grid and to neighboring Asian and Russian pricing grids, transmission and distribution fees, and retail tariffs. NDRC also charges additional penalties to those entities with above average electricity consumption or utilizing old machinery that consumes more energy.

In addition to these guidelines, the Energy State Council also provided some recommendations with regard to fuel efficiency in order to diminish dependency on foreign oils and fuel grades. For instance, the China Energy Administration requires energy efficiency practices for car engine technologies intended to save 23 million tons of carbon in 2020. This measure should save 212 million barrels of oil, equivalent to 25 million polluting cars. Mopeds, trucks, and motorcycles should also save twice as much through using alternative fuels by 2020. This encourages the development of low and zero-emission vehicles such as hybrid-electric vehicles. China aims to bypass the traditional applications of matured markets and to move directly to hybrid-electric vehicles. It also aims to export them and create a new market in order to conserve energy or at least to maintain global energy pricing in balance. To achieve that goal, the Chinese Ministry of Science and Technology has invested more than $100 million over 5 years in the development of advanced hybrid-electric and fuel cell vehicles. Since 2002, China has been undergoing a growing demand for vehicles, leaving bicycles behind and employing more modern technologies at cheaper costs than mature counterparts. This phenomenon is also drastically changing the daily lives of those Chinese citizens who are soon to be the leading global middle class. Evident since 1993 the phenomenon has been more obvious since 2002 with new car sales rising by 25 per cent on average each year, up by 38 per cent in 2003 alone.

Research and development of new machinery continues, and so do reforms, fiscal policies, and tariff legislation. The Chinese Energy State Council formulated a new taxing and fiscal framework, and the Chinese government is leveraging rising oil imports by adding intermediary tax spread interests. To achieve this goal, the China Automotive Technology and Research Center (CATARC) proposed recommendations on tax and fiscal policies for promoting cleaner and more efficient vehicle technologies on 21 and 22 March 2005 in Beijing. The Ministry of Finance, the National Development and Reform Commission (NDRC), and the State Environmental

Protection Administration are currently developing further reforms towards that goal. The Ministry of Finance (MOF) and NDRC also created a tax and developed fiscal policy reforms to promote energy efficiency and renewable energy. The State Council Development Research Center also recommended various energy pricing reforms, taxes, fiscal and investment policies to promote energy efficiency and renewable energy developments. Additionally, the Chinese State Council also promulgated new rules to favor transportion by buses, trains and other forms of public transportation to limit per capita energy consumption as well as pollution. Ultra-modern, dedicated-lane buses, moving station-to-station may move people with subway efficiency at 5 per cent of subway costs. Beijing created a 15-kilometer Bus Rapid Transit (BRT) in 2004. Kunming also created a 39.5-kilometer system.

The State Environment Protection Administration (SEPA) promulgated new national emission standards on 27 April 2004. SEPA also imposes tariffs and penalties for light and heavy duty vehicles and noise limits for motorcycles and mopeds. Demand-side management (DSM) also created energy saving programs in Eastern China. For instance, Beijing imposes off-peak power tariff and other policies to promote the efficient use of electricity. Hebei Province also imposes electricity wiring fees that finance a fund of about $10 million per year for energy efficiency projects. Jiangsu Province also invested $5 million in 2002 to leverage over $70 million in investments financed by about 70 industrial and commercial enterprises. Generation performance standards (GPS) cap emissions on a kilowatt-hour production basis, requiring older and dirtier power plants to limit pollution. These plants are also penalized for emissions. Such programs are promulgated in Zhejiang, Jiangsu, Shandong, and Shanxi provinces. Another program is the mandatory market share (MMS), which demands that Chinese citizens use a minimum energy percentage from renewable energies. Fujian and Sichuan regions have begun to apply the MMS program. CRED has required that 14 per cent MMS is allocated for Fujian Province and 10 per cent MMS for Sichuan Province by 2015. CRED and NDRC also require that 10 per cent is allocated in MMS by 2020.

According to the China Sustainable Energy Program, the Center for Renewable Energy Development (CRED) works in association with the Energy Research Institute (ERI) of the National Development and Reform Commission (NDRC). CRED and NDRC write renewable energy regulatory policies. CRED has worked with the World Bank, the United Nations Development Program, and the Global Environment Facility. Since 2001, CRED has been attempting to establish a renewable energy law in China with the US Center for Resource Solution. CRED's participation helped the Environment Protection and resources Conservation Committee of the National People's Congress (NPC) to establish the National Renewable Energy Law in August 2003. This enabled 15 per cent of all China's energy to come from renewable energy by 2020. CRED assists regions such as

Gansu, Jiangsu, Jilin, and Fujian provinces to implement the National Renewable Energy Law. For example, CRED partnered with Fujian Energy Research Society and Sichuan University, and the Global Environment Facility to create 200 Megawatts of wind energy in Fujian, at a cost of $200 million through the MMS program.

7.3 ENERGY LAW FOR RENEWABLES AND ALTERNATIVES

According to the China Sustainable Energy Program, the National People's Congress (NPC) Legal Affairs Committee, Environment and Resources Committee, Legislative Affairs Committee, National Development Reform Commission, Ministry of Finance, and the State Council's Legal Affairs Office in April 2005 implemented the Renewable Energy Law with regulations to come into effect by 1 January 2006. The renewable energy regulations impose fixed power tariffs and fees. CRED, the Energy Bureau and the Department of Pricing of NDRC define pricing legislation to monitor and control China's rising energy demand. New regulations require China to allocate a targeted volume for renewable energy. The National Development and Reform Commission (NDRC) created the 2020 Renewable Energy Development Plan. The framework requires that 10 per cent of electricity is generated by renewables by 2020 and that renewable technologies employ wind and biomass. The Chinese Sustainable Energy Program also speeds up the development of demand-side energy efficiency and renewable energy programs across most Chinese provinces. Table 7.1 provides a description of wind and solar energy policies' regulatory framework by regions within China.

According to the Chinese government and the NESP, China's target for renewable energy amounts to an additional 90 to 100 GW of capacity by 2020, which includes 60 to 70 GW of small-scale hydropower, 20 GW of wind power, 1 GW of biomass-fired electricity, and residual rises in solar, geothermal, ocean and tidal energy. In addition, the Chinese plan also meets two-thirds of the renewable energy target with small-scale hydropower. The Chinese plan for wind power is to produce 20 GW of wind power by 2020, or roughly 1 GW per year. Zhengming et al. (2000) provide a description of the evolution of Chinese energy policy (Table 7.2).

Table 7.3 gives an overview of alternative energy resources in 2002.

7.4 THE CHINESE GOVERNMENT'S 10-YEAR GUIDELINES FOR ENERGY POLICY

A National Energy Research Scientific Center (ERI-LBNL) study (2005) contributed to the discussion of efficient energy use within a context of domestic economic balance, the avoidance of inflationary pressure on national

Table 7.1 Incentives for wind/solar energy development

Inner Mongolia	
Current status	14.5 MW of large wind power generators, 18.5 MW of small wind power generation; household photovoltaic (PV) systems dominated by 10–20 Wp.
Subsidy policies	RMB 25 million to users in 1986–90; RMB 200 for each 100 W wind power generator purchased or each 16 W PV unit from financial budget. Annual subsidies of RMB 300,000 for R&D activities; working capital provided by local authorities for establishment of extension station in 56 countries.
Taxation policies	3% VAT surtax on wind power generation; income tax relief for 2 years; 10.69–14.43 yuan for VAT surtax on PV units of 16 Wp–21.6 Wp.
Pricing policies	Tariff calculated on repayment of principal and interest. 713 yuan/MWh including VAT in 1995 and 609 yuan/MWh not including VAT; the difference is shared by grid and subscribers with 200 yuan/MWh by grid and rest by subscribers in the form of subsidies 2.5 yuan/kWh.
Loan policies	RMB 400 million for wind power by State Economic and Trade Commission; Danish government loans for wind power generation.
Other	Land use policies: Land tax collected on the land actually occupied; 5-year income tax holiday for occupying arable land; 10-year land tax holiday for occupying unused land.
Xinjiang	
Current status	16.7 MW of large wind power generation; 8000 household PV systems dominated by 10–20 Wp.
Subsidy policies	50–200 yuan for PV unit and small wind power generation unit purchased; 1 million yuan for R&D; working capital for extension stations; 300 yuan subsidies for PV users.
Taxation policies	2-year tax holidays, 3-year tax relief and 5-year 15% income tax for foreign invested or joint venture with an operational life of 10 years. VAT holiday for products export. Monthly collected VAT and surtax at 17% and 10% respectively. Seasonal collected income tax at 15–33%; import tariff and VAT at 12% and 17% respectively, with duty free for international donation.
Pricing policies	Tariff calculated on repayment of principal and interest. 698 yuan/MWh including VAT in 1995. The regular grid tariff is 118 yuan/MWh in the area; the difference is shared by grid and subscribers. The added 2 cents/kWh with 0.5 cent for difference with rest borne by grid.

(*Continued*)

Table 7.1 (Continued)

Gansu	
Current status	5000 household PV units mainly of 20 Wp.
Subsidy policies	300 yuan for each PV unit purchased; subsidies for R&D; support for extension station.
Taxation policies	Policies similar to that of Xinjiang in taxation on PV system with only exception for monthly collected value added tax on non-donated PV system.
Loan policies	Guarantee by local government for household PV system to secure a loan with interest rate at 3%. The subsidies came from additional tariff by 3 yuan/MWh and 20% interest subsidized from financial budget.
Qinghai	
Subsidy policies	300 yuan for each PV unit purchased, RMB 500,000 for R&D; working capital for extension stations.
Pricing policies	Addition tariff 2 yuan/MWh, with some of the revenues used to finance installation PV system.
Northeast grid	
Current status	6 MW of wind generation.
Taxation policies	6% VAT on wind generation. No tax relief for high power tariff (900 yuan/MWh) in Henshan, Liaonin. 6% for Donggang, Liaonin (100 yuan/MWh).
Pricing policies	Tariff calculated on repayment of principal and interest.
Loan policies	Discount loans for wind power.
Other	Land use fee paid on the area actually occupied by wind power generation with preferential treatment as foreign invested businesses.
Gaungdong	
Current status	11.7 MW of wind power generation.
Taxation policies	VAT collected at 20 yuan/MWh and 15% for income tax.
Pricing policies	Tariff calculated on repayment of principal and interest; grid tariff 770 yuan/MWh with difference shared by subscribers.
Zhejiang	
Current status	1 MW of wind power generation.
Pricing policies	Tariff calculated on repayment of principal and interest.

Sources: Zhengming et al. (2000); *Draft Final Report on Financial Policies Promoting China's Renewable Energy Development by World Bank and China.*

Table 7.2 Typical renewable energy developmental phases

Phase I	*Research, development and demonstration of new technologies*
	• Resource assessment
	• Technology development and demonstration
Phase II	*Wholesale market development*
	• Establishment of stable wholesale market rules and processes
	• Establishment of companies willing and able to undertake resource development (usually IPPs)
	• Establishment of manufacturing facilities
	• Development of a financing framework
Phase III	*Cost reduction*
	• Project experience (multiple projects by individual companies)
	• Increased manufacturing volume
	• Development of related infrastructure and service companies to support the technology
	• Standardized product
Phase IV	*Price reduction*
	• Assured volume/long-term opportunities
	• Competitive market
Phase V	*Retail market development*
	• Clear market rules
	• Four or more individual companies
	• Public education/information

Source: Zhengming et al. (2000).

prices, and international security, through the monitoring of energy pricing efficiency to avoid future conflicts over pricing inefficiencies, and the adequate sharing of natural resources. Part of the Chinese government's 10-year plan objectives is the monitoring of pricing mechanisms and adjustments between actual tangible commodities and the trading in speculative negotiated prices among international entities. The government's strategic guidelines aim to set relative prices, such as rising renewables and energy prices, or decreasing prices of natural gas. Part of the plan is to maintain market efficiencies and ensure that commodities deliverables are not subject to liquidity gaps domestically or internationally that would endanger the stability of ramified financial markets and commodities exchange or trading entities. The Chinese 10-year guidelines also suggest the establishment of an energy ministry, with adequate administrative powers to coordinate all

Table 7.3 Overview of technically exploitable potential and currently installed capacity of renewable energy for power generation in China, 2002

Renewable energy source	Technically exploitable potential	Installed capacity (MW)
Grid connected RE power production		
Bagasse fired power generation	21.2 m ton bagasse	807
Wind energy	253 GW[a]	440
Large and small hydropower	265 GW	90,200
Geothermal	5.8 GW	30
Tidal energy	110 GW	5.9
Decentralized RE power production		
Solar PV		43
Wind energy		25
Mini-hydro (<10 kW)	80 GW	154

Note: a. Based on an evaluation of winds at 10 meters above ground.
Source: Zhengmin et al. (2000).

entities – private, governmental, institutional, international corporations, exchanges, and financial reserves. Wiser et al. (2005) compared energy policies as shown in Table 7.4.

As of 2005, the 10-year energy policy focused on the following objectives:

- *Energy saving and conservation* The 2005 energy demand of 10,000 yuan GDP should be reduced to 2.2 tons of coal. The guidelines also target the accumulation of alternative reserves to 340 million tons of standard coal with a recommended yearly energy conservation ratio of approximately 4.5 per cent. Fuel oil savings and substitution ought to be around 16 million tons and 5 million tons, respectively. Additionally, the Chinese energy demand per unit for most usable energy products, such as oil and natural gas, ought to be monitored and reduced, based on 'needs' not on 'wants', to avoid creating inflationary pricing trends based on 'waste' and 'systematic consumerism'. China's 2005 energy consumption per ton of steel for medium and large iron and steel enterprises is expected to be less than 0.8 tons of standard coal. Coal demand for thermal power generation production is to be decreased to 380 g standard coal/KWh. Additional guidelines to reduce consumption also involve ammonia, down to 37 Giga-joule, and cement and glass, down by 20 per cent. In addition to these savings standards, the energy conservation ratio in 2005 for public works, administrative, industrial and residential should reach 50 per cent.

Table 7.4 Comparing the policies

Policy objective	RPS	Feed-in	Tendering
Incentives for cost and price minimization	Policy creates incentives for generators to lower RE prices in order to compete for contracts; does not inherently reduce costs of generation except those related to technology learning and efficiencies of scale	Few inherent incentives to minimize market prices for renewable energy, though there are likely to be project cost reductions related to technology learning and manufacturing volume	Policy creates significant competitive pressures for price minimization that will be linked to cost minimization where there is sufficient competition, technology learning and manufacturing volume
Ability to maintain targets for renewable energy	Purchase obligation can be effective at meeting RE targets provided RPS is well designed	Ability of feed-in tariff to help government meet RE targets is variable depending on host of factors	Ability of tendering policy to help government meet RE targets is variable depending on host of factors
Assurance of resource diversity	Diversity possible with bands and tiers, but has administrative drawbacks	Can successfully stimulate a more diverse set of resources by setting one price that many technologies can meet or setting a separate price for each technology band	Diversity possible with bands, but as with targets, policy does not guarantee that projects will be built
Sustainable market for power	All three policies build markets for RE power – RPS may be more technically & politically sustainable	Can be vulnerable to political 'tinkering' and if viewed as 'subsidy' makes it less economically and politically sustainable	Tends to be tied to a resource planning process that can make it more politically vulnerable if planning out of favor
Political viability	Depends on circumstances – unclear in China	Depends on circumstances – unclear in China	Depends on circumstances – unclear in China

(*Continued*)

Table 7.4 (Continued)

Policy objective	RPS	Feed-in	Tendering
Local industry development	Needs companion policies to ensure local development	Feed-in tariffs can create local manufacturing and development infrastructure benefits	Will favor least cost generation over local industry development; benefits established industry
Compatibility with electricity industry and regulatory structure	RPS is compatible with industry and regulatory structure in China, though appropriate phase-in and enforcement are important	Compatible with existing regulatory and industry structure but current tariff structure needs fixing to work	Tendering is compatible with industry and regulatory structure in China and can be used by utilities in conjunction with an RPS.
Policy stability	Provides less certainty than feed-in tariffs, must be carefully designed	Provides high degree of certainty and stability	Can provide high degree of certainty and stability, but only if well designed
Competitive parity	Creates competitive parity as the same standard applies to utilities and developers equally	Parity achieved only if cost sharing mechanisms are established that spread the costs broadly	Policy favors established market players over new market entrants and can allow market manipulation by existing companies
Integration of renewable energy supplies	Creates incentives for full integration and barrier reduction	Fewer incentives than under RPS to reduce institutional barriers	Neutral – doesn't help reduce institutional barriers
Simplicity	More challenging policy to design and administer, and more complex contractual and development process for generators as compared to feed-in	Most simple design, administration, enforcement, contractual, and development simplicity	More complex than feed-in laws, because requires the development of a system to raise money for the incremental costs of RE.

Source: Wiser et al. (2005).

- *Energy usage efficiency* In 2005 industrial energy usage efficiency targeted an output of 40 billion yuan. Additionally, recycling and waste reutilization are intended to save 55 billion yuan. Industrial waste usage ratio is targeted at a 60 per cent limit for 10 years.

Since 2002, China has overhauled its entire domestic energy program. Between 2002 and 2006, it has reformulated its energy policy program with the aim of quadrupling GDP by 2020. At beginning of 2002, the China Sustainable Energy Program aimed to bring together 11 central government ministries and research institutes to develop a National Energy Plan for 2004–20. This consists of the analysis of five-year energy scenarios. The resulting study states that aggressive energy efficiency and renewable energy policies will contribute to a decrease in coal consumption by one billion tons by 2020. It also shows that fossil fuel consumption will rise by 90 per cent. The study also demonstrates that carbon emissions will diminish by 40 per cent and that Chinese per capita carbon emissions will decrease by less than one ton by 2020, which represents at least 33 per cent of OECD countries and 93 per cent of the global average.

The following policies and measures applied as of February 2006:

- According to CNPC and NESP, the government's primary goal is to improve the legal framework to conscientiously apply the energy conservation law, new measures and regulations, and to guide and standardize the activities of energy utilization. Government's new policies focus on the organization of the formulation of *Provisions for the Administration of Petroleum Conservation* and *Provisions for the Administration of Energy Efficiency Identification*.
- Renewable Resources Recycling and Utilization Law.
- Provisions for the Administration of Tailing Comprehensive Utilization.
- Provisions for the Administration of the Recovery and Utilization of Waste.
- Circular of the State Council on Approving and Transmitting Proposals of Departments including the State Economic and Trade Commission on Further Implementing Resources Comprehensive Utilization.
- Revise and improve List of Resources Comprehensive Utilization.
- Improve preferential policies on tax reduction and exemption.
- Implement energy efficiency policies.
- Standardize and regulate market of energy-saving products.
- Improve energy efficiency standards for energy-using products.

- Include energy efficiency standards for key industrial energy-consuming tools such as industrial apparels, electric engines, fans, air conditioners, pumps and transformers, and domestic appliances, lighting tools, and devices consuming intensive energy.
- Implement adequate technology and energy-saving product certification and energy efficiency identification systems.
- Install an energy-saving product certification program.
- Adopt the international intercertification system for certified engines and products.
- Create an energy efficiency identification system with codes to apply foreign energy regulations that save energy and are energy efficient.
- Implement international guidelines identification systems and create products compliant with international energy standards.
- Scale up the Chinese industrial infrastructure and product mix to develop tertiary industry with advanced new technologies with low energy consumption.
- Upgrade traditional industrial infrastructures with advanced high and new technologies.
- Invest in new techniques that do not pollute and are not harmful to the environment.
- Develop and implement clean coal technologies.
- Expand natural gas investments and projects.
- Develop new energy resources and renewables.
- Promote efficient and clean energy utilization.
- Promote technological advances and improve technological levels of energy conservation and resources.
- Establish and accelerate technological energy innovations and inventions and develop technologies.
- Promote education and research.
- Enhance the industrialization of scientific and technological improvements.
- Establish a new taxation framework to give tax benefits to energy saving technology investments, to restrain excessive consumption, and to benefit companies that implement energy saving programs.
- Impose heavy taxes on products and equipment with high energy consumption and heavy pollution, and implement the obsolescence of products with high energy consumption.

- Enhance energy pricing structures to further deepen energy prices and install adequate reforms. (This 2006 energy price forming mechanism to establish an energy price forecast system is new. The pricing of commodities in China had previously been performed separately from financial markets. The appearance of financial engineering derivatives trading in the Chinese financial markets in November 2005 forced the Chinese government to study energy prices more closely to value its tangible assets used as collateral against those intangible financial derivatives and to evaluate the pricing adequacy of its internal regime vis-à-vis outsider traders investing in the Chinese system. The fact that China has been a net importer of oil since 1993 forced China to cooperate with international investment banks and to allow speculative instruments in its systems. However, it is harmful to overspeculate on non-deliverables without having collateral assets as back up for potential defaulting securities and to create valuation gaps between the actual commodities reserves used as collateral goods for derivatives tradings on commodities.)
- Promote financial investments to favor policy-related banks and give preferential loan terms.
- Study new global market economies to connect to the international energy pricing grid and smooth policy regulation and pricing accordingly with energy transportation costs.
- Install a broad market-based energy savings program that promotes the popularization and application of new technologies, techniques and equipment.
- Promote information services, marketing campaigns, publicity, education, and training, to create a society aware of energy saving.
- Develop information services and create an intelligent network system.
- Implement an annual nationwide energy-saving publicity week.
- Create a centralized ministry to govern international partnerships and internal Chinese commodities firms and to enhance policy-making.

According to Fan et al. (2005), China's national energy policy focuses on transparent operational rules. The evolution of Chinese energy policies began with its entrance to the World Trade Organization and continues with the inclusion of Chinese financial derivatives within the global financial markets. Chinese policy trends are now diversifying in various ways, primarily to ensure future global demand and geopolitical security. The Chinese energy authorities encourage energy firms to create international partnerships, new products, technologies, qualitative grades, and regulations, although, Chinese energy policy remains non-transparent and lacks operational rules and compliance (Fan et al. 2005). Since 1997, Chinese energy policies have included more transparency as part of the framework for partnerships with international entities. They also demonstrate various ways in which China has responded to the rising demand for energy resources. According to Sabina Sequeira (2006), however, Chinese policies still lack internal coordination, operational transparency and compliance between the various legal entities developing energy policies. According to her 2006 research, there is still a 'gap' between policy-making and the regulatory framework implementing and ensuring policies' efficiency and effectiveness.

From 1990 to 2000, energy policies were produced by:

- State Planning Commissions
- State Economic and Trade Commission
- State Science and Technology Commission
- Ministry of Finance
- State Environmental Protection Agency

According to Andrews-Speed (2004), in 2003, the National Development Reform Committee Energy Bureau (NDRC) replaced the State Commissions, becoming the major Chinese government delegate in energy affairs. The National Energy Leading Group (NELG) is responsible for coordinating energy policy. Andrews-Speed (2004) asserts that industry experts would prefer to see China have a ministry of energy to centralize policy-making and regulations. Until recently the Chinese government has attempted to remain independent as regards energy policy, concentrating on fulfilling its own rising needs. However, China's national oil policies are now connected with foreign exchange valuations and will eventually be directly connected to global resource availability and proportional market sharing and economic conditions.

CHAPTER 8

Chinese Oil Pricing and Global Market Integration

As seen in previous chapters, Chinese oil pricing mainly depends on supply and demand, but is also affected by external factors, such as global historical oil prices, geopolitical uncertainties, alternative energies and financial market speculation. The analysis of oil prices no longer depends only on supply and demand, but also on arbitrary political decisions by OPEC, on operational technological and infrastructural flaws between interconnected financial systems, on financial speculation on commodities prices and volatilities, which create an element of uncertainty, and on geopolitical factors, local regulatory policies and transportation and freight capabilities. So, now that China has integrated its financial and commodities industries to global markets, we have to add more variables into the equation of global oil pricings and their impact on various economies and policies. Inflationary oil prices, coupled with global warming and increasing emissions, have promoted the use of new technologies, such as hydrogen vehicles or models of car which use a water additive to make them cleaner and cheaper to run. But as yet we have advanced only tentatively towards these alternatives because oil is still the main, often the only, revenue source of African and Middle Eastern countries that have never reformed as independent democracies and the interdependencies on oil provides a valuable political weapon in dealing with these countries. However, if significantly larger numbers of consumers are competing for oil – such as in India and China (given population versus up-scaling globalization consuming average ratios) – it is foreseeable that oil reserves will no longer be sufficient as a sole dependency trading tool and revenue source. With this in mind, the impact of energy pricing on global economic equations also changes.

Energy costs as a share of GDP fell from 8 per cent in 1973 to only 2 per cent in 2005. Western economies rely more on services now than in 1973. Also, higher oil prices forced individuals to become more energy efficient by using alternatives. Increased Chinese demand contributed to the rise in commodity prices, especially since 2001. Chinese demand alone is the primary variable leading to the assertion that global prices will rise exponentially. Yet, as the main global manufacturing hub of the world China needs 50 per cent more energy and commodities. So, China incurs rising costs not only because of its new consuming middle class but because it needs to sustain its manufacturing activities. This double embedded premium within Chinese oil prices is countered by global citizens who purchase cheaper Chinese goods and can afford to pay an extra premium in rising oil prices. Most 'passive' European middle classes already suffer slowly eroding economic purchasing power in relation to day to day goods (food, health, insurance) as they export less and become dependent on expensive overvalued currencies. Yet, barrels are priced in dollars which reduces this margin slightly. In addition, Europeans pay an added premium in the form of a tax on oil. It is also worth noting that while the European middle classes are enjoying special, if diminishing rights, that other global competitors do not share, they pay inflationary premiums that will eventually reduce their economic freedom and democratic values. The new global economy's trade-offs offer costs and benefits, and the key to leveraging these changes is to select the newest and best opportunities that justify resource efficiency, high productivity, capital wealth creation, savings and adequate usages (based on needs not wants – unlike consumerism).

Since it joined the WTO (2001), China has changed its international trading terms to favor countries exporting commodities and importing manufactured goods, such as Australia, while moving the terms of trade against countries importing commodities and exporting manufactured goods, such as Bangladesh. In fact, China has already started to exploit those cheaper forex Third World countries to leverage wealth creation in technology, science, and advanced innovative industries. China has multiplied international partnerships with countries that have natural resources, and subsequently has already leveraged with technologies to subcontract heavy manufacturing work to poorer countries whose foreign currencies have not been readjusted or revalued against the dollar or other currencies. This is notably the case for countries rich in oil in Africa (Guinea, Kenya, Nigeria, Chad, and others) and also in some cases in South America, for example, Venezuela. Yet, it seems that despite rising inflationary pressures on day to day goods, since the end of the 1990s interest rate policies have not shown any correlation with rising oil prices. The rise of oil prices in the late 1990s was significantly affected by speculative investment, notably from the proliferation of hedge funds, private speculators trading on their own accounts, commodities pool operators (CPOs), and commodities trading advisors (CTAs) (Figures 8.1–8.4).

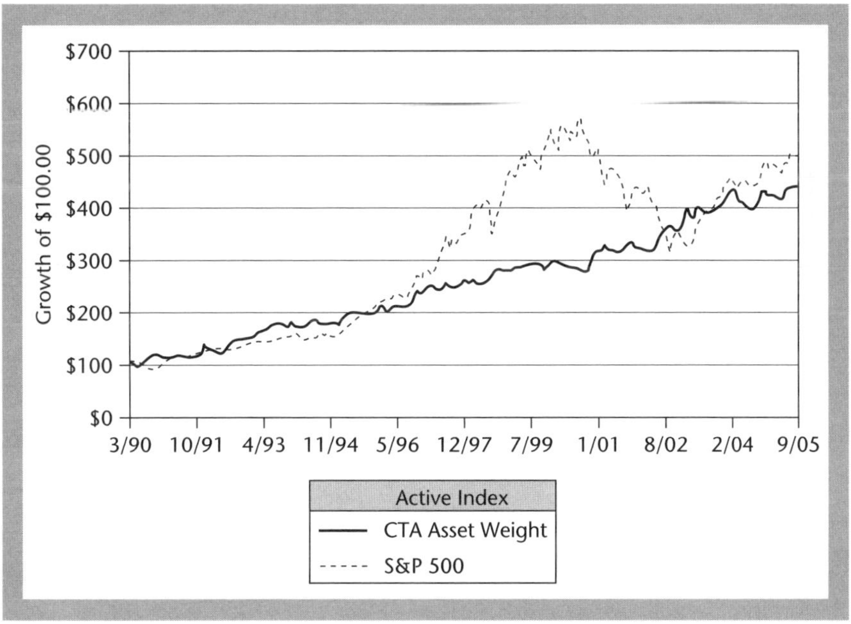

Figure 8.1 CISDM CTA asset weighted index
Note: See note to Figure 8.4.
Source: © CISDM.

As the figures in this chapter show, energy traded investments have correlated rising returns with traditional indices, except that physically traded instruments correlated more closely with the supply curve of energy and plateau at saturation levels. In 2005, physical traded instruments diverge from previously correlated functions and are no longer correlated to financial speculative instruments. Physicals use actual tangible asset valuations as collateral contingency in case of defaults.

In 2002, it is more difficult to determine what component of the oil price increases comes from geopolitical events, a global unquantifiable supply shortage, financial speculations on commodities prices distorting 'real' asset quantifications and delivery, corruption, or non-transparent bureaucratic decisions by OPEC. But each of these elements certainly has an impact on the actual barrel price at various points on the Chinese grid. A random internet study indicates that 'lower interest rates signify that the inflationary impact of higher oil prices is small'. The bond markets do not expect higher oil prices to trigger an inflationary spiral similar to that of the 1970s. Bond markets in the late 1990s were more reflective of debt levels and tended to alleviate correlating factors with commodities. Now, asset classes are immune from each other primarily as a result of greater technological, infrastructural and operational risks affecting the correlations of

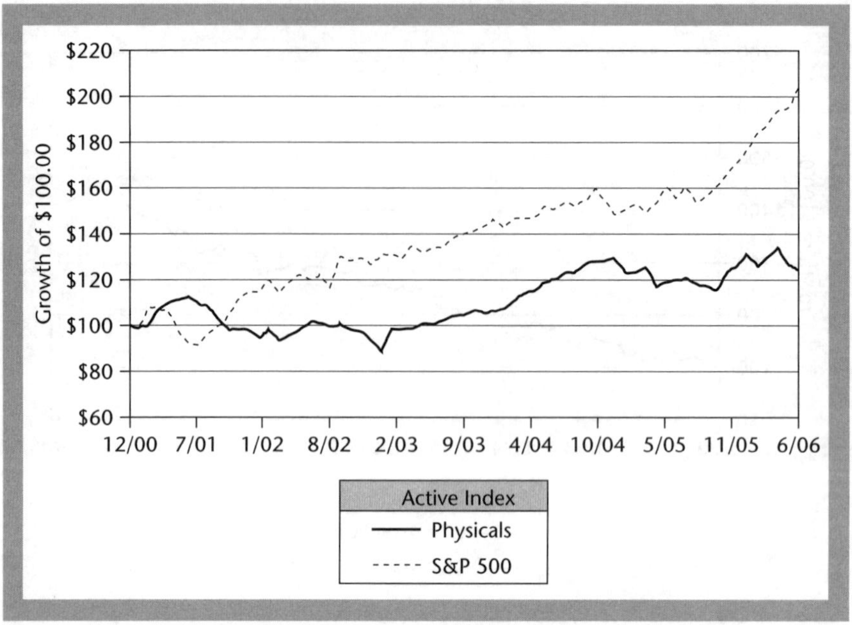

Figure 8.2 CISDM CTA asset weighted physicals index
Note: See note to Figure 8.4.
Source: © CISDM.

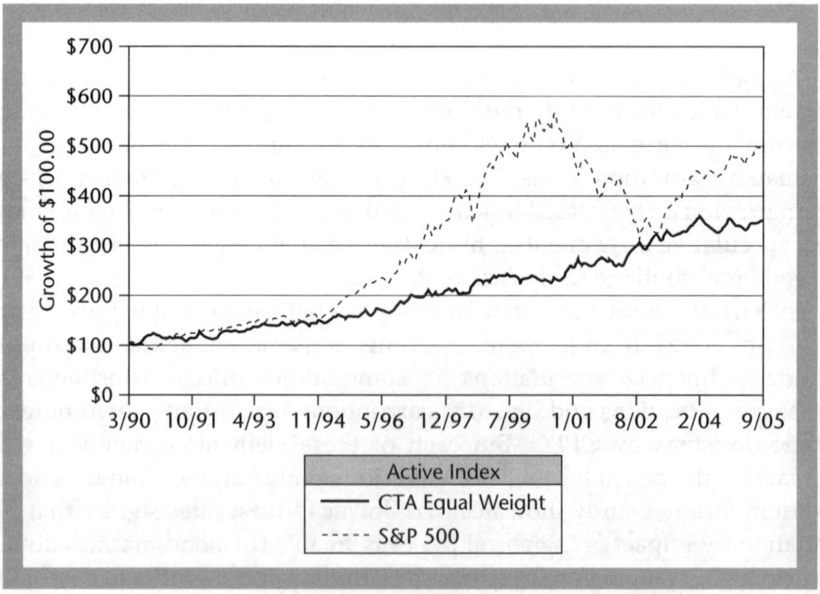

Figure 8.3 CISDM CTA equal weighted index
Note: See note to Figure 8.4.
Source: © CISDM.

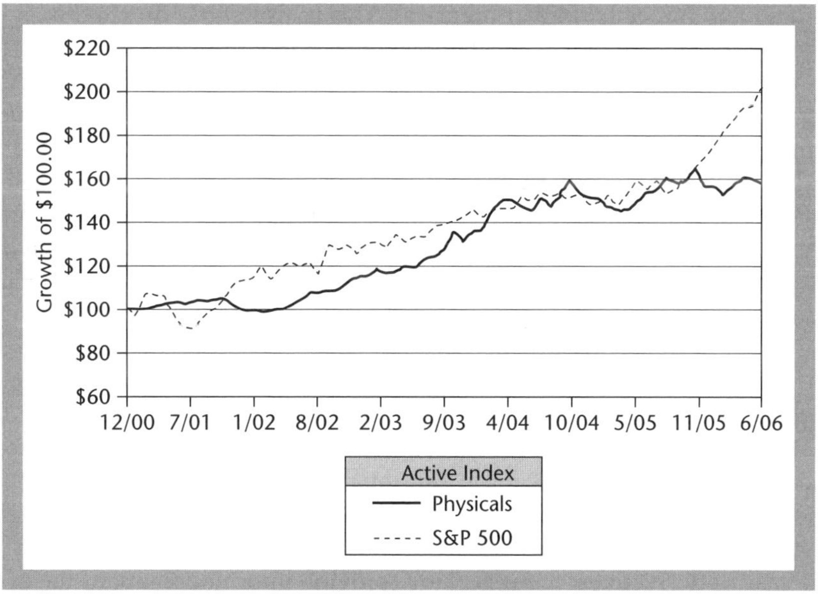

Figure 8.4 CISDM CTA equal weighted physicals index
Notes: The CISDM CTA asset weighted physicals index reflects the dollar weighted performance of physicals commodity trading advisors (CTAs) reporting to the CISDM Hedge Fund/CTA Database. Physicals CTAs trade a wide variety of OTC and exchanged traded commodity futures/options within the energy, agricultural and metals complex and use a wide variety of trading models including those based on systematic models, fundamental economic data and/or individual trader's beliefs. In order to be included in the physicals index universe, a physicals CTA must have at least $500,000 under management and at least a 12-month track record. The index goes back historically to January 2001.
Source: © CISDM.

those products. But by how much? No one knows. Insurance companies have problems in calculating correlations and thus overprice premiums to insure for all risks in their models. Another major difference between the 1970s and the 1990s comes from the immunization or 'wall' between financial trading markets and the economy, fiscal and monetary policies. There is less and less correlation between Chinese economic growth in all sectors (outperforming yearly growth rates of 20 per cent) with a yearly GDP of ±2 per cent barely deviating from the 10 per cent mark since the beginning of 2000. So, in effect, rising oil prices have little direct impact on the economy.

Energy costs are less significant to GDP than service economies and the deflationary impact of greater manufactured exports from China in conjunction with less accommodating monetary policies which reduce the inflationary impact of higher oil prices. This is also partly due to the transition period during which substantial global economies such as the United States or Western Europe are undergoing a 'soft' landing, marked by a depletion in real estate, and gradual rising interest rates from overvalued tangible assets facing new 'hard' landing economies such as China and India. They also have been subject to rampant inflationary rising interest rates due to overheating producing and capital generating economies. There are now global distortions between supply and demand, which are both inelastic. Speculative trading in correlation to geopolitical events added a speculative premium to the price of oil globally, especially since the fall of Enron. Consequently, this no longer reflects accurately the relationship between supply and demand. An embedded element of error is increasingly accounted for in the pricing sub-layers of oil, creating more inefficient global markets. The added speculative element also comes from increases in financial banking institutions trading commodities and other energy derivatives, which are extremely lucrative businesses offering instruments through which to disguise unexplainable profits, especially in the case of derivatives. Figures below provide information about the evolution of energy reserves, current account balance correlations with energy exports and bank deposits.

From 1999 onwards, additional variables, such as direct and portfolio investments, change the initial models. However, the International Monetary Fund should be more exponentially shaped with added errors from operational, infrastructural and geopolitical risks. While numbers become more accurate in market and credit risks, the level of errors remains as important as before, if not more so, due to added new types of risks (Figure 8.5).

Figure 8.6 represents the impact of energy exports on current account balances, surpluses and their correlations on financial instruments' traded volumes and valuations. As in Figure 8.5, there are added embedded variables, such as operational infrastructural risks, that are not taken into account in the exponentially rising functions. These operational infrastructural risks are within the accurate accounting of financial instrument valuations in financial institutions. These risks have risen exponentially since 2002.

Additional financial institutions in global energy markets created an extra volatility in the pricing of oil from overtrading products without adequate infrastructures. This is also partly why the price of oil increased by 70 per cent between December 2003 and 2006. Prior to 2001 global oil prices were broadly correlated, but in the intervening years, oil pricing functions from various regions have no longer been so closely correlated and have begun to diverge. Some oils are going to be more expensive than others due to their vulnerabilities to geopolitical tensions and transportation risks and costs which add a premium. Yet evidence also shows that global oil prices remain correlated to some degree and that all exhibit a common similitude toward an exponential rise (see Figure 8.7).

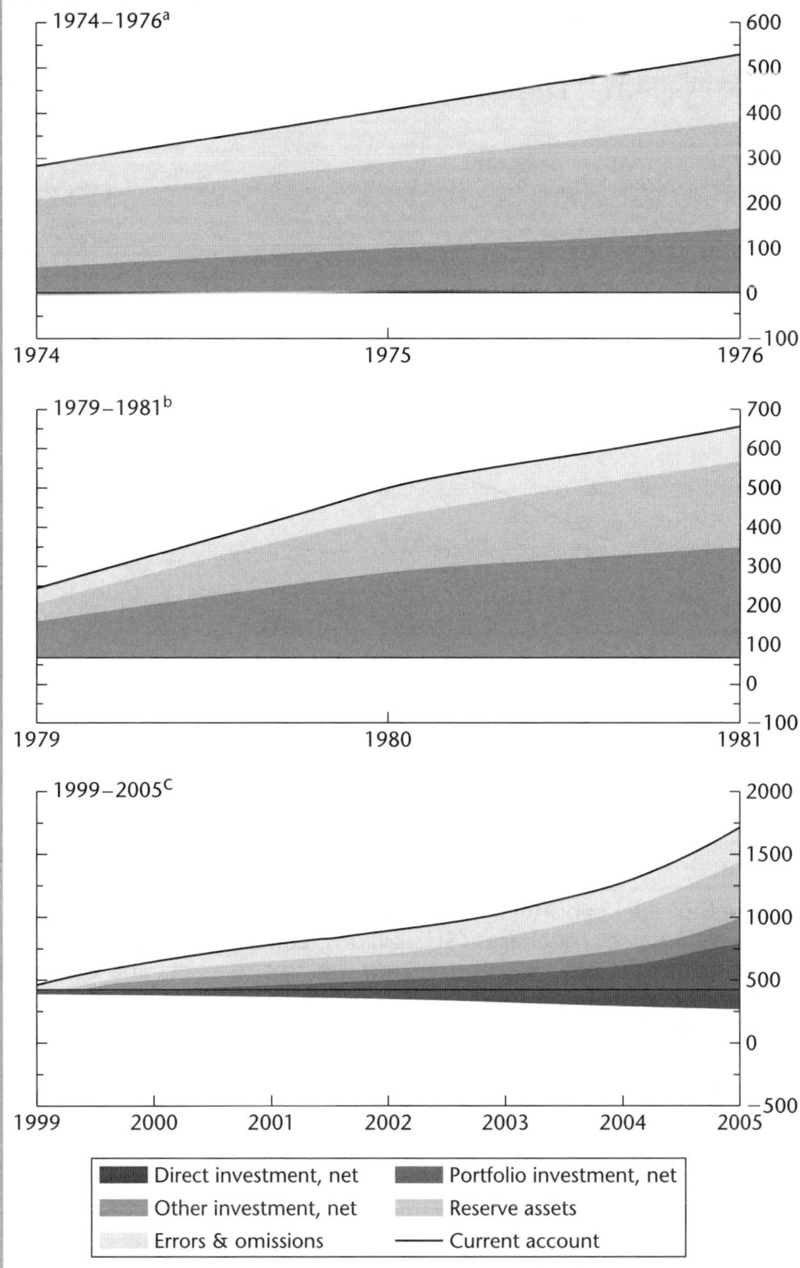

Figure 8.5 Fuel exporters' cumulative current account balances and capital flows (US$ billions, accumulative)
(Note and sources given overleaf)

Global oil pricing curves have risen exponentially from 2002 to June 2006. From then to January 2007, prices declined, primarily due to interventions by OPEC, which altered supplies in order to soften overinflated global prices. Brent and WTI pricing curves reflect this (see Figures 8.8 and 8.9).

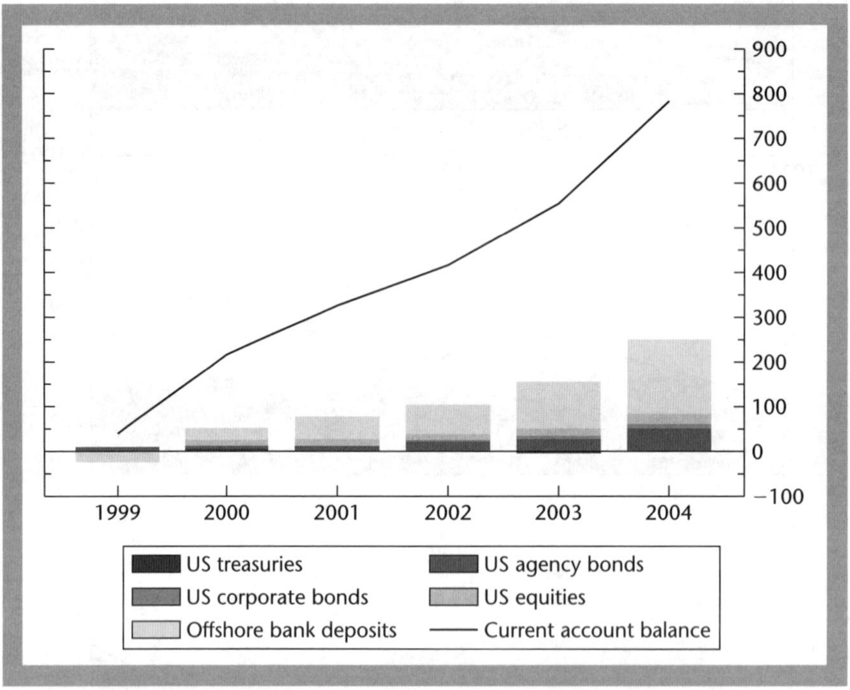

Figure 8.6 Fuel exporters' cumulative current account balances and identified asset purchases ($US billions, cumulative since 1999)
Note: In contrast to the 1970s, tracking the precise assets and countries into which oil revenues have been invested over the past few years is difficult. Identified purchases only account for a small share of current account surpluses.
Sources: Bank for International Settlements: Treasury International Capital System; IMF staff calculations.

Figure 8.5 continued
Note: Current account surpluses in the 1970s were associated with significant increases in official reserves and bank deposits. During the past few years, there has been relatively little accumulation of bank deposits, while portfolio investment flows have been sizable.
Source: IMF staff calculations: a. cumulative, starting from 1974; b. cumulative, starting from 1979; c. cumulative, starting from 1999.

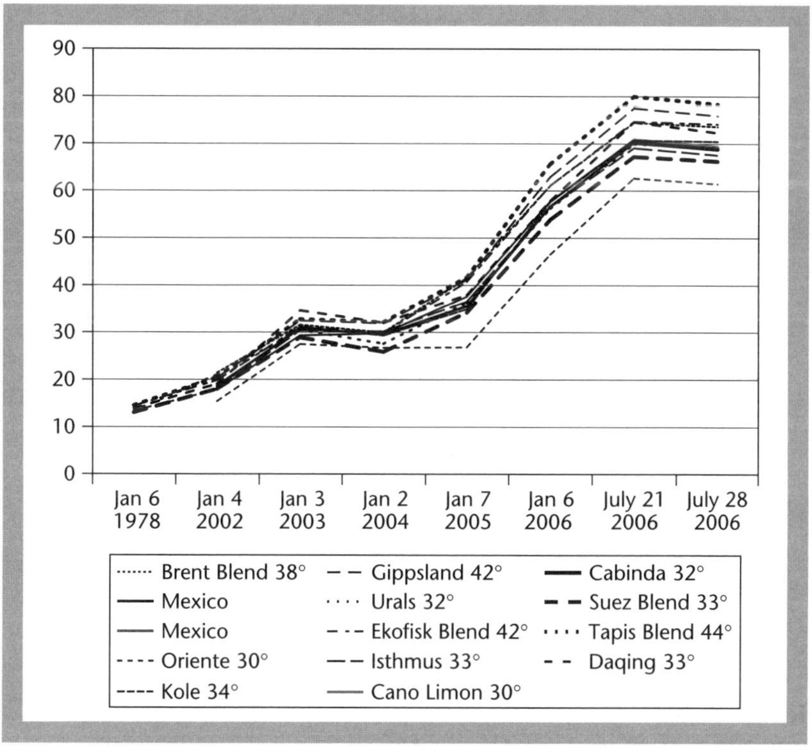

Figure 8.7 Barrel prices in non-OPEC production centers (US$)
Source: Bloomberg Markets June 2006 commodities prices.

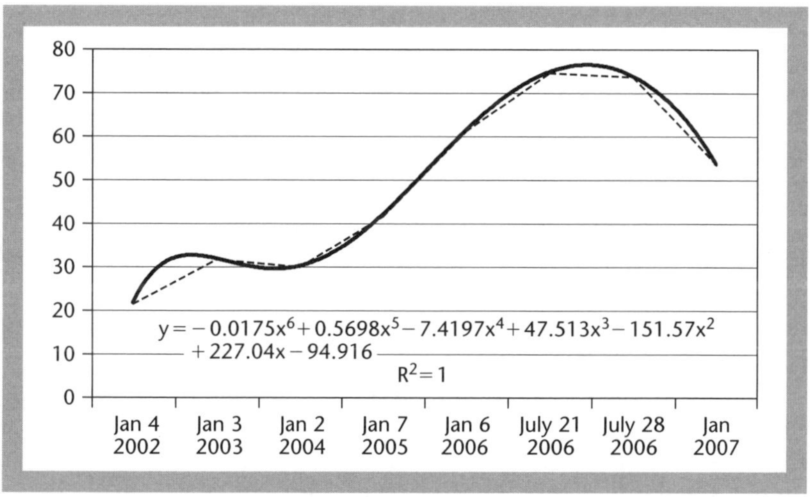

Figure 8.8 Brent pricing curve, 2002–07 (US$ per barrel)
Source: Yahoo finance 2007.

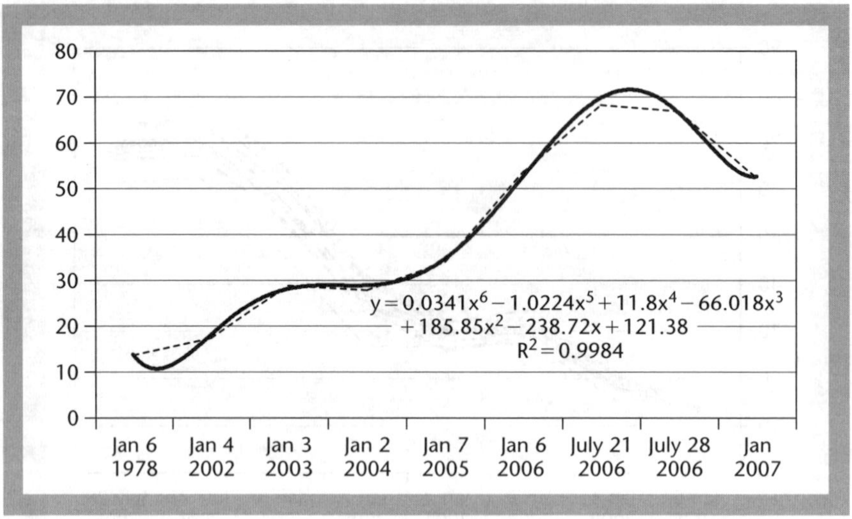

Figure 8.9 West Texas Intermediate pricing curve, 1978–2007 (US$ per barrel)
Source: Yahoo finance 2007.

All main oil curves were modelled for various geopolitical points to evaluate pricing forecasts.

Geopolitical changes in the supply or demand for oil cause dramatic shifts in the oil pricing structure. Since the beginning of the twenty-first century, higher oil prices have encouraged more drilling and further exploration, and spurred technological progress. Inflationary oil price trends also increased the demand for oil substitutes and encouraged oil conservation. Historical oil prices increased in line with traditional global indices on the broader scale of a century. Yet, what is most significant about historical financial markets and commodities is the major shift between the plateaus that traditional indices have undergone since 2002 while commodities prices rose exponentially, flirting with vertical limit lines (except between June 2006 and January 2007). The correlation between traditional financial indices and oil markets is also marked and consistent throughout various infrastructural economic shifts as seen in the analysis below. Real reserves have been substituted by 'artificial intangible' investments, such as derivatives and financial engineering products, thus adding a falsifying intangible layer in the pricing structure of oil.

According to the ChinaBidding report of 2 February 2007, the Chinese Ministry of Commerce declared on 30 January 2007, that China's crude and refined oil market would open to the world in 2007. As part of its WTO required standards, China opened retail and wholesale oil markets on 11 December 2004, and 11 December 2006 respectively. Investors from all countries now have the opportunity to access China's markets with money, technology and services. Domestic civilian-run enterprises and private companies can also access China's oil market. As the oil market grows, China should both open it and improve management and domestic regulatory policies.

8.1 OIL FINANCING

Oil exporting nations secure oil revenues from capital flows to offshore investments and structured financial corporations (notably hedge funds), which are unregulated. Part of this financial capital goes into lost investments or eroded evaporated assets and part of it remains under the control of hedge fund managers. Guizot (2007) shows capital flow depletion and erosion resulting from malpractice, misappropriation, misrepresentation, mispricing and insufficient compliance in this industry. Operational risk failures are even more obvious in hedge funds trading foreign exchange and commodities (CTA and CPOs). These market abuses are more exaggerated when transparency is low. When those financing shells appear in the Chinese and Indian markets, there will be no immediate substitutes to enforce compliance, and global market pricing could soon become anarchic. Spiralling corruption in global financial markets could outrun corporate governance standards and increase with rising economic competition to gain market shares. This is partly why the Bank for International Settlements cannot identify or trail 70 per cent of oil exporting nations' combined revenue surpluses since 1999. The US Treasury data indicates that oil exporting nations made net purchases and deposits of approximately $270 billion in US instruments between June 2003 and the end of 2005. Figure 8.10 indicates that, as opposed to the twentieth century, the oil pricing structure is now being operationally readjusted as a function of the wealth redistribution between the widening deficits of mature economies versus emerging democracies occurring trade surpluses. Since the beginning of the twenty-first century, the value of oil imports, goods and services, and of oil exports has risen exponentially. Between 2000 and 2005 the value of oil exports from OPEC countries (excluding Iraq and Indonesia) nearly tripled, reaching more than US$500 billion in 2006 (see figures 8.11 and 8.12).

However, the International Energy Administration estimates that $270 billion accounts for approximately 25–33 per cent of total oil exporting investment revenues. Investments are now being scattered throughout the globe in various innovative ways through corporate shareholding purchasing or offshore unregulated investment structures. Various OPEC countries are affiliated with those unregulated shells and own minority interest shares. For instance, Kuwait Investment Authority is the largest shareholder in DaimlerChrysler AG with a 7 per cent stake worth $3.7 billion.

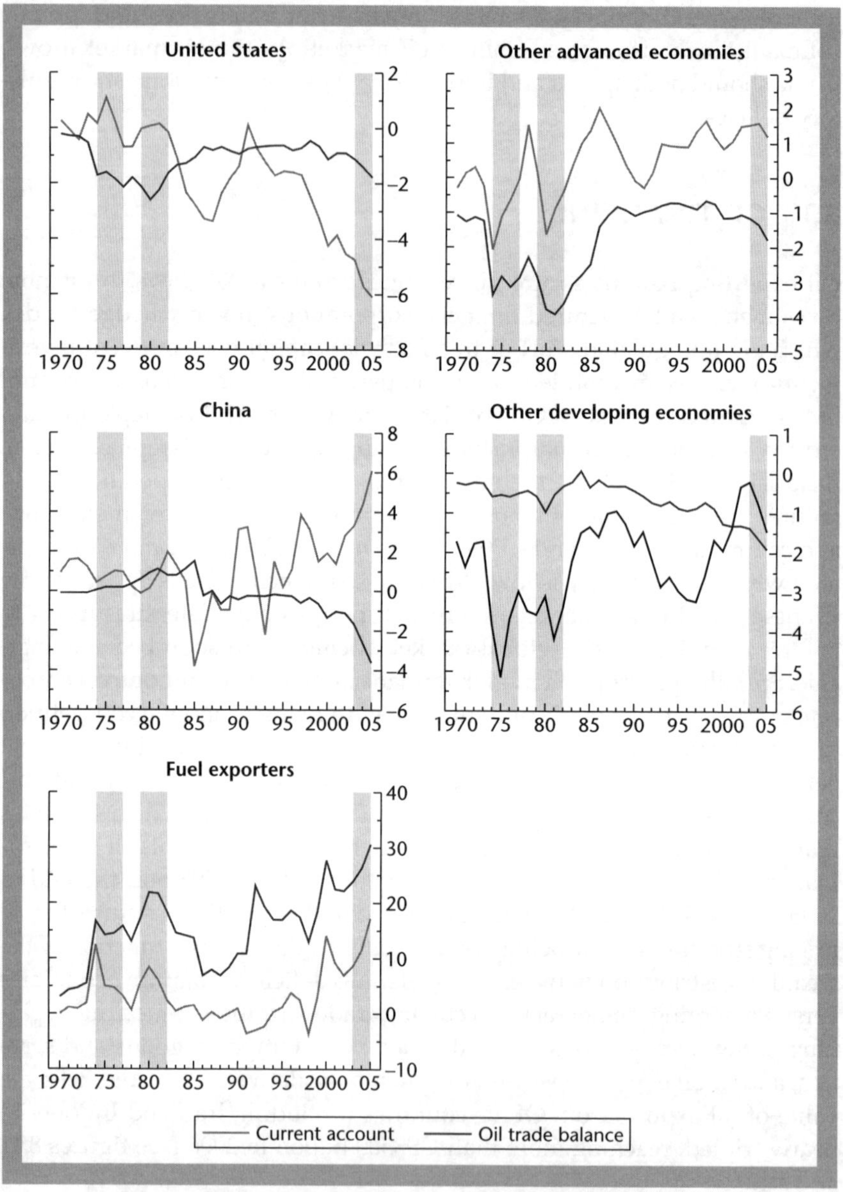

Figure 8.10 Current account and oil trade balances

Note: In the 1970s, large external deficits financed by the recycling of petrodollars were concentrated in oil-importing developing countries. In recent years, the oil price shock has instead contributed to a widening US current account deficit and has redistributed current account surpluses from other advanced economies and emerging Asia toward fuel exporters.

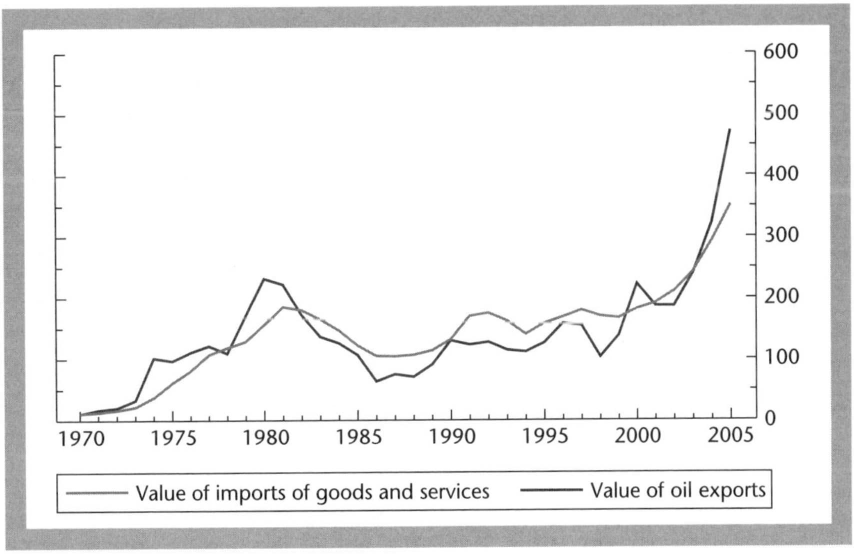

Figure 8.11 OPEC imports and oil exports ($US billions)
Notes: OPEC-9, excluding Iraq and Indonesia; data for United Arab Emirates start from 1971.
Sources: World Integrated Trade Solution: OECD; IMF staff calculations.

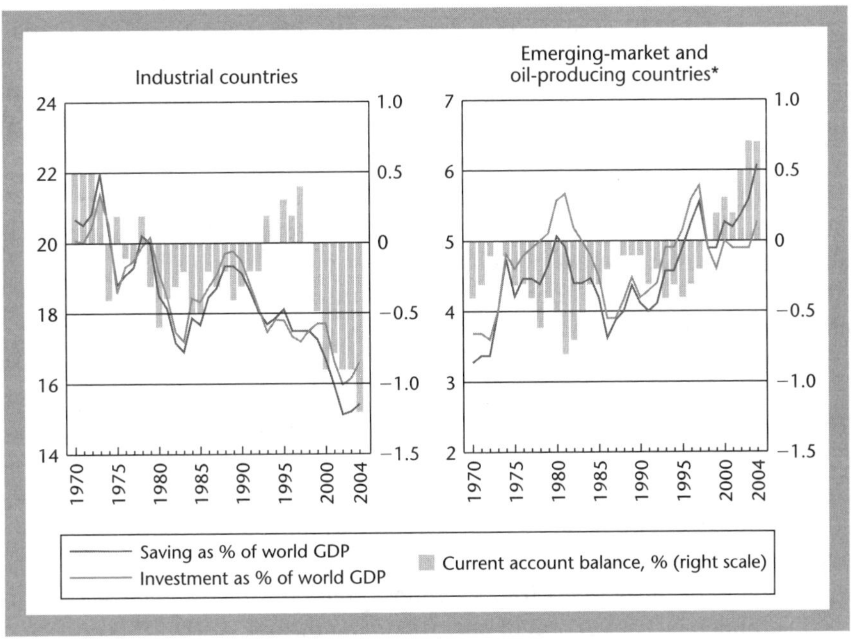

Figure 8.12 The red and the black
Note: * Includes Norway.
Sources: OECD; World Bank; IMF.

Table 8.1 Marginal propensity to import out of oil revenues[a]

	1973–1974	1973–1975	1978–1980	1978–1981	2003–2005
GCC[b]	0.08	0.34	0.18	0.25	0.15
OPEC[c]	0.14	0.52	0.24	0.42	0.24
Iran, I.R. of	0.17	0.68	0.35	0.24	0.37
Saudi Arabia	0.01	0.32	0.27	0.39	0.26
Venezuela	0.18	0.65	−0.15	0.01	0.46
Major non-OPEC[d]	...	...	...	...	0.31
Russia	0.77	1.37	0.76	1.08	0.20
Norway	...	...	0.18	−0.30	−0.13
Mexico	...	...	...	...	0.78

Notes: a. Defined as (change in imports net of non-oil exports. Investment income, and transfers)/(change in oil exports); b. The Cooperation Council of the Arab States of the Gulf (GCG) includes Bahrain, Kuwait, Oman, Qatar, Saudi Arabia, and the United Arab Emirates; c. OPEC-9 excluding Iraq and Indonesia, data for the United Arab Emirates start from 1971; d. Major non-OPEC includes Angola, Canada, Kazakhstan, Mexico, Norway, Oman, and Russia.

Sources: World Integrated Trade Solution: OECD: *World Economic Outlook*; IMF staff calculations.

Dubai Investment Group, an affiliate of Dubai Holding LLC, joined New York-based The Milestone Group and bought 21,000 apartments in Dubai. Additionally, Russia – a non-OPEC country – aims to use $22.3 billion of its $262.9 billion in reserves to repay debts owed to official creditors. Yet, Table 8.1 shows that oil-rich countries have reduced their own strategic oil revenue utilization since the beginning of the twenty-first century.

Norway, the world's third-biggest oil exporter, has a government pension fund that is funded by oil revenues and assets of $234 billion. It is already clear that owners of debts are backing their debts with revenues from commodities. The financing of the countries incurring deficits will be as expensive as the financing of imported oil and energy.

According to the International Energy Administration, China needs to accumulate $2.3 trillion by 2030 to secure its domestic energy consumption. This represents one-third of its foreign currency reserves valuations, financed primarily by US Treasuries until the introduction of new currencies onto world trading exchanges. This implies that China – upon the re-evaluations of its monetary policy – will also make energy imports more expensive and will have to create and issue its own bonds and debts. The introduction of derivatives to the Chinese markets since 2005 is accelerating the monetary conversion of new financing means. And so it is expected that the price of oil in China will be directly linked to the rise of its interest rates and will be implicitly more expensive beyond 2010. Most research centers do not think that insufficient resources will be an issue by 2030 and

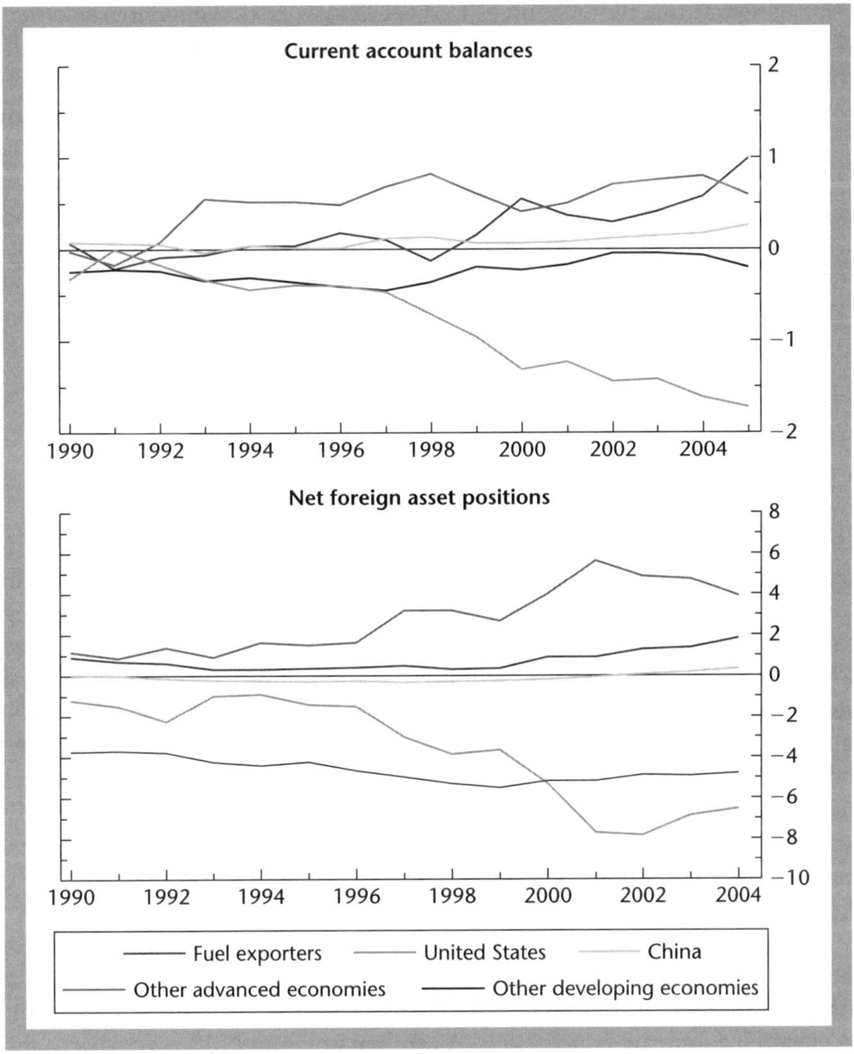

Figure 8.13 Current account balances and net foreign asset positions (% of world GDP)

Note: Large global external imbalances emerged starting around 1996. In particular, the United States is now running an unprecedented current account deficit, with fuel exporters emerging as the main counterparts. Also, the United States is by far the world's largest net debtor. As a group, other advanced economies remain the largest creditors; fuel exporters' net foreign assets, while growing, remain relatively small.

Sources: IMF staff calculations; Lane and Milesi-Ferretti (2006).

that there are sufficient resources to meet demands; however, it is also perceived that apprehensions of shortage might realistically surge by then, thus causing oil prices to rise above normal historical ($45–$55 a barrel) ranges. Lane and Milesi-Ferretti (2006) demonstrate the current account balances and net foreign asset positions of the US, China, and emerging countries with those of exporting countries as a percentage of global GDP (Figure 8.13).

In the meantime, drastic changes in political, economic, and environmental circumstances could also improve or worsen oil supply and demand balances, global market efficiencies, and trading equilibriums (Tables 8.2–8.4 and Figure 8.14).

8.2 NATIONAL AND INTERNATIONAL MONETARY POLICIES

Chinese government oil policies are directly linked to international monetary valuations and national debt financing. Oil prices are correlated to depreciating US dollar values against those currencies corresponding to national surpluses and rising yuan/renminbi valuations. Monetary forex, despite current values, will be realigned on actual tangible natural resources in the future. Guizot (2007) explains how China's $1 trillion forex monetary reserve was accumulated and how it is financed by US Treasury bills in depletion and currently being artificially substituted by other types of speculative investments such as derivatives. Yet, what does not get substituted is the tangible collateralizations of asset valuations created by real capital wealth but highly contaminated by injections of bad loans and artificial debts.

Thus increasing debt in order to invest in new projects and foster economic growth might deplete Chinese wealth faster than expected, reducing its country risk accreditation to the level of those of mature markets depleted by abnormal debt levels. As shown in Figure 8.15, oil prices correlate with global interest rates. Thus, oil prices rose exponentially in correlation with global average interest rate risks. The premium absorbed into pricing per barrel equates to a reduction of the daily purchasing power of the global consumer, thus forcing consumers to borrow to purchase everyday goods.

Real rates of interest declined (Figure 8.16) since they are immune to the inflationary factor which is not included in the figure. Figure 8.17 provides a more thorough description of the operational risk variations between various interest rate policies. The pricing of oil is inversely correlated to US interest rate policies and is a direct function of the US deficit gap. This balancing act is also reflected in the less obvious changes in interest rate risk policies across various countries.

Table 8.2 Increase in fuel exporters' net oil exports ($US billions, constant 2005)[a]

	CPI deflated	Trade price deflated[b]	Per cent of world GDP[c]	Per cent of own GDP[c]	Per cent of world private capital flows[c]	Per cent of world stock market capitalization[c]
1973–81	436	289	1.9	48.9	78.6	7.6
1973–76	239	139	1.1	27.8	58.4	5.5
1978–81	218	174	0.8	14.5	39.3	4.5
2002–05	437	382	1.2	33.2	37.3	1.6

Notes: a. All values deflated by US CPI, except where otherwise noted; b. Trade price deflated figure is calculated using a trade-weighted average of the G-7 non-oil export price deflator; c. World GDP, own GDP, private capital flows, and stock market capitalization are all computed for the first year of the relevant period (except for private capital flows and stock market capitalization during 1973–76 and 1978–81, when the final year of the relevant period was used instead, reflecting limited data availability). Private capital flows are defined as the sum of net direct investment, portfolio investment, and other investment, from the balance of payments. Russia is excluded from all calculations in the 'Percent of own GDP' column, since it was not a market economy during 1973–81.

Sources: IMF staff calculations, *World Economic Outlook*, *International Financial Statistics*; World Bank, Financial Structure and Economic Development Database.

Table 8.3 Change in net oil exports, 2002–05

	Billions of constant 2005 US dollars[a]	**Per cent of world GDP**[b]	**Per cent of own GDP**[b]
Fuel exporters[c]	437	1.24	33.2
United States	−124	−0.35	−1.1
Other advanced economies[d]	−198	−0.56	−1.3
China	−53	−0.15	−3.8
Other developing countries[e]	−53	−0.15	−1.2

Notes: a. All values deflated by US CPI; b. both world GDP and own GDP are computed for 2002; c. includes all the countries in the *World Economic Outlook* group of fuel exporters, with the addition of Kazakhstan and Norway; d. includes all the countries in the *World Economic Outlook* group of advanced economies, except for the United States; e. includes all other countries.

Source: IMF staff calculations.

The fact that oil prices are primarily quoted in dollars and that the dollar is depreciating in the face of global rising interest rates on main national debts is causing an imbalance between the actual price of oil in the US compared to some of the counterparts experiencing a national surplus with strengthening of their domestic monetary exchange values. Figure 8.18 shows the exponential rise of US Treasuries owned by foreign entities. Commodities are also subject to speculation inflating the real value of the underlying deliverable tangible assets.

On 1 August 2006, the Global Association of Risk Professionals and Risk Center released a paper about oil excesses being priced with national debts. Oil-producing nations are challenging the Asian central banks as the biggest source of cash in world financial markets, and the result may be higher US borrowing costs. The current account surplus of countries such as Kuwait and Norway is projected to widen to $311 billion in 2006 from $242 billion in 2005, according to an International Monetary Fund report in April 2006. Asian central banks tend to invest their surpluses in US Treasury securities which help finance the US current account deficit. US Treasury securities used to be considered as guaranteed instruments, but with the depreciation of the dollar and rising interest rates, the debt owed back on home equity loans, personal credit cards, educational loans and car leases are huge, thus depleting further the value of US Treasuries. This gradual depletion effect is a result of the aggressive monetary policies employed to force service industries to keep the economy afloat as service jobs replace manufacturing jobs which are being offshored into cheaper, more cost efficient countries. Figure 8.19 depicts falling US economic growth with depleted US Treasury yields.

Table 8.4 Composition of merchandise imports (% of imports of given importing region sourced from given exporting region)

	Exporting region					
	Fuel exporters[a]	United States	Other advanced economies[b]	China	Other developing countries[c]	Total
Importing region – 2004						
Fuel exporters[a]	–	8.4	59.0	7.6	25.0	100
United States	8.3	–	54.0	13.8	23.9	100
Other advanced economies[b]	19.5	25.8	–	21.3	33.4	100
China	9.2	8.6	65.3	–	17.0	100
Other developing countries[c]	13.3	19.5	59.3	8.0	–	100
Importing region – change between 1981 and 2004[d]						
Fuel exporters[a]	–	−5.7	−9.0	6.6	8.0	–
United States	−11.4	–	−4.9	13.0	3.2	–
Other advanced economies[b]	−19.6	−8.7	–	18.1	10.1	–
China	8.5	2.6	−14.7	–	3.6	–
Other developing countries[c]	−9.2	−4.8	7.8	6.3	–	–

Notes: a. This group is as defined in the text; b. This group includes all the countries in the *World Economic Outlook* group of advanced economies, except for the United States; c. This group includes all other countries; d. Percentage point difference in the share of imports between 1981 and 2004 (i.e., a positive number indicates an increase since 1981). The year 1981 is the earliest date available with data coverage comparable to 2004.

Source: IMF, *Direction of Trade Statistics.*

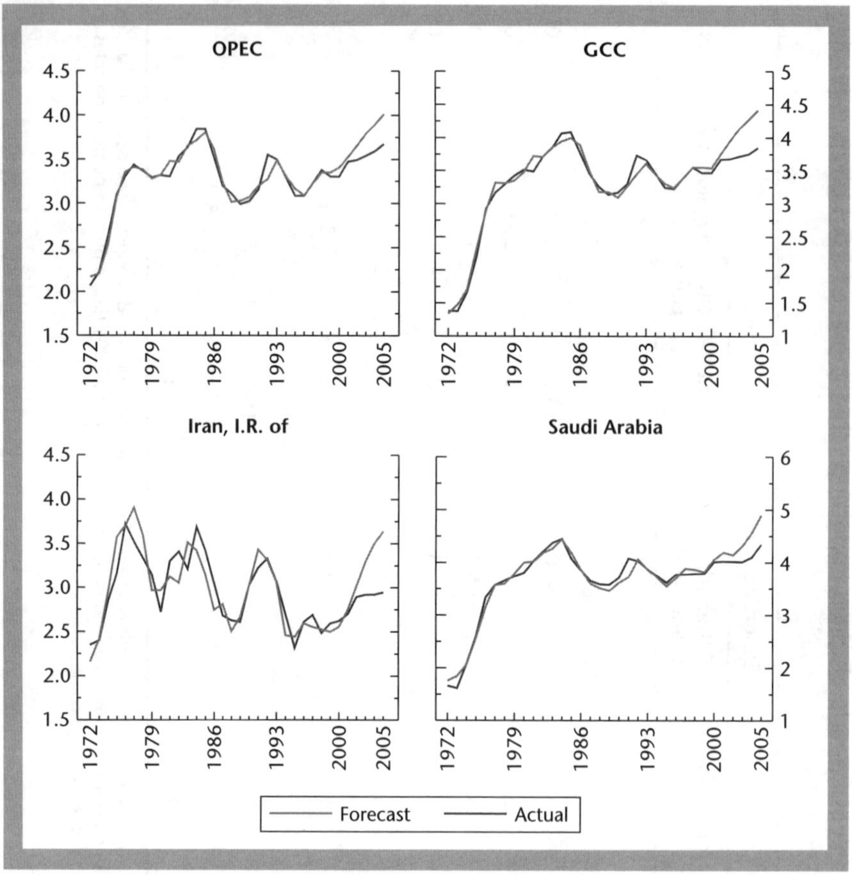

Figure 8.14 Dynamic forecasts for real imports (log of billions of 2000 US dollars)
Source: IMF staff estimates.

Service jobs will also eventually be substituted by more cost efficient and productive machinery. What has maintained the strong levels of the dollar is the buying of US Treasuries by Asian foreign banks in order to keep exports high. While China is increasing its financial infrastructures and is gradually readjusting the yuan according to domestic economic growth potential, the liquidity for ownership of short-term capital in dollars is rising among various international competing players such as Japan, Europe, the Middle East, and Asia (Figure 8.20).

Ultimately, it is the over-consumerism of Western societies that lost then economic control over the so-called new democracies. To keep Asian currencies weaker and exports competitive, Asian central banks used revenues received from selling goods to the US and Western Europe to buy low-yielding, dollar-denominated Treasuries. Figure 8.21 shows a change

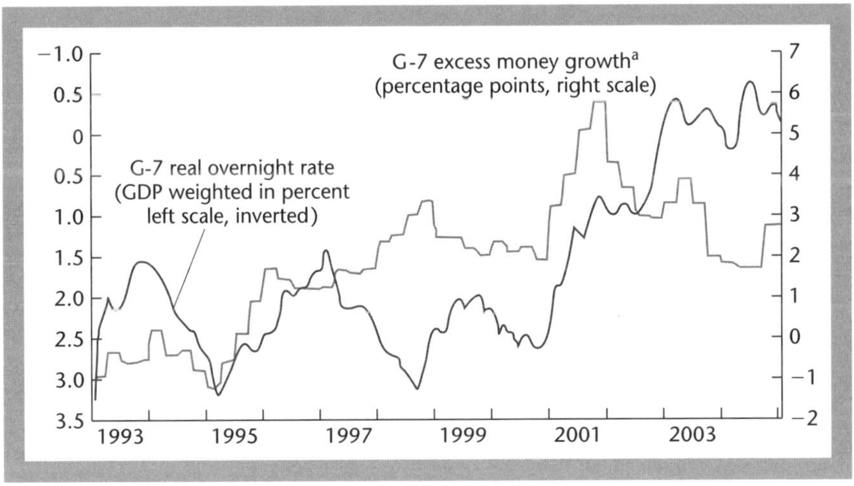

Figure 8.15 Global real interest rates and excess liquidity
Note: a. Excess G-7 monetary growth is defined as the difference between broad money growth and estimates of demand in each of the G-7 countries, weighted by their respective GDP.
Sources: OECD; IMF staff estimates.

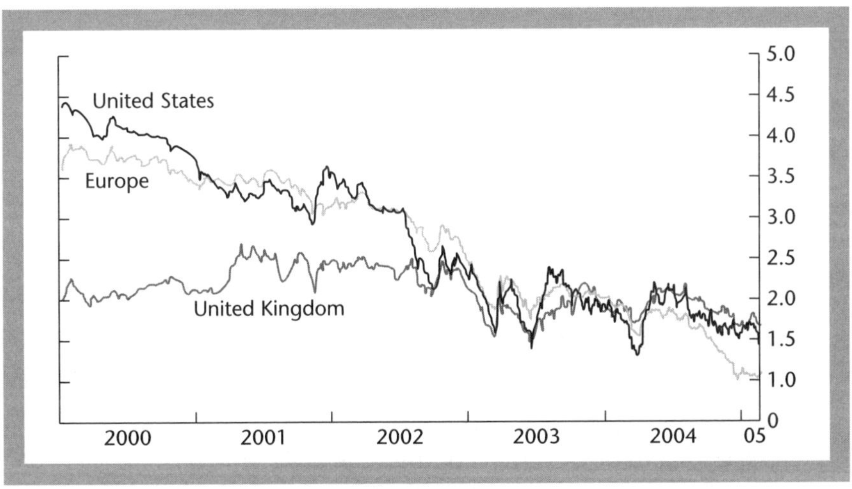

Figure 8.16 Real yields on inflation-indexed bonds (%)
Source: Bloomberg LP.

in Treasury yields prior to and after an increase in the overnight Federal Reserve rate.

By selling their own currencies to buy dollars, Asian central banks kept the US currency stronger than it should have been. This cycle has been

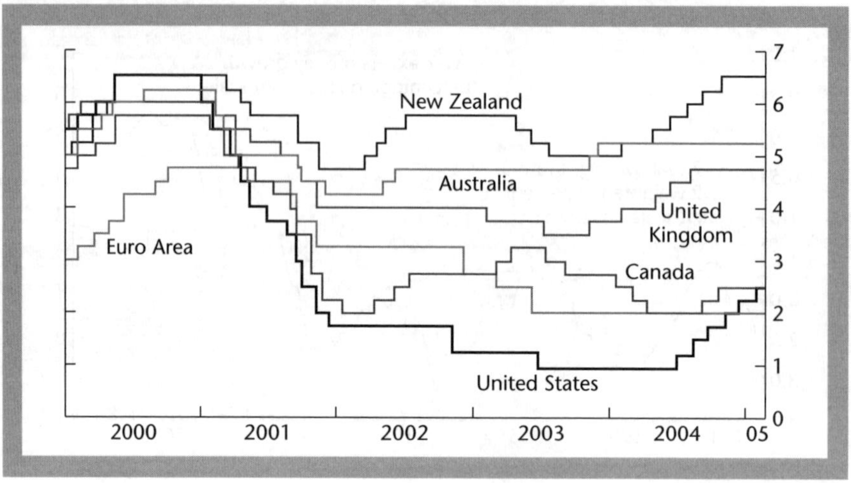

Figure 8.17 Policy rate (%)
Source: Bloomberg LP.

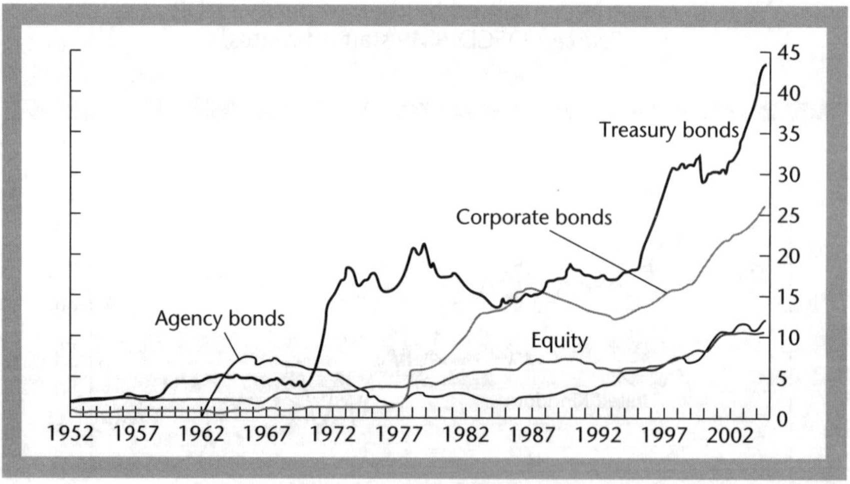

Figure 8.18 Foreign ownership of US securities (% outstanding)
Sources: US Board of Governors of the Federal Reserve system, *Flows of Funds Accounts of or the United States*; IMF staff estimates.

maintained for almost a decade. This mechanism fueled more purchases of Asian goods. Consequently, the significant buying of US Treasuries maintained US interest rates at an artificially low level causing American consumers to take out more home equity loans, car leasings, educational loans, and credit debt card. At the corporate level, financial banking institutions are currently restructuring debts with structured and derivatives instruments (Figure 8.22).

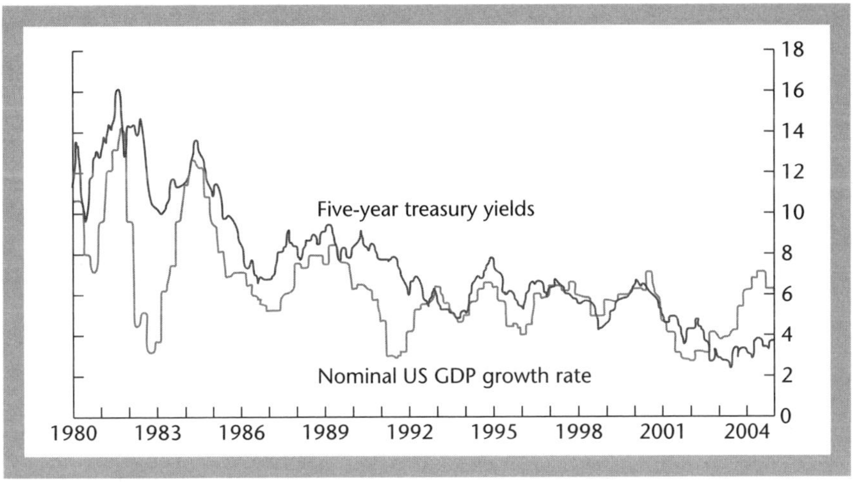

Figure 8.19 US economic growth and Treasury yields (%)
Source: Bloomberg LP.

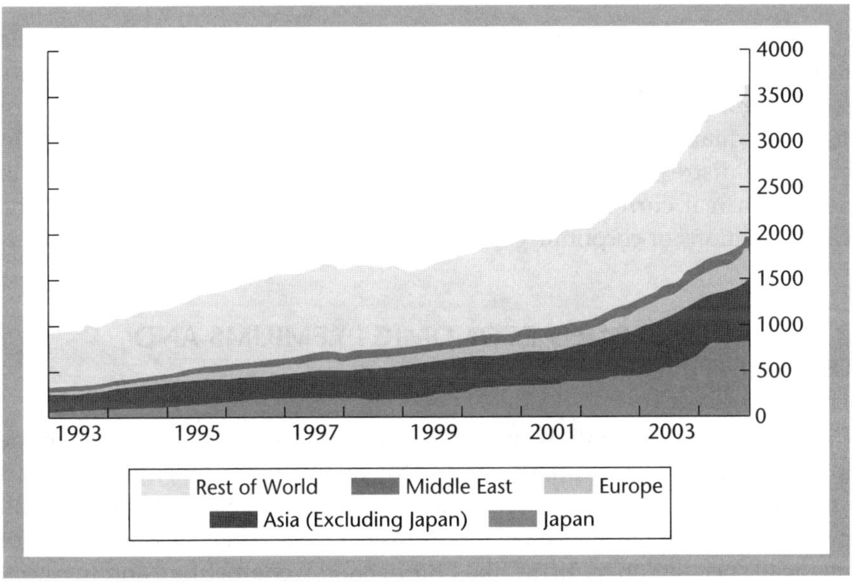

Figure 8.20 Rising global liquidity: international reserve accumulation ($US billions)
Sources: IMF, Bank of International Settlements, bis.org; Treasury International Capital System.

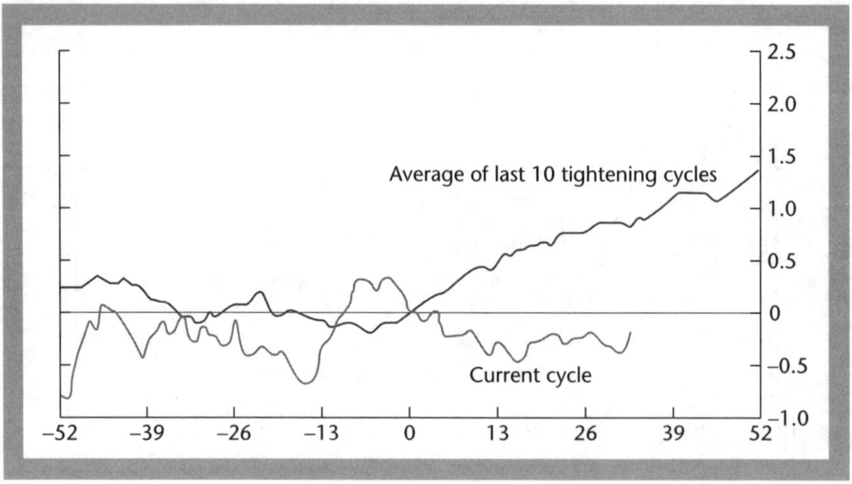

Figure 8.21 US tightening cycles: movement in 10-year Treasury yields (in weeks before and after first Fed rate increase, % change)
Sources: J.P. Morgan Chase & Co.; IMF staff estimates.

The ownership of commodities and energy will be used in the foreseeable future as a contribution to the credit quality of the country. Commodities are tangible goods that may be used as collateral assets to negotiate the country risk accreditations and government or sovereign bonds credit spreads. In economic distress, countries use tangible goods, such as commodities, to trade as a real currency of exchange in negotiating political deals at the United Nations or corporate commercial contracts in global affairs.

8.3 NET EMBEDDED ECONOMIC PREMIUMS AND DISCOUNTING EFFECTS

Oil prices rose exponentially in the 1860s, 1910s and 1970s, declining in the intermediate periods primarily due to the fact that oil supply and demand are more elastic in the long run, thus smoothing or stretching long-term oil prices to more stable levels. Oil prices rose in 1999 principally as a result of financial speculation by firms like Enron, energy competitors and financial institutions. Now, all financial firms and hedge funds trade commodities and energy derivatives. Additionally, commodities pool operators (CPOs) and commodities trading advisors (CTAs), as well as derivative futures managers, have proliferated, contributing to the rising prices of energy products through speculation. For instance, Jim Rogers, co-founder of Quantum Fund with George Soros, reported at the Hedge Funds World Global Opportunities 2005 conference that oil prices could run to $150 by 2015. My own estimates are not too far from his predictions. These

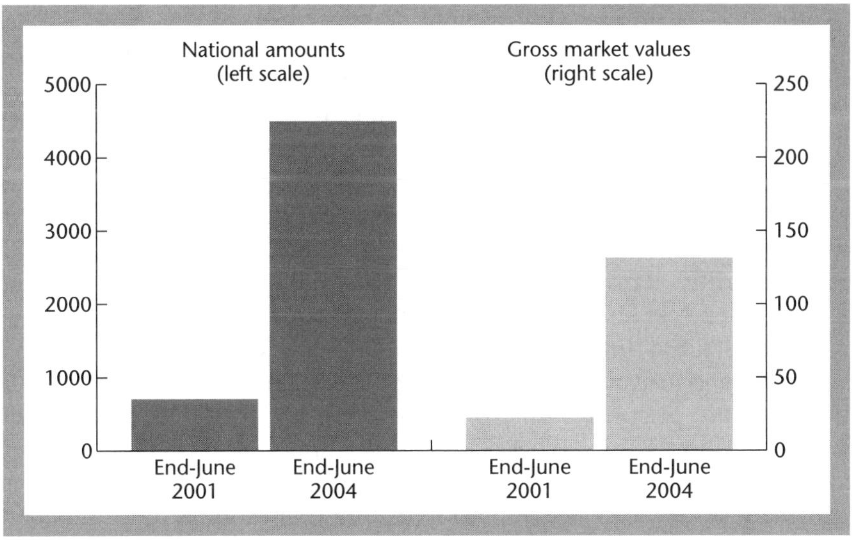

Figure 8.22 Credit derivatives market (amounts outstanding, $US billions)
Source: Bank for International Settlements, bis.org.

assumptions are based on the assumption that once Chinese firms get into the trading of financial derivatives and commodities, the valuations of new Asian equities firms will pump up global oil prices and institutional trading activities at international exchange levels to stimulate added speculation from new democratic markets. In addition to China, India is also allowing hedge funds into its financial markets, creating an added element in the speculation of oil prices through financial trading. This embedded 'invisible' premium may be estimated as follows. As of the beginning of 2006, the (oil price) reference was around $55 a barrel, $78 by mid-2006, back down to $50 by mid-January 2007 and up to $75 by end July 2007. Emerging new democracies and new exchanges facilitate liquidity and trading as well as volatility. Figure 8.23 depicts the evolution of the actual value of real oil prices since 1970. The embedded rise in oil prices since 2002 is principally due to the speculative premium from overtrading on energy derivatives markets by large financial firms that have replaced Enron. The increased oil pricing since 2000 also coincides with rising volatility, speculative investments, Chinese demand, and geopolitical uncertainties after 2003.

To adjust for inflationary value, the average price of global oil curves was $23 through the twentieth century, taking the 2004 price as a historical reference. While the real price of oil varied from $10 to $20 until the early 1970, it spiked to $38 in 1973. Prices increased five-fold from $15.5 in 1973 to $83 in 1981 – still under the 2004 base pricing reference. Historical trends then show that prices fell to $13 in November 1998. With adjusted inflationary factors, this was the lowest price since the Second World War. Since 1998 oil prices have risen from $11 to $73 as of mid-2006. Note that the financial engineering derivatives did not come into play until 1973.

The application of financial engineering derivatives into commodities and energy appeared in the mid-1980s and by the 1990s heavy speculation contributed to the exponential premium. Yet, despite the fact that oil prices reached a historical peak in mid-2006, oil prices remained cheaper than in 1981, under the base level of 2004, without the inflation factors discounted back. In reality, oil prices would have to rise by at least 50 per cent from the actual price level of $73 to get to the real levels of 1981.

It is difficult to differentiate actual inflationary factors from speculative influences and to evaluate the importance of each in the extra layer between 1981 and 2006 prices. Yet, 2006 oil prices were 50 per cent higher than the average oil prices since the 1973 shock. As Jim Rogers (2005) has noted, most high peaks in oil commodities have been followed by shocks, but unusually since the beginning of the twenty-first century, oil prices have risen continuously and record six consecutive years of cumulative gains. Unprecedently, the same trends have been recorded in the hedge fund industry, while traditional indices levelled off in 2005.

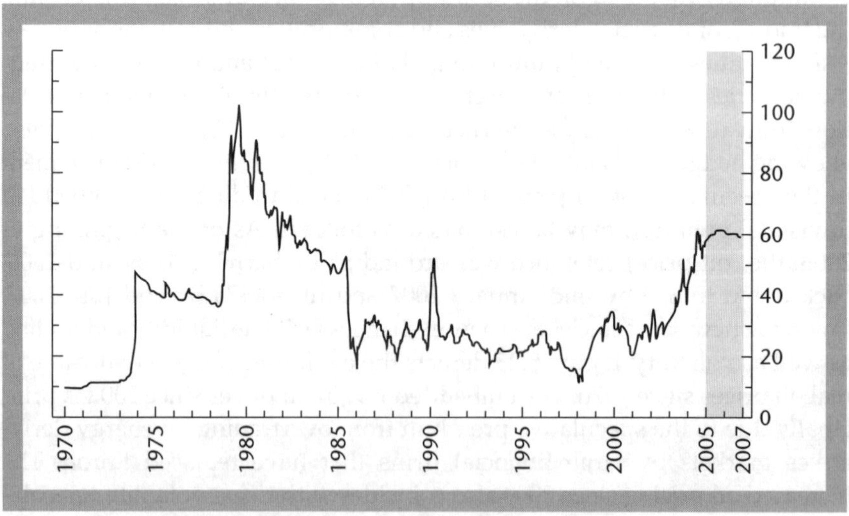

Figure 8.23 Real oil prices (2005 $US per barrel)
Note: Energy prices started to increases in 1999, with a sharp rise after 2003. This upsurge is to a large extent driven by growing demand in advanced and emerging economies, as well as by expectations of future market tightness. However, current and expected future real oil prices are still significantly below their value in the late 1970s and early 1980s.
Source: IMF, 2006.

In addition to being connected to foreign exchange volatilities and economic levels, oil prices are also directly linked to geopolitical risks in the Middle East where most of the world's proven reserves are concentrated. The price of oil also includes a 'terrorism premium' embedded in its actual pricing structure.

However, these factors contributing towards higher oil prices are compensated by the heavier usage of information technology and services which are less dependent on oil but more on alternative energies. Thus oil also has a minimizing effect on economic statistics such as GDP. As we saw earlier, the oil price represented 2 per cent of global GDP in 2005 while it accounted for 8 per cent in 1973.

8.4 FOREIGN EXCHANGE VOLATILITY FACTOR IN OIL PRICING VALUATIONS

An added variable in the pricing of oil is foreign exchange volatility with respect to the base price per barrel. Thus, in addition to financial trading volatilities, supply, demand, operational technological infrastructural risks, and political risks, the volatility of the embedded foreign exchange that is subject to the debt or surplus of a country also affects oil prices. Thus, although the price of oil is set in US dollars, the cost of oil in Europe, Japan or China also depends on forex volatility and is subject to a speculative premium embedded in the price of oil. The price of oil rose exponentially, especially from 2002 to 2006, as depicted in Figure 8.24.

In correlation to price increases, historical spreads also widened significantly from 2004 to 2006 as shown in Figures 8.25 and 8.26.

As a consequence of significant infrastructural operational shifts within global economies and trading activities, historical implied volatility significantly altered also from December 2005 (Figure 8.27). This produces further opportunities for dislocations of the underlying pricing structure as more and more arbitrage trades exacerbate volatility and expose underlying market economic efficiencies.

European currencies appreciated relative to the US dollar in the 1970s, reducing the economic impact of higher oil prices. The price per barrel rose from about € 3 at the beginning of 1973 to € 25 by the summer of 1980. Similar volatility shifts occurred again from 2002 to 2006. Although the US price for oil peaked in February 1981 at $38, the decline in the value of European currencies meant that the cost of oil continued to rise until 1984. Oil rose from € 32 in February 1981 to € 40 in September 1984. The price of oil increased by 20 per cent in euros during those three years, but fell by 30 per cent in US dollars. The price declined to around € 11 per barrel by March 1986 and remained under € 20 until 1999. The strengthening of the euro against the dollar had a mitigating effect on the eurozone compared to the US, while the American deficit

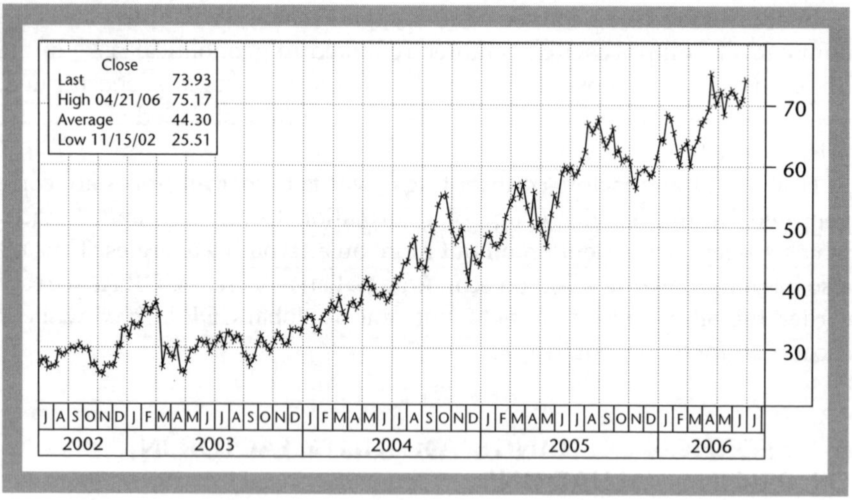

Figure 8.24 Daily oil prices, June 2002–06 ($US)

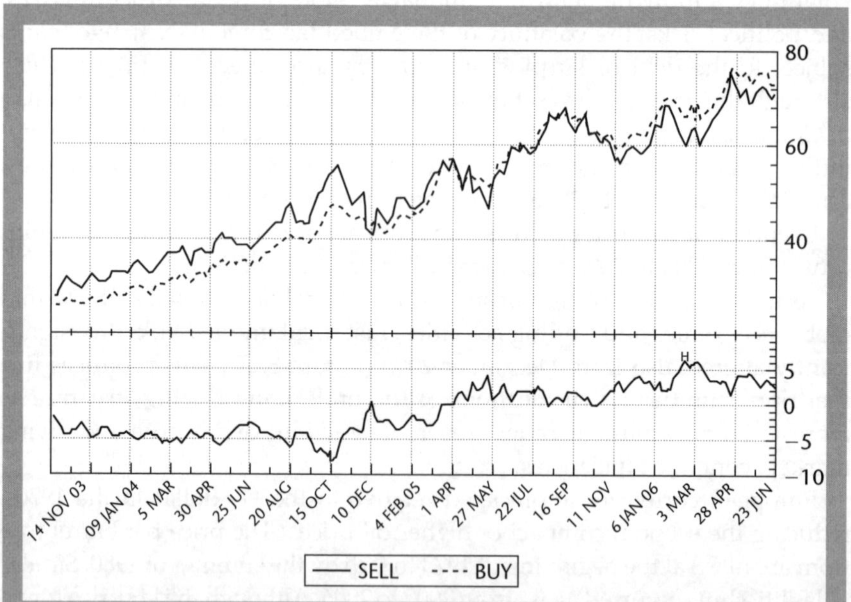

Figure 8.25 Historical price spread for oil 2003–06 (US$)
Note: The wider the spread, the sooner mature markets financial indicators will plateau, hedged by 'new' markets. Commodities, including energy, are most subject to speculative trading, causing markets to widen further and become inefficient.

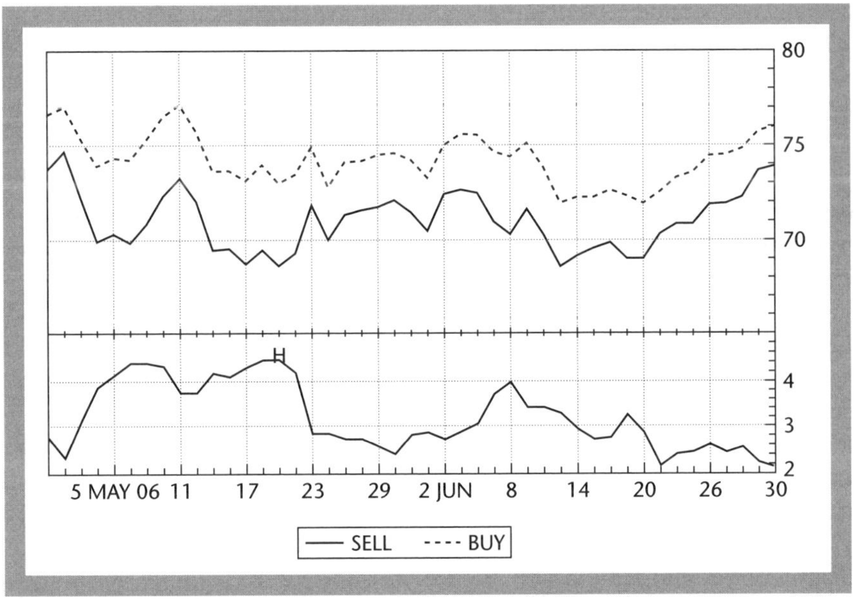

Figure 8.26 Historical price spread for oil, May–June 2006

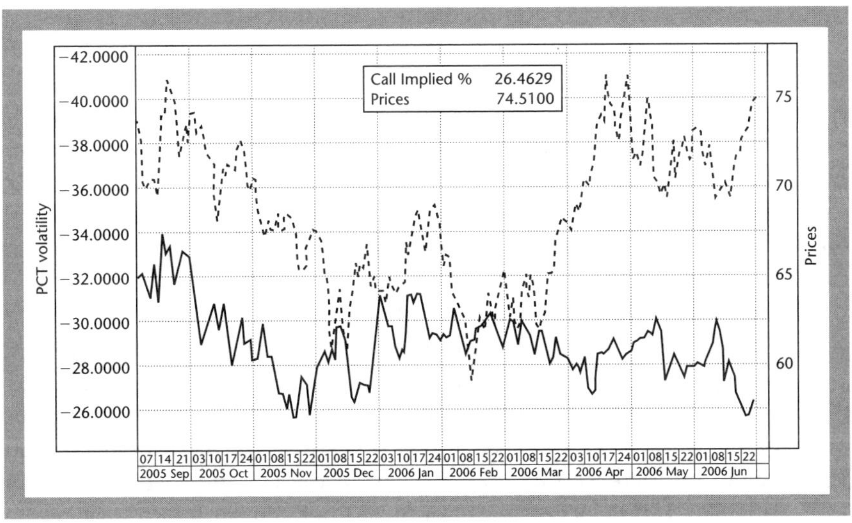

Figure 8.27 Historical implied volatility
Note: As of end 2005, oil prices encountered a significant pricing shift, adding a $30 premium in the price from December 2005 to June 2006.

actually accelerated. The dollar currency depreciation implies that it takes an additional $1 trillion to finance the deficit every 20 months. The US has been a net creditor since 1988, and profited from this situation until 2000. The price of oil per barrel reached € 38.50 or $33.60 in November 2000, and climbed up to € 44.20 or $56.37 in October 2004, and to $73 mid-2006. Basically, the barrel price increased by at least 70 per cent after November 2000 in dollars, but by only 15 per cent in euros, simply as a result of the serious unexpected currency depreciations on the dollar side. But the dollar depreciation did not come only from its 'own' twin deficits, part of it is related to the overwhelming internal debt in the US, and another part of it is implicitly due to the much higher productivity levels in new democracies, not adjusting currencies accordingly, and thus implicitly financing wealth artificially. Consequently, currency volatility also depreciates internal liquidity and the qualitative ratings of government sovereign bonds. The Chinese yuan was fixed against other currencies until mid-2005 when it started floating against a basket of other currencies. It was institutionally traded prior to that date at a fixed rate of 8.33 yuan to the dollar. It is now available at floating adjusting rates, and is traded institutionally and individually (figures 8.28 and 8.29).

This monetary depreciation or erosion of indebted countries' currencies in deficit also revalues actual commodities prices, which rose after 2002. The rise of commodities prices also implied more speculation, most particularly in hedge funds and large financial institutions, exacerbating derivatives trading activities in energy and commodities instruments. Consequently, commodities prices also suffered from an increase in volatility as shown in figures 8.30 and 8.31.

The logical scenario also implies that this over-abused volatility created a depreciation or devaluation of the actual natural pricing of the underlying asset class.

While in 2006 oil prices have been at their highest since the 1980s in both the US and in the eurozone, Japanese manufacturing still benefits from forex volatilities as shown in Figures 8.32 and 8.33.

A barrel of oil in Japan cost around 1000 yen in 1972 and rose to over 9000 yen in October 1982. When the yen strengthened, Japanese oil decreased, to 1750 by July 1986 or by 75 per cent from its 1972 level. Japanese manufacturing was at its highest productivity levels in the 1980s, partly as a result of lower oil prices, but also the economy was still in the inflationary period. The price of oil increased by 400 per cent in the US and Europe between 1973 and 1986, but it increased by only 150 per cent in Japan, principally because of the strengthening yen. Further strength in the yen meant that by the end of 1998, a barrel of oil cost only 1400 yen, only 40 per cent higher than in 1972. As of the end of 2005, the Japanese oil price was about 6000 yen, which is still lower than the yen price in 1982. Thus with foreign currency volatilities included in the model of global oil price fluctuations, the real value of oil takes a different shape.

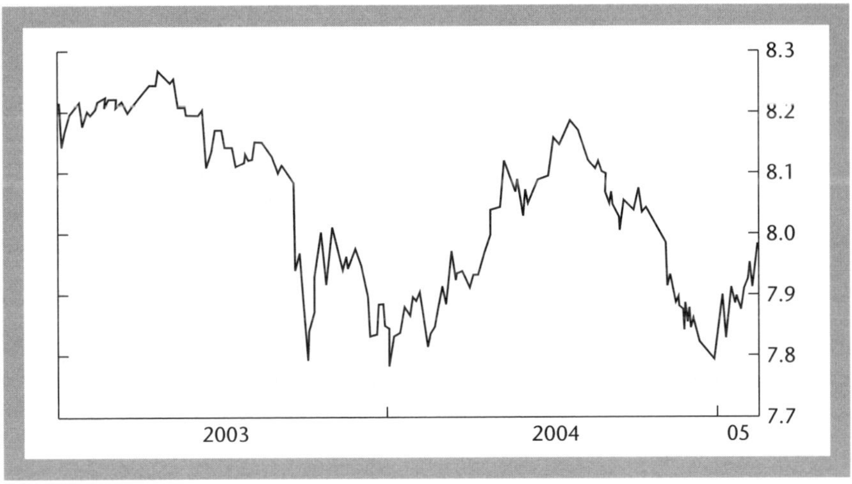

Figure 8.28 Chinese yuan 12-month forward rates (yuan per US dollar)
Source: Bloomberg LP.

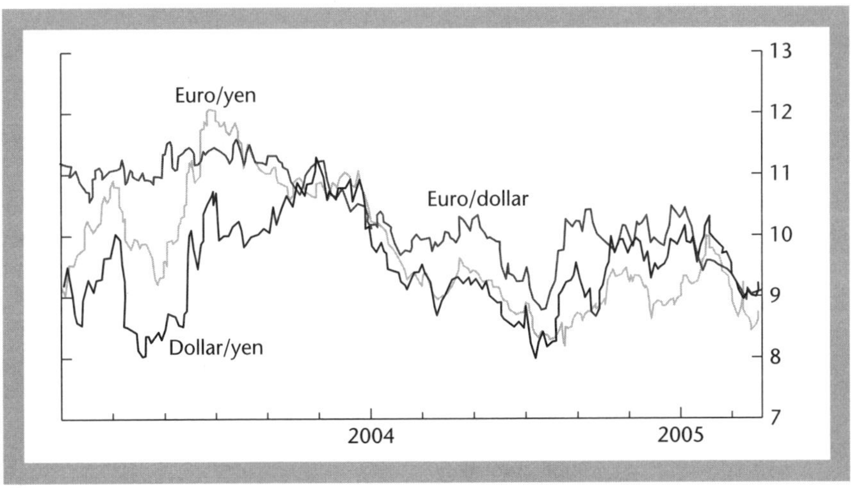

Figure 8.29 Currency volatilities (three-month forwards, %)
Note: The yuan ought to be part of global economic variables but has been avoided for many years because it was pegged or fixed. The yuan has been appreciating slowly since July 2005, but not significantly enough to alter global imbalances.
Source: Bloomberg LP.

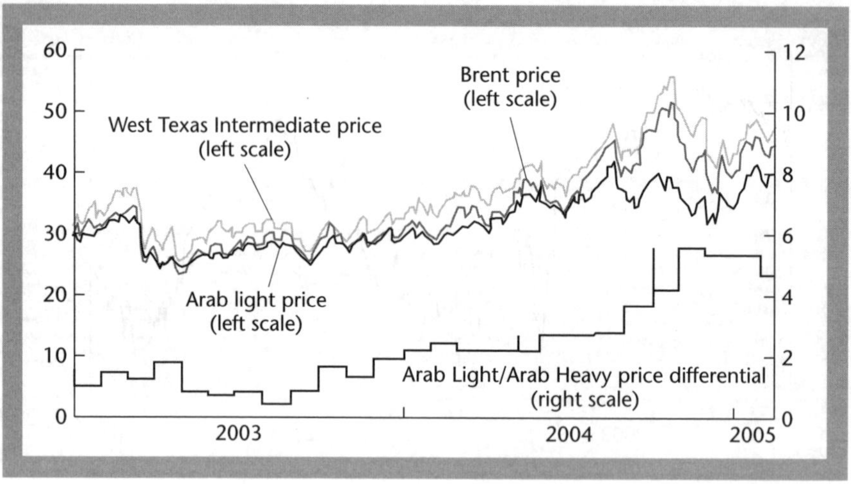

Figure 8.30 Arab Light/Heavy, Brent, and West Texas Intermediate crude oil pricing (US$ per barrel)
Source: Bloomberg LP.

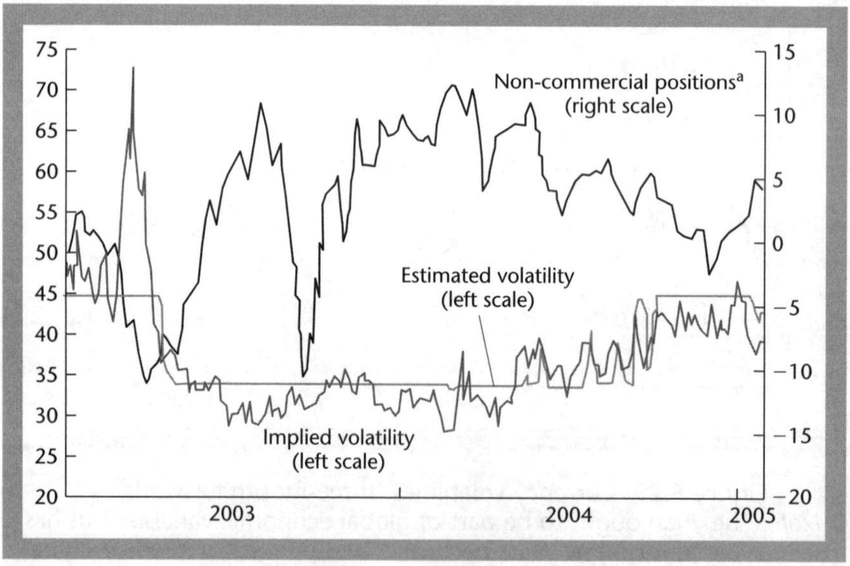

Figure 8.31 Implied volatility of oil futures prices and crude oil positions of non-commercial traders (%)
Note: a. Long positions as a percentage of total open interest.
Sources: Bloomberg LP; IMF staff estimates.

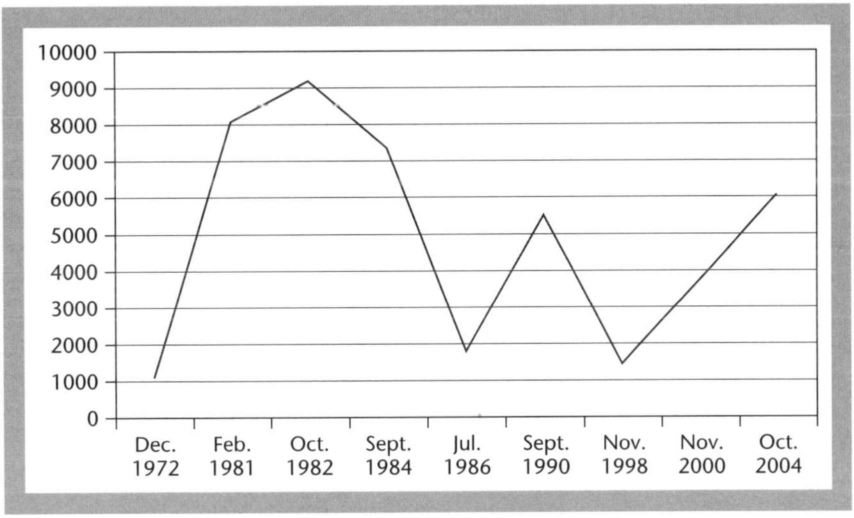

Figure 8.32 Real oil prices in Japan if discounting yen foreign exchange currency effect in barrel prices
Source: Yahoo finance data and author's estimates 2006.

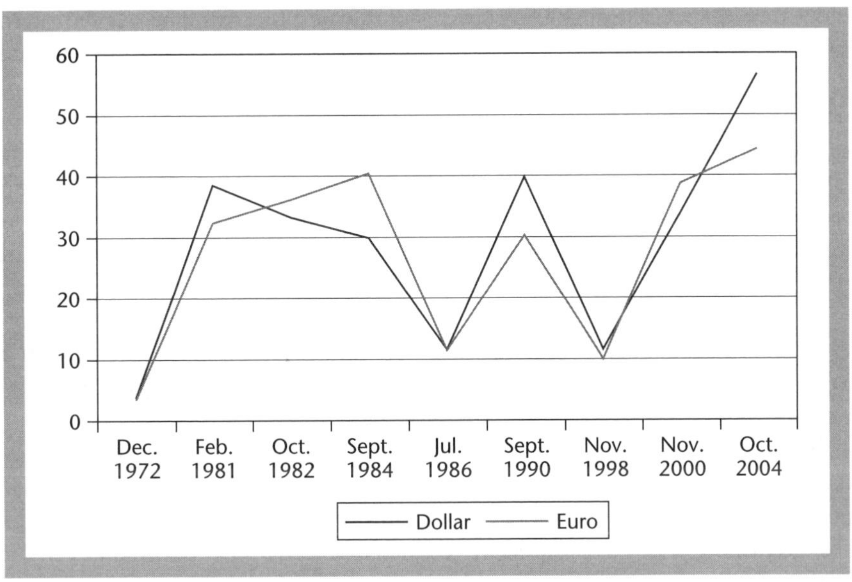

Figure 8.33 Real oil prices if discounting foreign currency volatility factors in dollar versus euro implicitly embedded in actual barrel prices
Source: Yahoo finance data and author's estimates 2006.

8.5 OIL VOLATILITY FACTOR

Global oil prices have been especially volatile over the last 25 years. Oil pricing volatility has also been subject to geopolitical and economic instabilities. High real prices exacerbate competition from large marginal sources of oil.

Despite the search for alternative energy, those substitutes are insufficient and too expensive to meet the rising demand for energy. In fact, the availability of alternative fuels, such as natural gas, hydro, power, wind, and solar are minimal compared to oil, causing prices to rise exponentially and with added volatility.

8.6 FUTURE OIL PRICING MODELS

The International Energy Administration (2007) predicted that oil prices from 2003 to 2006 would experience historical highs and that oil prices in the following ten to thirty years would not rise beyond what has already been exhibited. Many financial analysts also think that oil prices have peaked, and expect them to be lower in future than the peaks and exponential trends experienced between 2003 to 2006: 2004 oil prices rose by more than $9 per barrel; 2005 oil prices continued to exhibit an escalation to $15 per barrel; in 2006 the price neared $75 per barrel. The International Energy Administration reported in IEO2006 reference case that these exponential trends did not reflect longer term predictions and that prices are expected to fall, not exceeding $60 until 2030. From June 2006 to January 2007, the price per barrel of oil fell by almost $18. From record nominal high levels throughout 2006, oil prices in the reference case are predicted to decline gradually to $47 per barrel in 2014, then rise by about 1.2 per cent per year to $57 per barrel in 2030 with adjusted forex volatilities. According to the IEA's forecasts, oil demand increases substantially. Oil consumption is expected to rise between 2003 and 2030 and is forecast to be between a lowest level of 22 million barrels per day and a high point of 48 million barrels per day.

The model presented below is straightforward and concurs with the IEA forecasts of significant rise in consumption primarily due to increased geopolitical risks, new democracies' rising demands, and scarce resources shared among many more people. It does not consider the forex and inflationary factors because debt will be assumed and distributed between all nations by 2020, and debt will be taken into the forex and yield curves, netting the benefits from increased service jobs relying more on hydrogen or other alternatives substituting for oil. The model is a simplistic representation of the forecast supply and demand, given historical consumption and production levels in China and erasing most of the new assumptions and distorting variables netting each other out. Those variables are considered to have an insignificant impact in comparison to the rising demands for oil by the new democracies. In addition to these assumptions, the author also takes into account the entry of the Chinese market into the global derivatives and financial engineering markets which subject oil prices to added speculative premiums, not yet of huge significance in China as it only started trading derivatives in November 2005 (figures 8.34 and 8.35).

The resulting oil pricing curves are exponentially shaped as demand will 'unbalance' or create an inefficient market, aggravated in part by rising geopolitical tensions and financial speculation which no longer assumes that for each buyer there is a seller. (See figures 8.36–8.43.)

The slope of the convexely shaped curve since 2002 reflects a never previously seen historical pricing trend exacerbated by the rise of demand and scarcer resources disputed among many new globalized emerging nations. This would be true if and only if there was perfect matching and netting of assets collateralized valuations, but there is not. The rise of credit derivatives also reinforces this school of thought.

Another factor in the predicted inflationary oil prices is the integration of Chinese energy firms into the global trading and speculating scenes. According to ChinaBidding of 31 January 2007 in 'China Adopts New Pricing System for Oil Products', the Chinese State Development and Reform Commission reported that it was considering the trial operation of

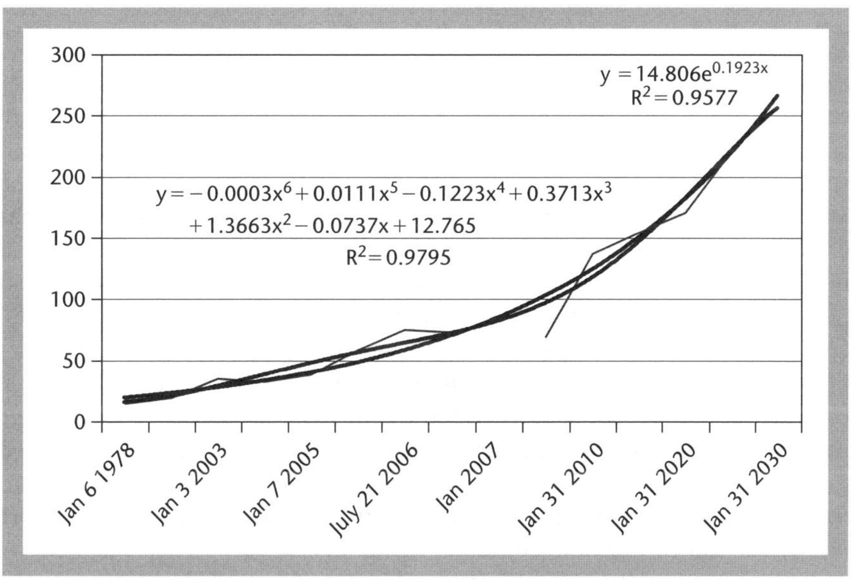

Figure 8.34 Oil pricing function of Chinese oil in $US equivalent (assuming no widening of economic and political conditions between China and the US)

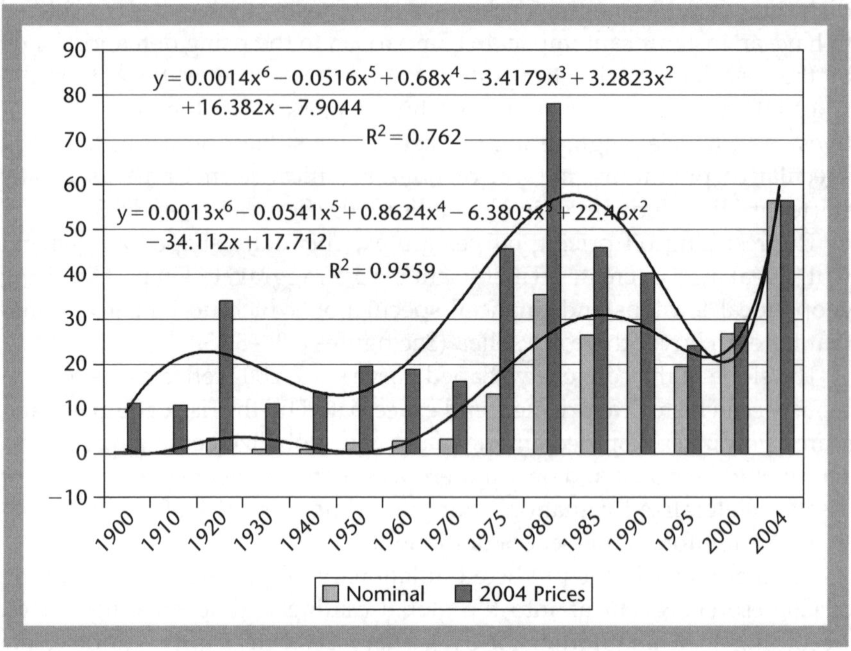

Figure 8.35 Historical oil prices ($/barrel)
Sources: Bloomberg LP and Yahoo finance 2004.

a new system to put an end to heavy losses in the oil processing and petrochemical industries. The new system means that domestic oil prices will become more closely linked to those on international markets, and consumers will have to bear bigger margin price rises on imported oil. The 'crude price plus cost' method was based on the Brent, Dubai and Minas crude oil prices, taking into account processing costs and possible opportunities for enterprises to profit. Sinopec, China's leading oil processing and petrochemical enterprise, relied on imports for three-quarters of the total crude it processed a year. It was highly sensitive to price hikes on international crude markets. It often bought crude at a high price, but sold oil products at a low price, as prescribed by the state to stabilize the domestic market and safeguard consumers' interests. Though the state subsidized the company substantially in 2005 and 2006, the move failed to stem losses for the oil refiner and petrochemicals producer. China needs to build more oil production entities to meet rising demand.

In the past regulations have imposed limits on energy prices (for example, the California power crises in 1999 and 2000) and the capping of energy prices has exaggerated inflation. This has forced private companies to trade in other countries or to create offshore corporations and this has created market

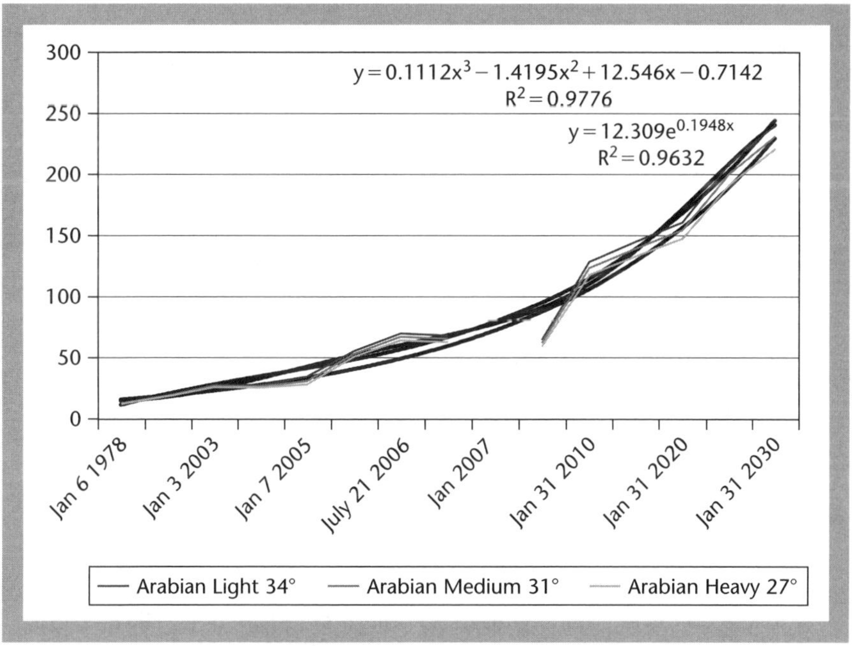

Figure 8.36 Saudi Arabian oil price approximations, 1978–2030 (estimations as 2006 in $/barrel)
Notes: Unkeyed curves indicate various grades of Saudi Arabian oils. Figure assumes no political change in the region.
Source: Author's estimations (2006).

illiquidity and inefficiencies. The same phenomena occur with financial corporations and hedge funds trading commodities and energy products.

Another fundamental cause of rising energy prices is that the high cost of higher grades of oil will rise substantially. A longer-term view demonstrates that oil pricing varies significantly from one geographical point to another, creating greater economic tension in adequate resource sharing. Some countries will be able to buy oil more cheaply than others because of their natural resources, proximity to exporting countries, and capital and or productivity abilities such as the Middle Eastern countries or China, while others will incur a higher pricing premium as a result of total dependency and a lack of compensating alternatives to the purchase of oil, such as the United States or Western European countries. Moreover, China is likely to incur higher oil prices as a result of import transportation costs, which will be the main variable affecting the spreads in oil pricing between various global geopolitical points in the future. With rising living standards, the Chinese will eventually own a car per household; the average cost of a car made and assembled in China is equivalent to € 5000 as of 2007.

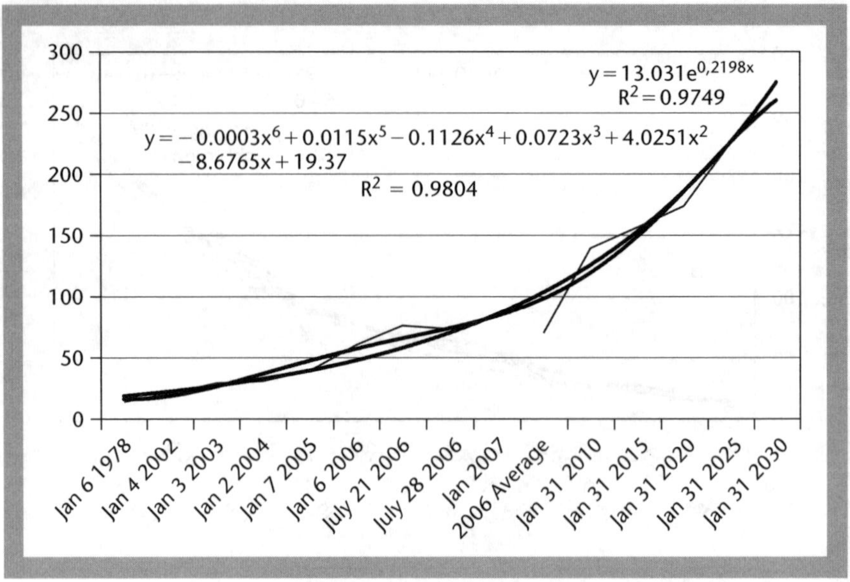

Figure 8.37 Abu Dhabi Muban 39° oil pricing function, 1978–2030 (approximations as of 2006)
Source: Author's estimations (2006).

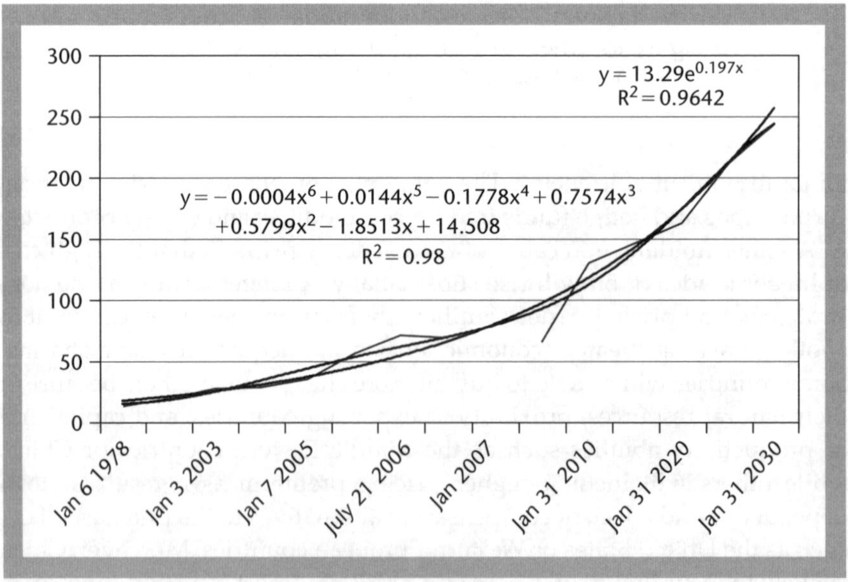

Figure 8.38 Dubai Fateh 32° oil pricing function, 1978–2030 (approximations as of 2006)
Source: Author's estimations (2006).

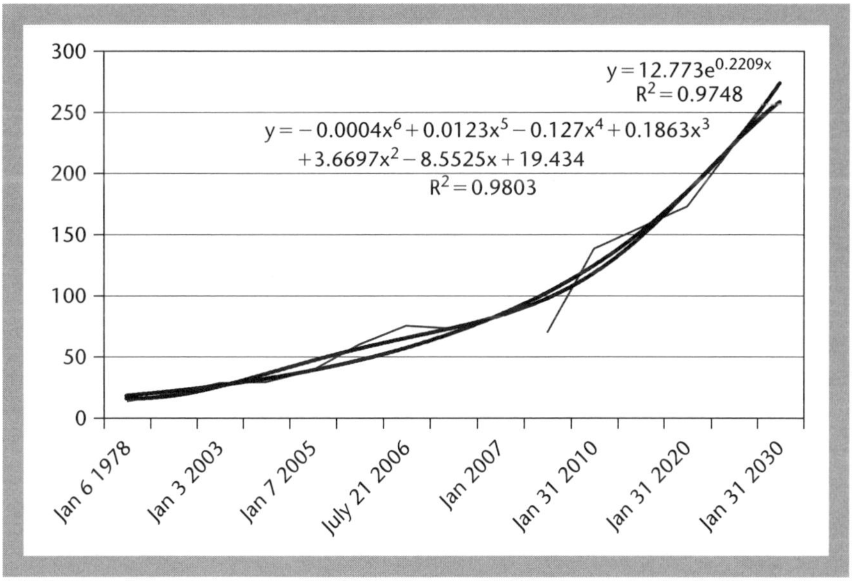

Figure 8.39 Qatar Dukhan 40° oil pricing function, 1978–2030 (approximations as of 2006)
Source: Author's estimations (2006).

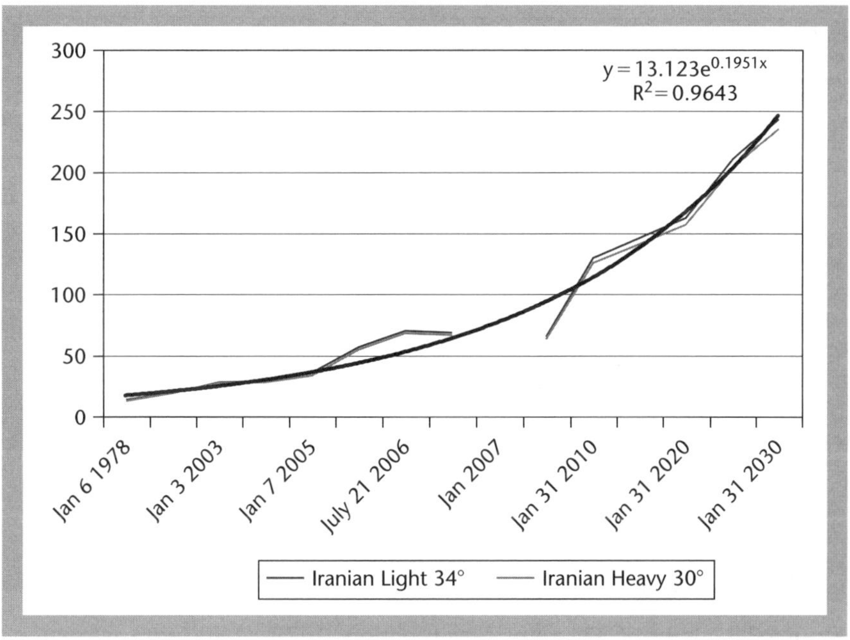

Figure 8.40 Iranian Light 34° and Heavy 30° oil pricing function, 1978–2030 (approximations as of 2006)
Source: Author's estimations (2006).

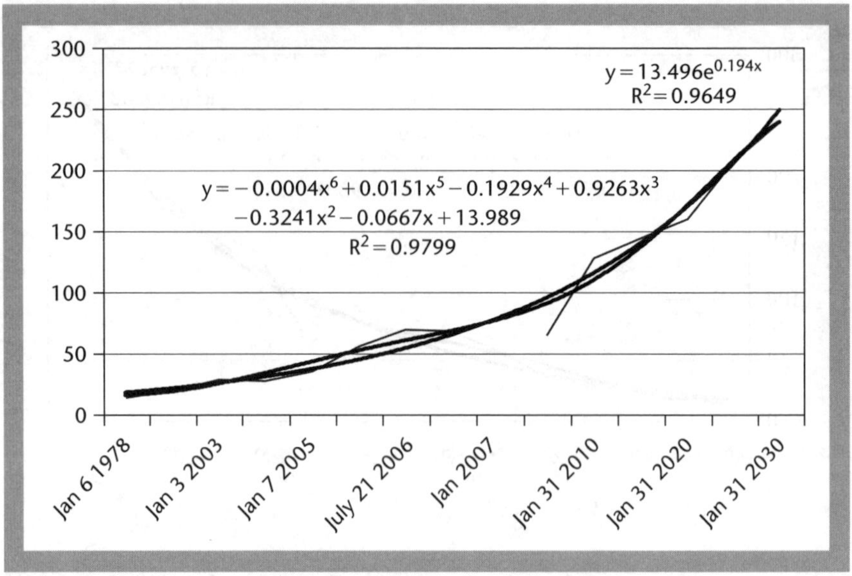

Figure 8.41 Iraq Kirkurk 36° oil pricing function, 1978–2030 (approximations as of 2006)
Source: Author's estimations (2006).

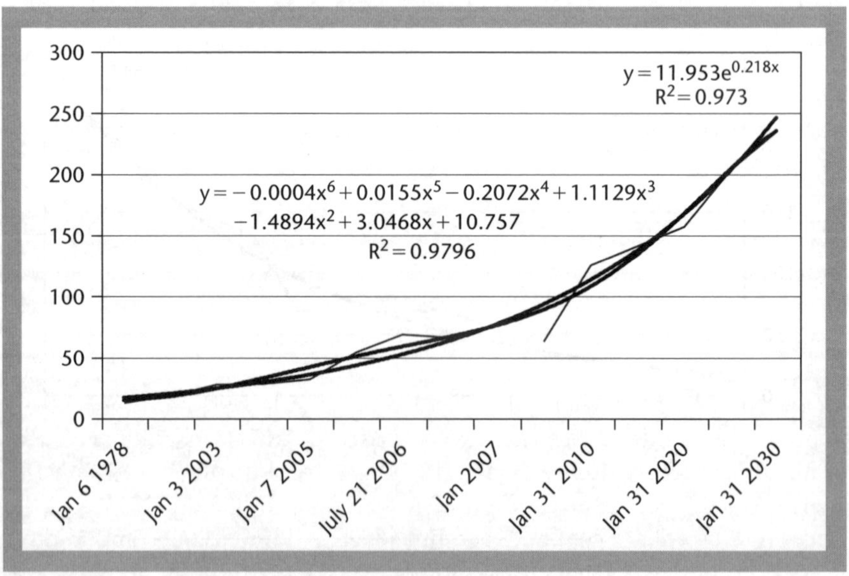

Figure 8.42 Kuwait 31° oil pricing function, 1978–2030 (approximations as of 2006)
Source: Author's estimations (2006).

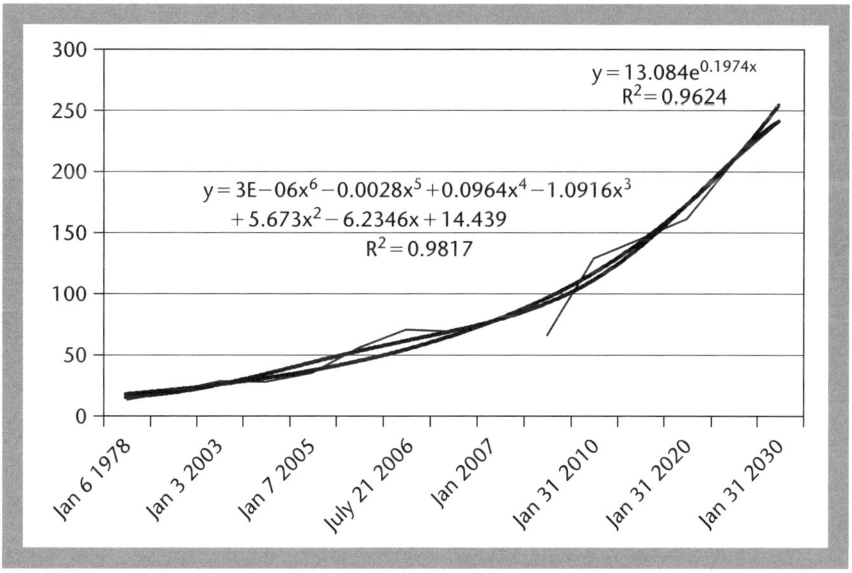

Figure 8.43 Neutral zone Khafji 28° oil pricing function, 1978–2030 (approximations as of 2006)
Source: Author's estimations (2006).

Figure 8.44 depicts the actual spreads and pricing spreads given various geographical points and world average.

8.7 CORRELATIONS WITH GOLD AND THE REAL ESTATE MARKET

Global energy and commodities prices are also correlated with gold prices and global real estate markets. According to the Global Association of Risk Professionals and Riskcenter's report of 15 August 2006, 'Rising Oil Prices Could Boster Gold', gold prices climbed 25 per cent in 2006, primarily due to a 22 per cent increase in oil prices. The gold price was $644.40 an ounce on 14 August 2006. When oil prices fell to $52.51, gold fell to $635.30 an ounce on 22 January 2006. The rise of oil directly affects inflationary trends in all major commodities. This inflationary trend also has an impact on energy costs, chemicals, labor, and industrial production. US consumer prices rose by 4.3 per cent between June 2005 and June 2006, compared to a 2.5 per cent increase over the previous year (Figure 8.45).

Along with gold, oil prices are correlated with real estate prices. While the United States' real estate market started a historical downturn in November 2005, new international emerging markets are surging. As shown in Figures 8.46–8.48, the trading of real estate investment vehicles in

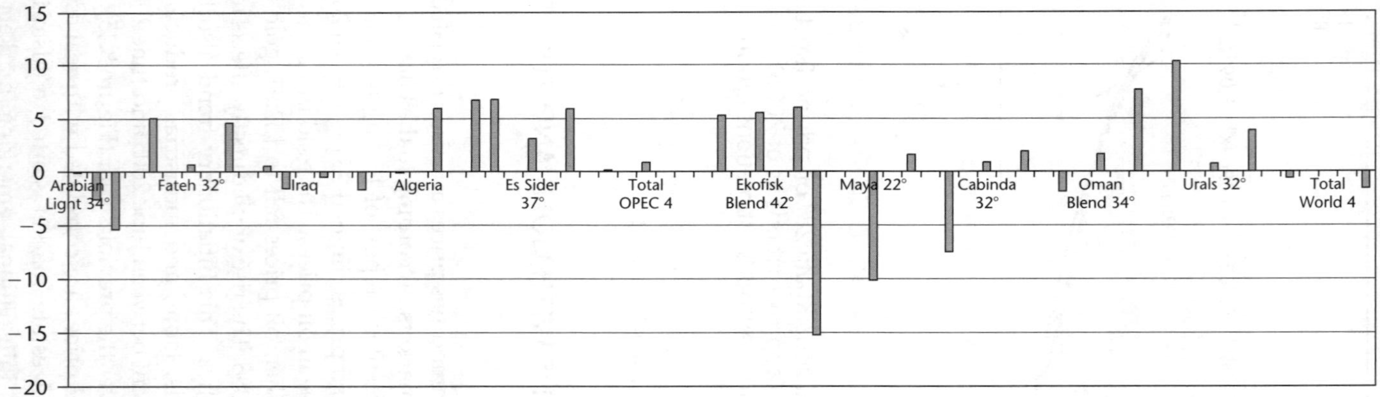

Figure 8.44 Pricing spreads between various geographical points and world average, 2006 ($64.2 barrel)
Source: Author's estimates and model (2006).

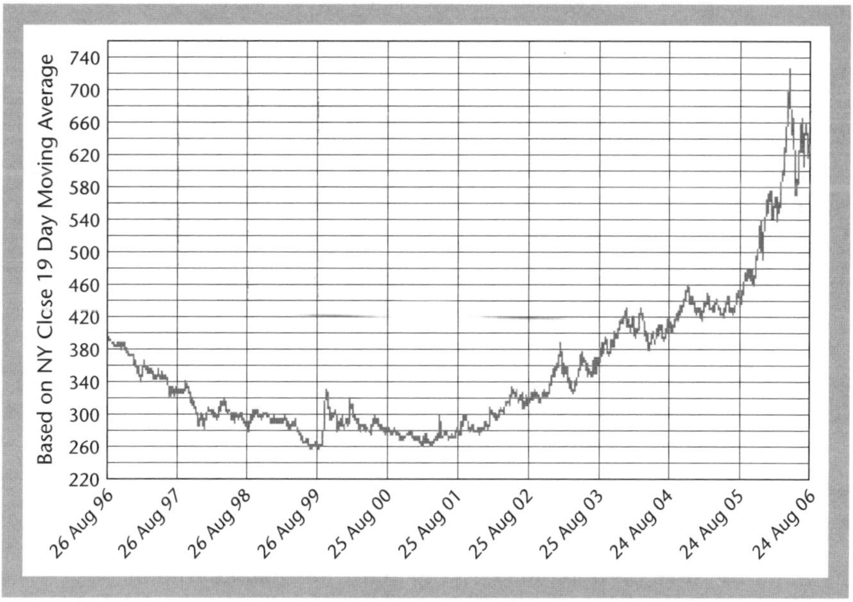

Figure 8.45 Ten-year gold prices ($US)
Notes: From 26 August 1996 to 24 August 2006; high $725 on 12 May 2006, low $252.80 on 20 July 1999.
Source: www.Kitco.com.

volume and asset value rose exponentially in correlation with global oil pricing trends.

The 1990s were a decade of indulgent speculation, highly destabilizing to the future of global financial markets and providing a poor basis for geopolitical balance. Transparency in pricing has been poor in large corporations, as has corporate governance.

8.8 OIL PRICING CORRELATION TO ALTERNATIVE ENERGY

Unlike at any other time in history, oil prices are not expected to decrease in the immediate future, not only because of the factors listed above but also due to alternative energy and renewables, which create barriers to entry, principally as a result of their inadequacies and inability to provide global market pricing efficiencies. We should note that alternative energies, used alone and unconnected to traditional energies, are irrelevant to China's emerging middle class, which is adding a significant share to the global consumption function. Utility barriers are considered by the Chinese national energy state council to define regulatory boundaries around wholesale and retail markets.

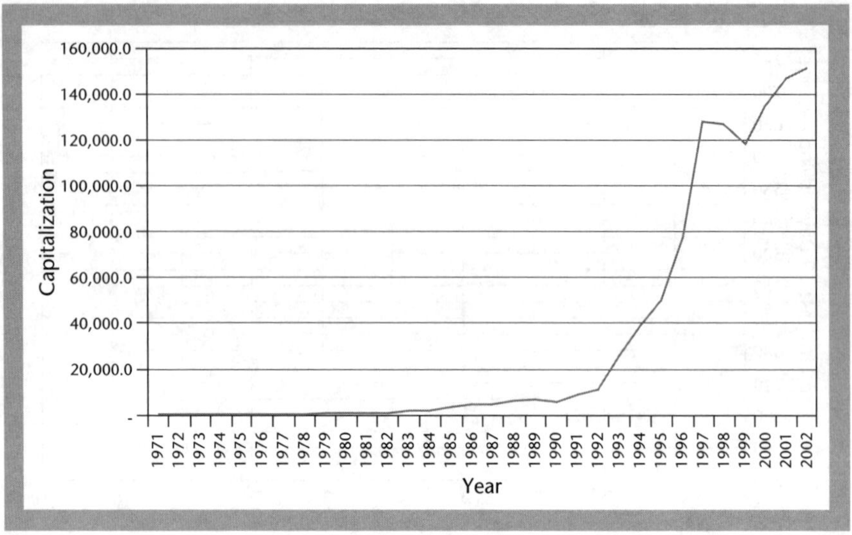

Figure 8.46 NAREIT equity REIT index capitalization
Note: NAREIT: National Association of Real Estate Investment Trusts.
Source: Yahoo finance 2002.

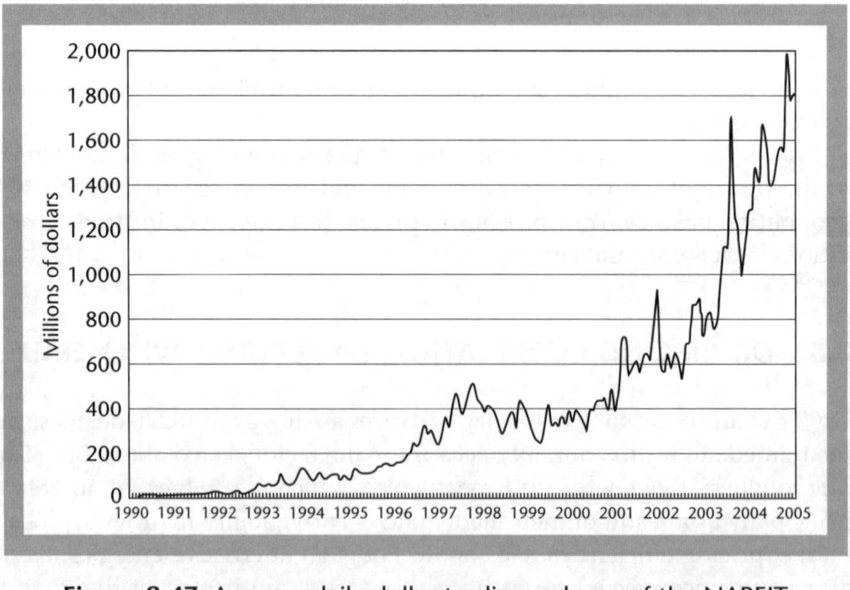

Figure 8.47 Average daily dollar trading volume of the NAREIT composite index, March 1990–October 2005
Source: © NAREIT.

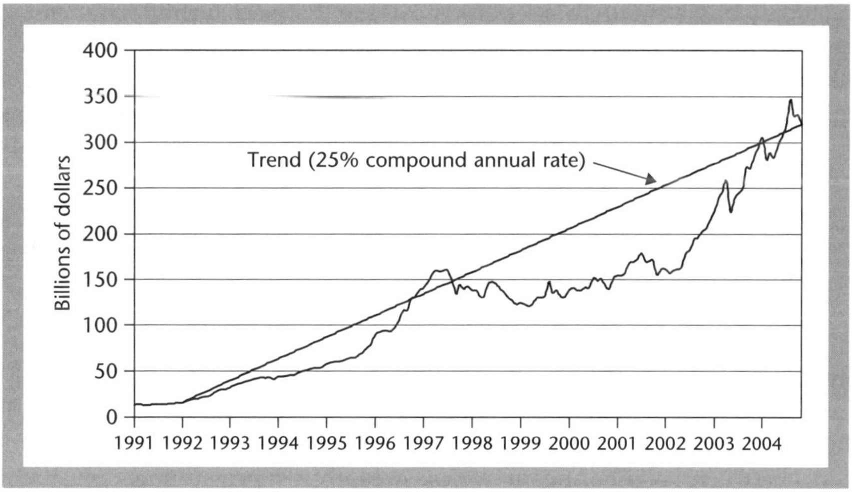

Figure 8.48 Equity market capitalization of US REITs
Notes: Figure does not include operating partnership units; data as of 31 October 2005.
Source: Uniplan, Inc.

Private Chinese firms are not invested significantly in renewable energy generating projects unless they are founded in partnerships or with foreign investors. This is primarily because renewable energies, taken individually, do not create sufficient resources to provide large-scale market pricing efficiencies. Renewable energies have to be bundled together as a complementary package in order to replace one individual traditional type of energy, such as oil or natural gas, and to be economically feasible and efficient. In addition to utility barriers, global regulating agencies put pressure on domestic corruption and red tape, such as regional regulations capping and ranging energy prices to provide governments with an added layer of revenues. National energy frameworks lack enforceable policies to favor the development of a globally independent power industry. The Chinese government – like all others, including the United States – has too many internal layers of bureaucratic regulation mainly imposing restrictions at national level rather than affecting international market policies. For instance, the US Congress passed the Public Utilities Regulatory Policy Act (PURPA) in 1978 to require utilities to buy power within specific pricing ranges and caps from independent power producers (IPPs). China arranged similar policies but does not enforce the utilities pricing floors and caps. Chinese utilities may also refuse to buy electricity from IPPs. Thus, international entities, such as the World Bank, can access special wind programs in Inner Mongolia at special negotiated prices. These prices add to the inefficiencies of the market between more regular dealers. Additionally, Chinese utilities may only buy a small number of kWh at the IPP price formula, then must agree to buy the rest of the output at the average system price. Very much like emerging markets which lack significant transparencies and corporate governance, China's separate negotiations of power purchase agreements for proposed generation projects have embedded added transaction costs and additional uncertainties.

As well as utility barriers and national government regulations, pricing policy is an additional disadvantage to the utilization of renewable energies. Chinese electricity prices are arbitrarily set, depending on consumption and production at various points of the domestic grid. Prices are then distributed with an added transportation function linked to the power pricing function. On a wider scale, national functions are being 'smoothed' in order to link them to international geopolitical global supply and demand. But these functions will be more effective if there are additional infrastructures between neighboring countries. Actually, Chinese utilities must still obtain national approval from the central government to raise prices. These national prices are still subject to local tariffs and quotas, and national government will finance infrastructures if and only if domestic budgets are well allocated and financed. Depending on the engineering of the national infrastructures, transportation costs create a more efficient system, thus facilitating new economic advantages for imports. Curiously, there are more barriers to entry the more local we get, so that the most backwards rural infrastructures are subject to the highest costs because of the extra layers of bureaucracy between international and national systems, regions, municipalities and even down to villages. So, intelligent governments ought to invest in modern adequate balanced infrastructures in order to be able to maintain future demand and enable international trading to facilitate improving globalization standards.

Another factor that allows renewables to remain as secondary energy resources is insufficient capital. Large Chinese firms do not have significant capital to invest into renewables, which are too expensive taken individually to meet demand levels. Not only are they insufficient in quantity but their scarcity also makes them extremely expensive. Added to that, there are extra levels of regulation to limit impractical utilizations. For instance, hydropower is 1.2 times the cost of coal, biogas 1.5 times, or wind power 1.7 times. Another barrier to expansion of use of renewables is the market itself. Renewables require far reaching investments in research and development. For instance, hydropower and solar water heating require advanced technologies that are more expensive to produce on a large scale. This is another reason why traditional energies have been more popular, despite the fact that global reserves are finite.

Since the late 1990s, China has focused primarily on renewable energies to meet rising demand in the manufacturing and industrial sectors. Yet, private companies incur a cost. With the exception of hydro-electricity, the power generation cost of renewable energy is much higher than the normal energy resources in China. For example, the initial investment in grid wind power is 8000 yuan/kw, unit power generation cost is 0.33 yuan/kWh, and the grid price including VAT is 0.52 yuan/kWh. The cost of power generation may climb to 40,000 yuan/kW, while the unit power generation cost is around 2.38 yuan/kWh. The unit investment cost of coal generating 300 thousand kW without the desulfurating equipment is 5000 yuan/kW, the unit power generation cost is only 0.21 yuan/kWh, the grid price including VAT is 0.33 yuan/kWh.

Renewable energies are much slower to develop because they require advanced technologies, regulatory approval and investment. In addition to the dependency on renewables pricing, oil prices also vary with regulations and new policies affecting other energies. For instance, the China Energy Research Society (CERS) provides a review on energy financing mechanisms. CERS reported results in the 'Study of the Investment Mechanism for Off-grid Electricity Generation Systems in Rural China'. In the report, CERS suggested that the government should increase investment in rural off-grid distributed generation systems. The Institute of Electric Engineering of the Chinese Academy of Sciences also developed similar policies in 2005.

8.9 THE PRICING OF OIL WILL BE AFFECTED BY FAST-PACED DEVELOPING PROJECTS

The pricing of oil in China will also be affected by new corporate valuation standards. This is partly due to the fact that China's domestic market is upgrading to international governance standards, and that there is new foreign competition entering and accessing the market. Thus, the price of domestic markets is subject to corporate governance changes and evolution, as well as alterations in internal financial valuations. According to ChinaBidding of 19 July 2006, China Petroleum and Chemical Corporation, or Sinopec, processed 5.3 per cent more crude oil in the first half of 2006 as China's economic growth boosted demand. Oil refining at Sinopec, Asia's largest oil refiner, rose to 71.7 million metric tons (527 million barrels). Crude oil production rose 3.1 per cent, and natural gas output rose by 20 per cent. Sinopec supplies about 75 per cent of the fuels sold in China, where the economy expanded 10.9 per cent in the first half of 2006, the fastest growth in more than a decade. The author's research proves that this figure is actually highly conservative and that a more realistic view demonstrates an overheating economy with annual growth rates close to 15–18 per cent per year since at least 2004. China's curbs on fuel prices, aimed at shielding consumers and manufacturers from higher energy bills, may cause Sinopec further refining losses because of record crude oil costs.

Sinopec shares fell in late July 2006, dropping 1.8 per cent to HK$4.15 (53 US cents). The stock gained 7.8 per cent in the first half of 2006, lagging behind PetroChina's 31 per cent advance. Oil output rose to 140.9 million barrels and natural gas production reached 126.2 billion cubic feet. Diesel output climbed 7.6 per cent to 28.3 million tons, while gasoline output declined 0.8 per cent to 11.2 million tons. Kerosene production dropped 6.2 per cent to 3.2 million tons. Sinopec's domestic sales of fuels including gasoline and diesel rose 7 per cent to 54.3 million tons in the first six months of 2006. Retail sales increased 19.5 per cent to 35.3 million tons. Production of ethylene, a chemical raw material used to make plastics, rose 24.5 per cent to 3 million tons. Sinopec's first-quarter profit fell 3.6 per cent to 9.29 billion yuan (US$1.16 billion) as it recorded an operating loss of 7.88 billion yuan (US$985 million) at its refining division, compared with a profit of 1.67 billion yuan (US$208.7 million) around April 2005. High oil prices have a negative impact on downstream companies like Sinopec. On 24 May 2006, China increased fuel prices for the second time that year to help refiners cover the rising cost of crude oil. Gasoline prices rose by 10.6 per cent , diesel prices by 12.3 per cent, and jet fuel by 10.3 per cent. Sinopec increased oil production at a faster pace in the first half of 2006 than bigger rival PetroChina. Oil output increased 1.8 per cent to 419.1 million barrels. Natural gas output surged 31 per cent to 684.7 billion cubic feet at China's biggest oil company. PetroChina's oil processing rose 3.5 per cent in the first half of 2006 to 392.6 million barrels. Crude oil prices in New York also climbed 32 per cent in 2005, reaching a record US$78.40 on 14 July 2006, the highest since the New York Mercantile Exchange began trading in 1983. Between 2007 and 2014 China may consume 5.5 per cent more crude oil as a result of rising living standards and higher car ownership (IEA 2007). Oil pricing has been subject to major volatilities throughout 2006, and according to the Energy International Administration, demand is expected to rise to 7.4 million barrels a day in 2007, or 390,000 barrels a day more than 2006.

In conclusion, China is challenged by its overheating domestic growth in all industry sectors. The Chinese market downturn at the end of March 2007 shows that it is not vulnerable to future corrections and that global markets are fully ramified to the Chinese economy. As a consequence of its rising living standards, China already faces, as of 2007, internal economic imbalances in the management of its resources versus the demand. The demand for oil has dramatically risen since the beginning of this century and so did its price which doubled from 2001 to 2007 and tripled from summer 2002 to summer 2006. It is unclear what premium of the global oil pricing structure comes from Chinese demand alone. It will be less transparent once the large Chinese energy companies become fully integrated into global commodities derivatives markets and engage actively in speculative trading with international commodities exchanges. Let us now review how the Chinese commodities corporations are preparing for this integration.

CHAPTER 9

Chinese Commodities Corporations

The three largest Chinese oil and gas firms are Sinopec, the China National Petroleum Corporation (CNPC), and China National Offshore Oil Corporation (CNOOC). The main difference between those firms and their foreign peers is that the Chinese companies are solely involved in the physical trading of natural resources as opposed to international peers which have already significantly developed their financial derivatives operations. Chinese companies are not focused on the trading of financial derivatives primarily because derivatives did not exist in China until 2005. Chinese commodities companies focused instead on developing foreign partnerships in the trading of physical goods that will serve them as collateral value and forms of payments when resources become scarce and in case of defaults by financial derivatives contracts. Soon enough, they will be able to monetarize their physical tangible goods at much higher prices, and thus gain value and catch up with financial engineering forms of payment.

This is also partly why Chinese companies are behind their international counterparts in investing in corporate governance and infrastructural transparencies. Corporate ethics, compliance and global regulatory mandates impose major costs that do not provide any competitive edge to the future of global resource sharing. And the fact of the matter is that these are ideal concepts that are costly and serve little purpose in the face of economic competition.Yet, since China's entry into the World Trade Organization in 2001, Chinese oil companies have partnered with foreign counterparts to share physical commodities and energy resources on a rotational basis. It is as of yet unclear to Chinese commodities companies what the effect of financial derivatives will be to financial statements, but it is most likely to occur in the same fashion as the integration of speculative financial instruments on competitive foreign peers.

The corporate capitalization of China did not occur until 2001, when China was preparing to become the most significant manufacturing hub in the world. Chinese oil companies were just starting to initiate capitalistic public offerings (IPOs) of stocks in 2000, raising billions of dollars in foreign and domestic capital. The Chinese system has been very astute and aggressive in teaching Chinese citizens to day trade with ebanking tools and advanced internet telephony. In fact, Chinese individual investors bought shares of the Bank of China's stocks and are becoming more financially educated and aware – more so than European consumers, relying on service industries to maintain their way of life. A 'relative injustice' contributing to this global imbalance is that, as of 2006, most day trading operations do not provide European investors with any Chinese financial products in euros, or very few. All proposed Chinese products are either in dollars or quoted as Chinese shares directly for knowledgeable traders. European middle-class investors are thus deprived of any opportunities for direct returns on investments from Chinese economic growth and are reduced to investments mainly in euros. It is, however, worth noting that most European politicians signed major trading corporate agreements with the Chinese government, selling out European know-how and resources and cashing in on those contracts to profit boards of directors and corporate management. Thus, the majority of European investors never saw a share in this dream. Europe has short working hours and a high currency. It is utopian to believe that Europe's economy runs on its exports and high productivity rates.

Since the start of the twenty-first century, it has become highly noticeable that major Chinese companies are adopting capitalist forms and are restructuring in a more Westernized style. Thus, CNPC reorganized in 2000 to separate assets and to create a subsidiary named PetroChina. CNPC floated its share with an initial public offering (IPO) of its minority interest shares on both the Hong Kong and New York stock exchanges in April 2000. The IPO created more than $3 billion, and British Petroleum now owns 20 per cent of minority shares. Sinopec also conducted its initial public offering in New York and Hong Kong in October 2000 and managed to get approximately $3.5 billion. In this deal, the minority interest share amounted to 15 per cent. ExxonMobil, British Petroleum, and Shell all contributed to about $ 2 billion of the initial public offering. CNOOC conducted its initial public offering with 27.5 per cent in interest in February 2001 after an earlier attempt in September 1999 had been abandoned. Shell bought a large block of shares valued at around $200 million. As of 2002, Chinese oil companies started to re-engineer to separate their business units into subsidiaries and different legal structured shells. CNPC created subsidiaries for drilling services and geological sample drilling work. CNOOC created oilfield services, such as the China Oilfield Service (COSL), which was listed on the Hong Kong stock exchange in November 2002. Let us review, in this chapter, the three main Chinese corporations: Sinopec, PetroChina and CNOOC. We will conclude with a review of smaller firms.

9.1 SINOPEC: CHINA PETROLEUM AND CHEMICAL CORPORATION

The following information is issued by Sinopec Corporation to investors, customers, regulators, media and press contacts.

China Petrochemical Corporation, the sole initiator of Sinopec, is a very large petroleum and petrochemical group re-engineered and restructured as the main Chinese state entity as of 1998. It is a state-owned company functioning as a state-authorized investment institution. The state holds the controlling share of the entity. Sinopec Group is one of China's largest oil producers with some of the largest revenues, with 832.5 billion yuan (US$104 billion) as of 2005 for sales, accounting for 14.25 per cent of the total revenue of the 500 companies. Three Chinese firms are among the top companies ranked with China National Petroleum Corporation (CNPC) and Sinochem Corporation. China Petroleum and Chemical Corporation, Sinopec, is a publicly listed company re-engineered in February 2000 by the China Petrochemical Corporation (Sinopec Group) according to the Company Law of the People's Republic of China. Sinopec's financial statements include 16.78 billion H shares as of October 2000 with 2.8 billion A shares floated on the Shanghai Stock Exchange in July 2001. Sinopec counts an approximate total number of shares of 86.702 billion. The state holds the majority control with 71.23 per cent of the shares in the legal name of Sinopec Group. Additionally, other domestic banks and minority asset management companies hold a major stake of 6.2 per cent of main equities, 19.35 per cent by H shares and 3.23 per cent by various private investors in China.

Sinopec describes itself as an integrated energy and chemical company, and its major business lines are oil and gas exploration and drilling throughout the world. Other business lines include renewables development, commodities production, oil refining, chemicals and fuel derivatives production, petrochemical production and usage, chemical derivatives such as chemical fibers, chemical fertilizers and others. In addition to producing chemicals, commodities and energy derivatives, the company also engineers oil reserves storage and builds pipelines for transportation of crude oil and natural gas within China and with neighboring countries. Sinopec is an expert in the import and export of crude oil, natural gas, refined oil products, petrochemicals, chemicals, and other commodities and technologies. Sinopec is also involved in the research, development and application of technology and information to advance in more efficient ways the production of energy derivatives and commodities in China to respond to rising demands. Sinopec is China's largest producer and supplier of oil products such as gasoline, diesel and jet fuel and major petrochemical alternatives such as petrochemical intermediates, synthetic resin, synthetic fiber monomers and polymers, synthetic fiber and chemical fertilizer.

Sinopec's proved recoverable oil reserves total 3294 million barrels and its proved recoverable natural gas reserve was approximately 2951.7 bcf as of the beginning of 2006. Sinopec's crude oil processing capacity totals 160 million tons with an ethylene production capacity of 5.395 million tons. Sinopec also owns 29,647 service stations around China, with 27,367 self-owned entrepreneurships. Additionally, Sinopec produced 39.27 million tons of crude oil and 6.3 billion cubic meters of natural gas as of end of 2005. In 2007 Sinopec progressed in marine exploration in Northeast Sichuan Province, in Western Tahe and Eastern China. In March 2007 output of crude oil and natural gas increased by 2 per cent and 10.1 per cent over 2006. Sinopec's refining processing volume of crude oil as of March 2007 increased by 5 per cent from March 2006 figures (see Table 9.1). Refining products also amount to 140 million tons, oil product output to 84.52 million tons including gasoline, kerosene and diesel. Sinopec also created 5.32 million tons of ethylene, 7.61 million tons of synthetic resin, 0.63 million tons of synthetic rubber, 6.72 million tons of synthetic fiber monomers and polymers, and 1.57 million tons of synthetic fiber. Moreover, Sinopec's annual oil products sales volume reached 105 million tons with retail amounting to 63.52 million tons.

During the restructuring of its corporate strategy, Sinopec re-engineered corporate governance and developed a management system with centralized decision-making, delegated managed networks and business operations handled by specialized business units. Sinopec counts more than 80 subsidiaries and branches with wholly-owned, equity holding, shareholding entities also engaged in oil and gas exploration and production, refining, chemicals, marketing, research and development and foreign trading partnerships. Sinopec has also participated in advanced technological developments and research to improve energy reliability diversification. In addition to major developments in research and development and advanced technologies, CNPC also enhanced and proliferated partnerships and deals throughout the world.

According to Ellen Silverman of riskcenters.com (6 July 2006) in 'Energy Risk China's Oil Giant Looking at $3 Billion Stake in Rosneft', CNPC purchased stakes of $3 billion worth of shares in the initial public offering of Russian state-owned oil group Rosneft which was issuing an initial public offering worth $11.6 billion in order to remodel and re-engineer its capital structure. The oil companies can stake one to two billion dollars and CNPC responded by offering three billion dollars. Rosneft is to sell shares of the companies by mid-July 2006 in Moscow and London while other oil companies such as Malaysian Rosneft are also staking into the IPO. Russian rival Yukos also disputed Rosneft's IPO claiming that it had stolen some of its

Table 9.1 Operating data for Sinopec, 2007, first quarter

Operational results		Unit	Three-month period ended 31 March		Changes over the same period of the preceding year (%)
			2007	2006	
Exploration and production					
Crude oil production		10 thousand tonnes	999.43	979.86	2.00
Natural gas production		Million cubic meters	1,996	1,813	10.07
Crude oil realized price		RMB/tonne	2582.70	3112.81	(17.03)
Natural gas realized price		RMB/000 cubic meters	799.27	745.10	7.27
Refining					
Refining throughput		10 thousand tonnes	3691.40	3516.88	4.96
Production of gasoline, diesel oil and kerosene		10 thousand tonnes	2164.94	2088.87	3.64
Of which:	Gasoline	10 thousand tonnes	587.24	553.40	6.11
	Diesel	10 thousand tonnes	1390.88	1381.87	0.65
	Kerosene	10 thousand tonnes	186.82	153.60	21.63
Light chemical feedstock		10 thousand tonnes	608.50	576.81	5.49
Light products yield		%	74.31	74.81	(0.50) percentage point
Refining yield		%	93.38	93.64	(0.26) percentage point
Marketing and distribution					
Total domestic sales of refined oil products		10 thousand tonnes	2746.60	2603.00	5.51
Of which:	Retail	10 thousand tonnes	1724.30	1666.80	3.45
	Distribution	10 thousand tonnes	469.80	491.50	(4.40)
	Wholesale	10 thousand tonnes	552.40	444.80	24.19
Total number of petrol stations		Stations	28885	29744	(2.89)
Of which:	Self-operated	Stations	28075	27464	2.22
	Franchised	Stations	810	2280	(64.47)
Annual throughput per station		Tonne/station	2457	2428	1.19

Source: Sinopec.

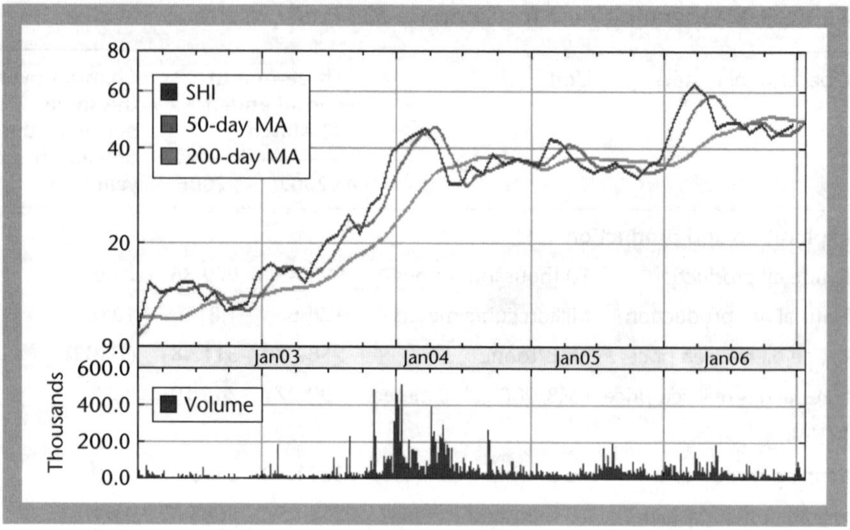

Figure 9.1 Sinopec's American Deposit Stock, as of 19 January 2007
Source: © 2007 Yahoo! Inc.

assets. China is expanding oil stakes throughout the world with contracts and partnerships, utilizing its forex surpluses to purchase as many commodities stakes as possible while its foreign exchange currency valuation is still low and reasonably stable. Deals are also particularly lucrative given that oil prices are conservatively anticipated to rise beyond $100 by 2010.

Sinopec's American Deposit Stock valuation nearly doubled between June 2006 and January 2007, from $20 to more than $40, mainly as a result of global increasing oil prices and rising international trading speculations (see Figure 9.1).

9.2 PETROCHINA

PetroChina Company Limited or PetroChina is a joint stock company, re-engineered under the Law of the People's Republic of China in 1999 as part of the restructuring of China National Petroleum Corporation (CNPC). CNPC invested in PetroChina the whole CNPC balance sheet involving the exploration and production, refining and marketing of oil, and its chemicals and natural gas businesses. PetroChina is China's largest producer of crude oil and natural gas, and is engaged in the exploration, development and production of crude oil and natural gas, refining, transportation, storage and marketing, including import and export of crude oil and petroleum products, production and sale of chemical products, and transportation, exploration and exploitation of natural gas.

PetroChina's financial statements in 2000 also had some stakes in the form of ADR American depositary shares and H shares. In addition to being listed, PetroChina also marked continuous performances with net profits of RMB 55.2 billion, RMB 45.5 billion and RMB 46.9 billion in 2000, 2001 and 2002.The company manages oil and natural gas exploration and production through 14 provincial subsidiaries throughout China: Daqing, Changqing, Liaohe, Xinjiang, Tarim, Southwest, Huabei, Dagang, Jilin, Qinghai, Turpan-Hami, Jidong, Yumen and Zhejinag. PetroChina provides energy and commodities resources to the company's operations in China's northern, northeastern, northwestern and southwestern regions. According to PetroChina's information page provided to investors, regulators, shareholders, media and press, the company is the main owner of China's total oil reserves with a 55 per cent share and of Chinese total natural gas reserves with approximately 60 per cent of the country's natural resources. Investor information also reveals that PetroChina owned proven oil reserves of 1.479 billion tons and natural gas reserves of 1.33 trillion cubic meters at the end of 2005. PetroChina's refined products, including gasoline, diesel, kerosene and lubricants, exceeded 40 per cent of the national total. During its major reorganization, PetroChina re-engineered the company's 26 oil refining subsidiaries, 22 refined products legal entities, and a chemical lubricant firm. The company's production centers and legal entities mainly operate along the northeastern and northwestern coasts of China where most of the natural resources are extracted to meet the demand of major urban centers such as Shanghai.

PetroChina owns the Dalian Marine Shipping Company to import commodities from international markets and started in 1999 to develop more marketing centers. The company installed approximately 430 wholesale distribution centers around China along the coastal areas, in urban city areas and along the Yangtze river, 1790 service stations privately franchised to entrepreneurs, 3600 wholly-owned service stations, and 1050 franchise service stations owned and operated by commercial third-parties.

9.2.1 Main products

PetroChina is primarily involved in the production of gasoline, kerosene, diesel, lubricants, light chemical oil, fuel oil, solvent oil, paraffin, bitumen, petroleum coke, liquefied gas, propylene and refinery benzene. Other refined products include naphtha, jet fuel and asphalt. It produces high grade oils such as 90-RON, 93-RON, 95-RON, 97-RON and 98-RON. PetroChina also produces various types of diesel derivatives, such as light diesel, diesel for urban vehicles, diesel for military use, Beijing local standard diesel and diesel for export. PetroChina's diesel products are graded as #10, #5, #0, #−10, #−20, #−35 and #−50. It is the country's main producer of kerosene, lamp kerosene, #3 jet fuel oil, JetA-1, JP-8, and DERD2494. As well as fuel derivatives, PetroChina also produces vehicle lubricating oils, industrial oils, synthetic lubricating grease and lubricants for metal processing. Petroleum bitumen for heavy-duty vehicles, modified asphalt, road bitumen, emulsified bitumen, pipeline protective asphalt, cable asphalt, waterproof asphalt, rubber asphalt, paint asphalt, and insulating asphalt are also among PetroChina's specialties. PetroChina wax products include semi-refined wax, fully-refined wax, paraffin scale wax, medicinal Vaseline, food-grade wax and electrical Vaseline. Fushun Petrochemical Company also benefits from Daqing crude oil with 25 per cent wax and Shenbei crude oil with 42 per cent.

PetroChina is also involved in the exploration and transportation of natural gas as the main sponsor of the west–east gas pipeline project to connect all Chinese urban areas within its grid and to ramify it to international energy providers. The west–east pipeline connects the gas field in Xinjiang's Tarim Basin and is ramified with branch lines and storage facilities to transport natural gas to the Yangtze River Delta area in China's Central and Eastern regions. It is also connected to other natural resource areas such as the Changqing.

A natural gas pipeline part-owned by PetroChina starts from Lunnan in Xinjiang and runs for almost 4000 kilometers through 66 counties in 10 regions and municipalities of Xinjiang, Gansu, Ningxia, Shaanxi, Shanxi, Henan, Anhui, Jiangsu, Shanghai and Zhejiang. PetroChina's operations started on 1 October 2004, and natural gas was commercialized by the end of 2004. The west–east natural gas pipeline gets energy resources from Tarim Basin that amount to approximately 8.39 trillion cubic meters of natural resource reserves. The Tarim Basin is considered the largest single source of natural gas in China, totalling approximately 38 trillion cubic meters. According to PetroChina's own information the company also produces chemical derivatives to maintain vehicles, cars, for engine construction, information technology parts, electronics, manufacturing devices, electrical appliances, household items, packaging, papers, textile, plastics, paints, footwear, agriculture and the furniture industry level.

PetroChina also creates synthetic resin derivatives such as polyethylene, polypropylene and polystyrene, as well as synthetic fibers including acrylon, dacron, pure terephthalic acid, ethylene glycol, acrylonitrile and polyester, and synthetic rubbers such as 1, 4 polybutadiene rubber, urea, organic epoxyethane, butanol, adipic acid, alkyl benzene, styrene, benzene, phenylamine, and paraxylene. PetroChina's inorganic products also include nitric acid, catalysts, and additives. PetroChina is the major Chinese producer of ethylene, synthetic rubber, synthetic fibers and fertilizer.

9.2.2 Productivity

PetroChina extracted 1119.5 billion cubic feet of natural gas in 2005 which represented an increase of 27.8 per cent over 2004. In 2005 its crude oil and natural gas production equated to approximately 1009.5 million barrels of oil, an increase of 5.1 per cent from 2004, and turnover reached approximately RMB 552,229 million, up by 39 per cent. Consequently, the company recorded a consolidated net profit of RMB 133,362 million, representing a substantial increase of 28.4 per cent from 2004. PetroChina improved efficiency and productivity levels by investing in new technologies and advanced materials to increase the capacity of its oilfields. The company also proliferated international partnerships to secure future rising demand. PetroChina's crude oil output in 2006 rose by 419 million barrels, an increase of 7.3 million barrels or 1.8 per cent. Natural gas production rose by 684.7 billion cubic feet, up by 161.1 billion cubic feet or 30.8 per cent. During 2006, PetroChina participated in the balancing of oil pricing to monitor its internal oil pricing developments based on supply and demand. In the first half of 2006 the company produced about 392 million barrels of crude oil, representing an increase of 13.20 million barrels, or 3.5 per cent over the first half of 2005. Gasoline production was 11.064 million tons and diesel 22.217 million tons, rises of 3.2 per cent and 2.6 per cent, respectively. To ensure that demand is met and to maintain oil pricing throughout the country, between January and June 2006 PetroChina opened 737 service stations to reach 17,952 customers, an additional 4.28 per cent from 2005. In 2006 PetroChina produced 1.003 million tons of ethylene, 0.155 million tons of synthetic rubber, and 1.525 million tons of synthetic resin, increases of 6.6 per cent, 9.2 per cent and 16.3 per cent respectively over 2005.

In 2006 PetroChina also extended the west–east gas pipeline, with the Huai–Wu line connecting to the Zhongxian–Wuhan stretch that constitutes approximately 90 per cent of the total. The west–east gas pipeline also connected via the Ji-Ning connection line to the second Shaanxi–Beijing gas pipeline. In addition to the main west–east pipeline project, PetroChina participated in the implementation of a line to facilitate the transportation of imported crude oil from Kazakhstan. PetroChina also established a crude oil pipeline between Shanshan in Xinjiang and Lanzhou in Gansu and a refined oil pipeline between Urumqi in Xinjiang and Lanzhou. In efforts to diversify and intensify its production, PetroChina multiplied its geopolitical partnerships and embarked on the production of new energy alternatives.

9.2.3 PetroChina's American Deposit Stock

PetroChina Company's ADS is valued at more than $100 as of January 2007 and gained more than 25 per cent since June 2006. This is partly due to gains from global oil prices (Figure 9.2).

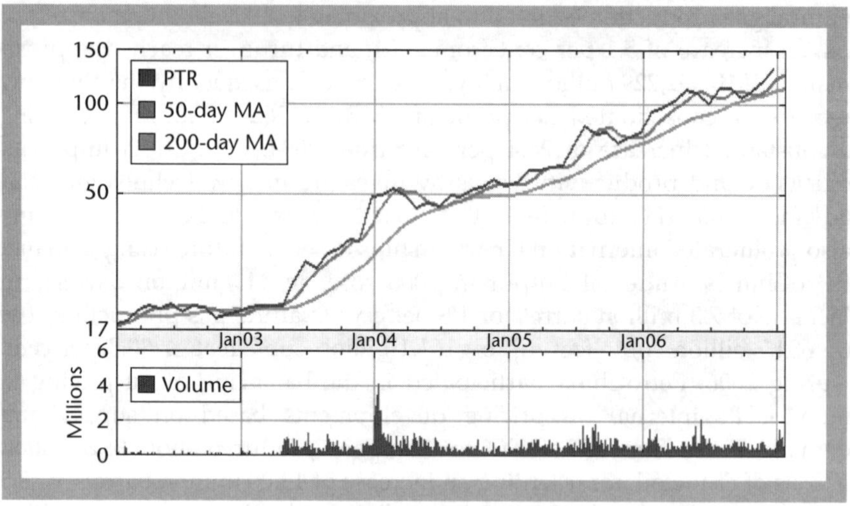

Figure 9.2 PetroChina Co. American Deposit Stock, as of 19 January 2007
Source: © 2007 Yahoo! Inc.

9.3 CNOOC

According to CNOOC's information provided to investors, shareholders, regulators, media and the press, it is a state-owned offshore and onshore petroleum company established in 1982. CNOOC is a Hong Kong incorporated public company which primarily explores, develops and produces crude oil and natural gas offshore in China. It is the main provider of oil and natural gas to most of China, and has net oil and gas reserves of approximately 2.36 billion barrels of oil equivalent (boe). The 2005 production averaged 424,108 boe per day. CNOOC currently expands strategic Chinese reserves through partnerships and new drilling investments, and its net proved reserves increased by 668 million boe in 1995 to 2.36 billion boe in 2005. CNOOC also raised average production from 107,600 boe per day in 1995 to 424,108 boe per day in 2005. The company has proliferated its operations through partnerships with multinationals in various offshore areas and through buying minority interest shares in foreign entities. CNOOC owns a number of subsidiaries: CNOOC Limited, CNOOC International Limited, OCGC Malacca, OCGC America Inc, OCGC Myannar, Malacca Petroleum, CNOOC China Limited, Tianjin Branch, Zhanjang Branch, Shenzhen Branch, Shanghai Branch, a 30 per cent interest in Shanghai Petroleum and Natural Gas Company Limited, and China Offshore Oil Singapore International Limited. Throughout its reorganization, CNOOC gained the following assets and liabilities: 37 production-sharing contracts, and one geophysical survey agreement, eight independent development

Table 9.2 Production of oils and gas by CNOOC, 2001–05

Production data	2001	2002	2003	2004	2005
Net production of crude and liquids (barrels/day)					
Bohai Bay	99978.0	127756.0	129506.0	134512.0	178840.0
Western South China Sea	41277.0	5691.0	60944.0	55873.0	49016.0
Eastern South China Sea	81404.0	73793.0	72981.0	96989.0	103741.0
East China Sea	3967.0	3223.0	2536.0	2121.0	1706.0
Overseas	2247.0	36944.0	40497.0	29941.0	23565.0
Total	228873.0	247407.0	306464.0	319436.0	356868.0
Net production of natural gas (mcf/day)					
Bohai Bay	46.2	47.1	47.1	47.7	49.1
Western South China Sea	139.0	142.3	127.8	215.2	229.6
Eastern South China Sea	0	0	0	0	0
East China Sea	9.8	12.4	14.2	17.1	18.3
Overseas	0	70.8	101.9	84.1	92.7
Total	195.0	272.6	291.0	364.1	389.7
Total net production (boe)	261379.0	346639.0	356729.0	382513.0	424108.0
Total net proved reserves at year end (mboes)					
Bohai Bay	1066	1092.3	1084.8	1092.3	1043.7
Western South China Sea	535.1	578.9	601.0	603.8	639.7
Eastern South China Sea	132.2	127.5	246.1	289.8	341.9
East China Sea	45.2	42.4	59.8	88.7	88.5
Overseas	8.4	174.7	136.8	155.5	249.1
Total	1787.1	2015.8	2128.5	2230.0	2362.6

Source: CNOOC (2006).

and production projects, the interest noted above in Shanghai Petroleum and Natural Gas Company Limited, rights to use facilities and rights to land exploration in Nanhai, Weizhou and the western part of Bohai Bay, and loans from, and swap agreements with, various Chinese and foreign banks. Table 9.2 and Figure 9.3 illustrate production of oils and gas by CNOOC.

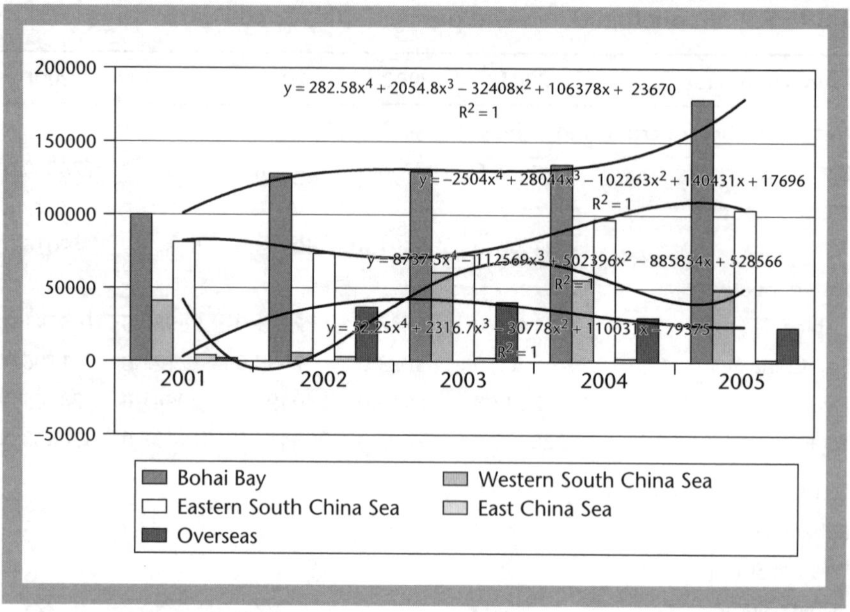

Figure 9.3 CNOOC net production of crude oil (barrels/day)
Source: CNOOC (2006).

CNOOC is investing in a number of major projects (Table 9.3), including Bohai Basin in northeastern China, 200 kilometers east of Beijing. Bohai is among the oldest and largest oil resources in China and is still exploitable. Its net proved reserves total 1044 million boe and account for 44 per cent of the total resources of CNOOC. Average daily net production is 87,021 boe. Bohai Basin includes Jingzhou 20-2 gas field, Chengbei, Qikou 18-1, Bozhong, Jinzhou 9-3, Suizhong 36-1, and the Jinghou 20-2 terminal. In addition to Bohai the western South China Sea, southwest of Hong Kong, is another of CNOOC's significant natural gas areas. CNOOC also owns the Yacheng 13-1 and Dongfang 1-1 natural gas fields which comprise a new production project which began in 2006. This region generates about 640 million boe of net proved reserves which represents 27 per cent of CNOOC's total. Its average daily net production accounted for 89,583 boe or 21 per cent of the company's total. CNOOC is also involved in the western South China Sea at Beibu Gulf, Yinggehai Basin and Qiongdongnan Basin. The eastern China Sea also totals 88 million boe of net proved reserves accounting for 4 per cent of CNOOC's total. Its daily net production amounted to 4751 boe or 1 per cent of CNOOC's total. The eastern South China Sea accounted, as of 2005, about 342 million boe of net proved reserves or 14 per cent. The company's average daily net production amounts to 103,741 boe or 24 per cent, shared among Xijiang 24-3, Lufeng 21-1, Lufeng 13-1-YL, Liuhua 11-1, and Huizhou 26-1. In addition to these domestic centers, CNOOC is also becoming involved in offshore production activities and in investing minority interests in foreign firms.

Table 9.3 Major properties under production and development by CNOOC

Name of block		Major oil and gas field	Oil (b/d) Gas (mcf/d)	Total (boe/day)	Net Reserves as on 31 December 2005 Total (million boe) Oil (million barrels) Gas (Bcf)
Bohal Bay					
Production					
		Jinzhou20-2,			
		Jinzhou9-3,		104.68	323.0
		Suizhong 36-1,			
		Luda4-2,	Oil 98,065		Oil 286.5
Liaoxi	1		Gas 40		Gas 219.1
	Sep-18	Chengbel		4.229	5.5
		QK18-1, QK18-2,		10.008	18.5
		QK17-2, QK17-3	Oil 8,951		Oil 15.9
Boxi			Gas 6		Gas 15.4
		Nanbao35-2,			
	May-36	Qinghuangdao32-		21.914	121.3
	11-May	Penglai19-3		7.422	132.5
		Bozhong34-2/4,			
		Bozhong28-1,		21.797	184.4
		Bozhong26-			
		2, Bozhong25-1,	Oil 21,289		Oil 165.2
Bonan		Bozhong25-1s	Gas 3		Gas 115.2
		Caofeidian11-1,			
		Caofeidian11-2,			
		Caofeidian11-3,			
	Apr-36	Caofeidian11-5		16.97	28.8
Development					
					90.7
		Jinzhou21-1/25-			Oil 38.9
Liaoxi		1S			Gas 310.3
		Qinhuangdao33-1,			
		Bozhong3-1,			
Bozhong		Bozhong3-2			9.8
		Caofeidian18-1,			
		Caofeidian18-2,			20.1
		QK18-9,			
		Bozhong13-1			Oil 9.0
Boxi				Gas 310.3	

(continued)

Table 9.3 (Continued)

Name of block		Major oil and gas field	Oil (b/d) Gas (mcf/d)	Total (boe/d)	Net Reserves as on 31 December 2005 Total (million boe) Oil (million barrels) Gas (Bcf)
	11-May	Penglai25-6			10.5
04/36&05/36		Caofeidian12-1, Caofeidian12-1S			13.8
Bonan		Bozhong34-1, Bozhong34-1S, Bozhong34-3/5			28.3
Liaodong		Luda27-2, Luda32 2			37.4
	Nov-19	Bozhong19-4, Bozhong26-2N			19.1 Oil 16.8 Gas 13.6
Bohai subtotal			Oil 178,840 Gas 49	187.021	1,043.7 Oil 920.2 Gas 740.7
Eastern South China Sea					
Production					
Huizhou14		Huizhou Oil Fields		18.117	22.0
16/19		Huizhou19-3 Huizhou19-2 Huizhou19-1		2.782	21.6
	15-Dec	Xijiang24-3		16.576	15.0
Xijiang24		Xijiang30-2		10.788	8.5
Huizhou31		Liuhua11-1		18.699	29.8
	16-May	Lufeng13-1, Lufeng13-2		3.513	20.6
Lufeng08		Lufeng22-1		2.382	2.4
15/34		Panyu4-2, Panyu5 1		30.885	29.7
Development					
Liuhua07		Panyu30-1, Liuhua19-5			96.7 Oil 3.0 Gas 562.0
Panyu33		Panyu34-1			30.7 Oil 0.6 Gas 180.8

Name of block	Major oil and gas field	Oil (b/d) Gas (mcf/d)	Total (boe/d)	Net Reserves as on 31 December 2005 Total (million boe) Oil (million barrels) Gas (Bcf)
Xijiang04	Xijiang23-1			47.2
15/34	Panyu11-6			2.6
				9.1
				Oil 2.2
Huizhou 14	Huizhou21-1 (G)			Gas 41.3
	Huizhou25-1,			
Huizhou 16	Huizhou25-3			6.1
			103.741	341.9
Eastern South		Oil 103,741		Oil 211.2
China Sea subtotal		Gas 0		Gas 784.2
Western South China Sea				
Production				
			25.864	46.4
		Oil 24,706		Oil 44.7
Yulin35	Weizhou Oil Fields	Gas 7		Gas 10.5
	Wenchang13-1,			
Yangjiang31	Wenchang13-2		23.077	30.1
			25.02	76.2
		Oil 1,027		Oil 4.5
Ledong01	Yacheng13-1	Gas 130		Gas 430.1
			15.622	223.1
		Oil 208		Oil 3.0
Changjiang25	Dongfang1-1	Gas 92		Gas 1,320.4
Development				
	Wenchang8-3,			
	Wenchang14-3,			123.6
	Wenchang15-1,			Oil 83.6
	Wenchang19-1,			
	Wenchang9-2,			Gas 239.7
Yangjiang31/32	Wenchang9-3,			
	Wenchang10-3			
				101.8
	Yacheng13-4,			
	Ledong22-1,			Oil 1.3
Ledong01	Ledong15-1			Gas 603.3

(continued)

Table 9.3 (Continued)

Name of block	Major oil and gas field	Oil (b/d) Gas (mcf/d)	Total (boe/d)	Net Reserves as on 31 December 2005 Total (million boe) Oil (million barrels) Gas (Bcf)
	Weizhou6-1, Weizhou11-1, Weizhou11-1, Weizhou11- Weizhou6-10,			
Yulin35	Weizhou12-8			38.5
Western South China Sea subtotal		Oil 49,016 Gas 230	89.583	639.7 Oil 205.7 Gas 2,604.0
East China Sea				
Production				
Pinghu	Pinghu Gas Field	Oil 1,706 Gas 18	4.751	9.1 Oil 3.0 Gas 36.5
Development				
Xihu Trough				
Canxue				9.4 Oil 5.1 Gas 25.5
Duanqiao				7.8 Oil 2.3 Gas 33.4
Chunxiao				32.0 Oil 3.8 Gas 169.0
Tianwaitian				6.1 Oil 0.5 Gas 33.7
Baoyunting				19.2 Oil 4.6 Gas 87.5
Wuyunting				4.7 Oil 1.9 Gas 16.6
			4.751	88.3

Name of block	Major oil and gas field	Oil (b/d) Gas (mcf/d)	Total (boe/d)	Net Reserves as on 31 December 2005 Total (million boe) Oil (million barrels) Gas (Bcf)
		Oil 1,706		Oil 21.2
East China Sea subtotal		Gas 18		Gas 402.2
			385.095	2,113.7
Offshore China subtotal		Oil 333,303		Oil 1,358.5
		Gas 297		Gas 4,531.0
Indonesia				
Production				
			39.013	123.2
		Oil 23,565		Oil 73.7
		Gas 93		Gas 296.9
Development				
Australia				
			12.851	125.9
		Oil 167		Oil 25.4
NWS	CNOOC Ltd.	Gas 76		Gas 603.0
			39.013	249.1
		Oil 23,565		Oil 99.1
Overseas Subtotal		Gas 93		Gas 899.9
			424.108	2.363
		Oil 356,868		Oil 1,457
Total		Gas 390		Gas 5,431

Source: CNOOC (2006).

For instance, CNOOC Limited partnered in 2006 with South Atlantic Petroleum Limited and owned a 45 per cent interest in the oilfield OML 130 in Nigeria for US$2.268 billion. Furthermore, CNOOC will acquire a 90 per cent interest in the PSC and a 45 per cent interest in OML 130. Moreover, CNOOC also signed a contractual agreement with the National Oil Company of Equatorial Guinea to produce from 2287 square kilometers in the south offshore Equatorial Guinea. To diversify its geopolitical interests, CNOOC also bought equity stakes with Kerr-McGee Australia Exploration to explore WA-301-P, WA-303-P, WA-304-P and WA-305-P in the Outer Browse Basin of Australia. CNOOC is also involved in Kenya with six PSCs, Block 1, Block 9, Block10A, L2, L3, and L4. The production centers are near Lamu, Anza, and Mandera within 115,343 square kilometers (see Figures 9.4–9.8).

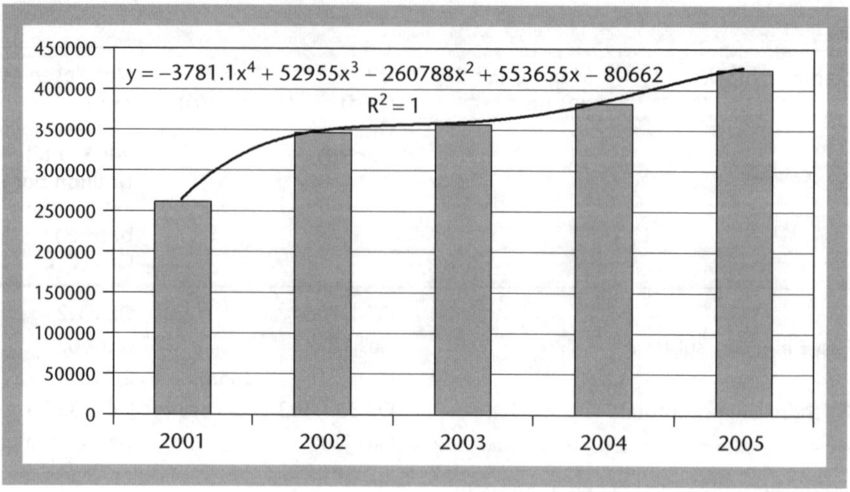

Figure 9.4 Total net production (boe) of natural gas in mcf/day
Source: CNOOC (2006).

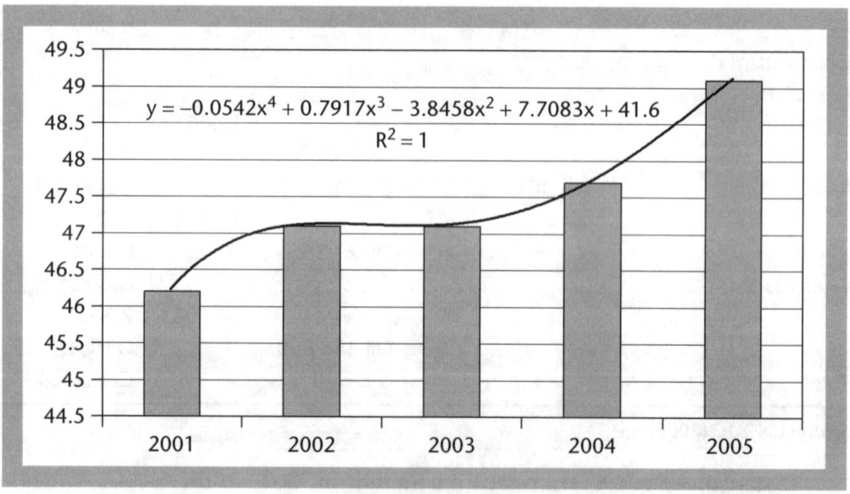

Figure 9.5 Bohai production of natural gas in mcf/day
Source: CNOOC (2006).

On 29 March 2007, CNOOC released its results for 2006. In this year, net income reached a historical high of RMB 30,926.9 million, or an increase of RMB 5603.8 million or 22.1 per cent from 2005. Earnings per share reached RMB 0.73. CNOOC's total revenue amounted to RMB 88,947 million, up 28 per cent from 2005. CNOOC generated revenue of RMB 67,828 million from oil and gas sales, an increase of RMB 14,410 million, or 27 per cent from 2005. CNOOC produced 167 million barrels of oil equivalent in 2006, or a

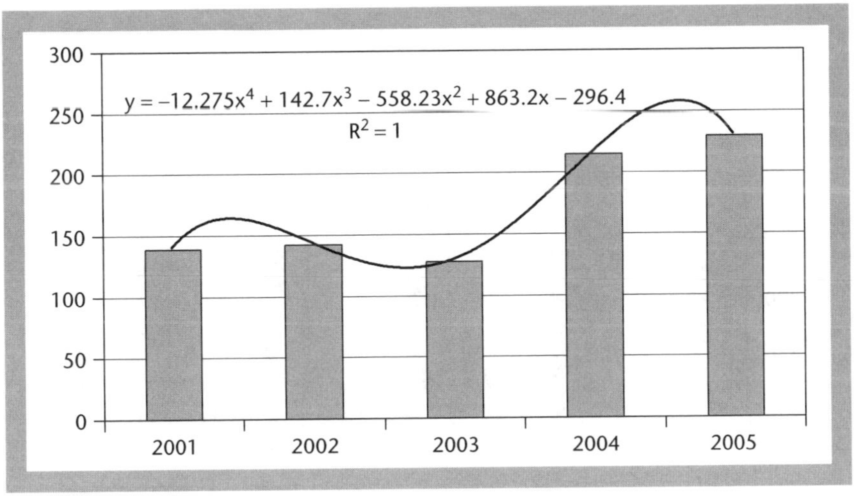

Figure 9.6 Western South China Sea net production of natural gas in mcf/day
Source: CNOOC (2006).

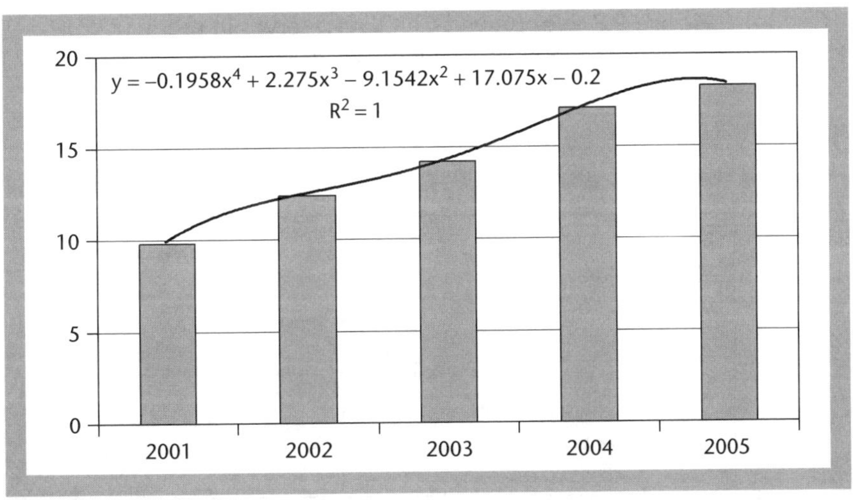

Figure 9.7 East China Sea net production of natural gas in mcf/day
Source: CNOOC (2006).

year on year increase of 7.9 per cent, including 136 million barrels of oil and 179 billion cubic feet of natural gas. In 2006, CNOOC discovered 10 new oil and gas resources, including the first deepwater discovery offshore China – Liwan 3-1, and 6. In 2006, seven new projects started producing new resources. Internationally, CNOOC acquired a 45 per cent interest in Nigerian OML 130 block and further extended exploration activities to Kenya, Equatorial Guinea, Australia and elsewhere.

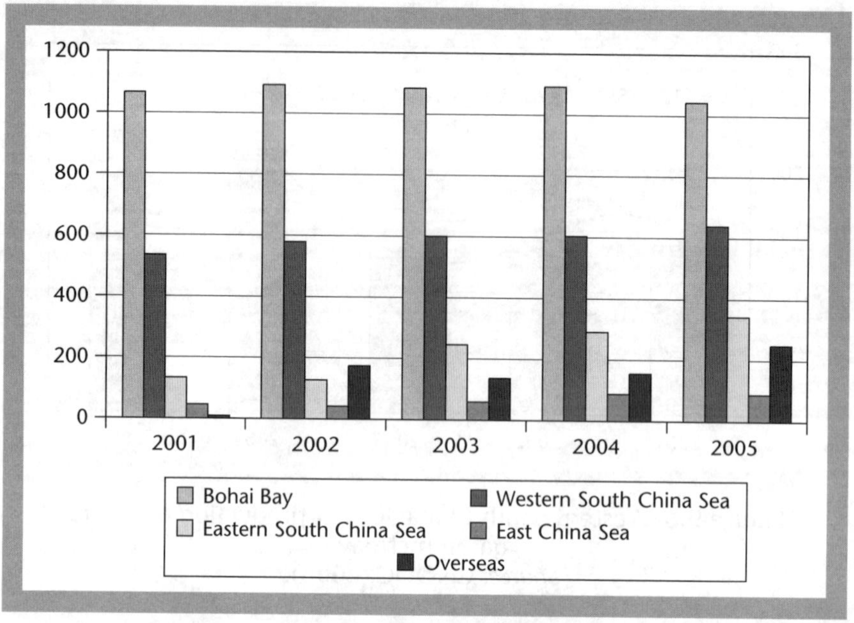

Figure 9.8 Net proved CNOOC reserves in mboe
Source: CNOOC (2006).

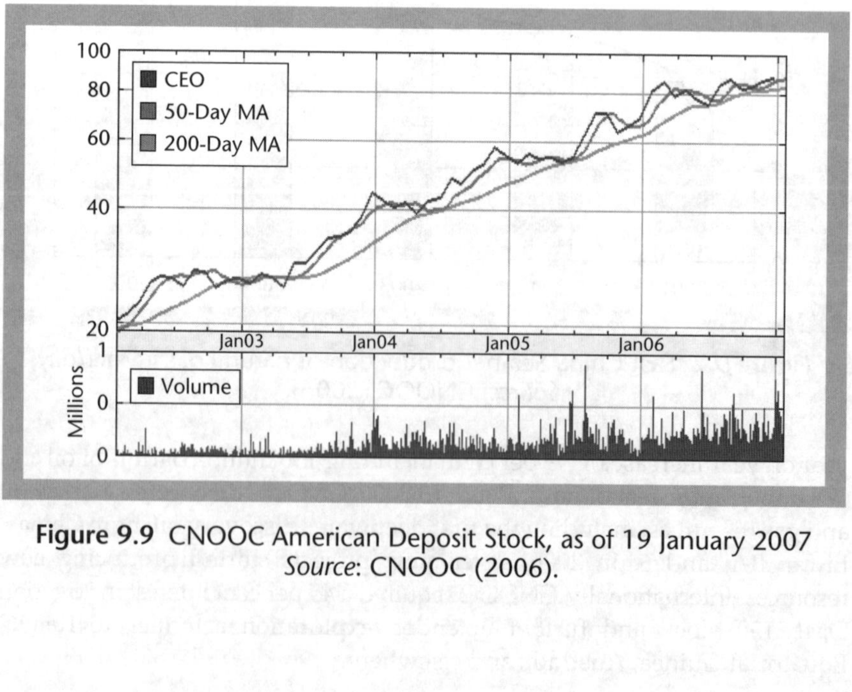

Figure 9.9 CNOOC American Deposit Stock, as of 19 January 2007
Source: CNOOC (2006).

Finally, CNOOC's American Deposit stock valuation nearly doubled from 2004 to January 2007 rising from $40 to more than $80 per share, mainly due to increasing global oil prices and financial trading activities (Figure 9.9).

9.3.1 The rise of smaller commodities firms and financial institutions

In *AsiaRisk* of July/August 2006, Pamela Tang reports on Hong Kong-based utility Towngas which has supplied natural gas for over 140 years and owns 65 per cent of the Hong Kong residential market share with 33 joint ventures. Towngas buys naphtha, which is a liquid hydrocarbon mixture, to refine into natural gas. Purchases are made in US dollars which engenders an implied forex risk embedded in the purchasing and financing of natural resources. China's growth is beyond the 10 per cent GDP line and all mainland purchases are currently performed in yuan or renminbi. Upon revaluations of the yuan against the dollar and the replacement of US Treasury bills by speculative derivative investments in yuan, the monetary repercussions on the individual pricing of natural resources will increase, causing inflationary trends on daily purchasing powers and thus leading to rising interest rates on global debts and making financing more expensive. The difference between monetary values in USD and yuan includes the discounted value of not having the same infrastructures as Western counterparts in alternative energies and energy efficiency. Though China has made major strides in recent years in energy efficiency, the gap between China and advanced world levels is still significant. In 1997, China's energy processing, conversion, transmission and end use efficiency was only 31.2 per cent, 10 percentage points lower than advanced international levels. Of 33 products in 11 industries, the energy consumption is 46 per cent higher than advanced international levels, consuming 230 Mtce more energy.

9.4 OTHER MARKET PARTICIPANTS AND COMPETITORS

Shanghai Zoom Intelligence Analyst reports the following list of industry participants in 'China Energy Topic Players Daily' (No. 157, Volume I-157) on 16 February 2006. Participants and competitors of the main Chinese energy corporations are Adnoc, Amerada Hess, Anadarko, Apache, Bashneft, BG Group, British Petroleum, BPCL, BHP Billiton, Burlington, Canadian NR, Cepsa, Chevron Texaco, Conoco Phillips, Cosino, CPC Taiwan, Devon Energy, Dominion, Eco Petrol, EGPC, Encana, ENI, EOG Resources, Exxon Mobil, Fortum, Gazprom, Husky, Energy Intelligence

Kazmunaigas, Idemitsu, INA, INOC, IOC, Kerr-McGee, KPC, Libya NOC, Lukoil, Marathon, Maerksh, Murphy oil, Nexen, NIOC, Nippon Oil, NNPC, Noble Energy, Norsk Hydro, Occidental, OMV, ONGC, PDO, PDV, Pemex, Pertamina, Petrobras, Petro-Canada, Petroecuador, PetroKazakhstan, Petronas, Pioneer, Pogo, Premcor, PTT, Qatar Petroleum, Reliance, Repsol YPF, Rosneft, Royal Dutsch Shell, RWE-DEA, Santos, Sasol, Saudi-Aranco, Sibneft, Sidanco, SKCorporation, Socar, Sonangol, Sonatrach, Statoil, Suncor, Sunoco, Surgutneftegaz, Syrian PetroleumCo, Talisman, Taftneft, Tesoro, TNK, Total, TPAO, Unocal, Valero Energy, WinterShall, Woodside, XTO, and Yukos.

CHAPTER 10

Alternative Energies and Renewables

The past fifty years have been marked by two major oil crises which severely affected the global economy and created a greater awareness among nations of the need to develop renewable energy. Because of the global capacity limitation of oil and natural gas – which is even more obvious with the emerging democracies – alternative and renewable energies will gradually have to substitute for traditional energy products. China is exploring the use of alternative energies and renewables – in spite of the additional costs – as a way of tackling the problems of polluting emissions and global warming. In 2001 renewable energies amounted to 2 billion tons standard coal equivalence (tce), or about 13.5 per cent of the total supply of primary energy. Worldwide, alternatives are composed of 10.8 per cent combustible biomass energy, 2.2 per cent hydropower, and the remainder geothermal, solar, wind, and tide energy. Wind energy is the fastest growing and most widely-used alternative. Wind power engines reached 32 kW in 2002, and Photovoltaic generation reached 560,000 kW, with a total actual installed capacity of 2.2 GW. Global installed biomass generation capacity rose by 15 GW, and biomass liquid fuel surpassed 20 million tons. Global solar water heaters increased to 100 million square meters representing about 50 per cent of the total area available in China and supplying about 14 Mtce. Further advances are being made in geothermal and tidal energy.

In 2002, the Chinese Energy State Council indicated that the situation concerning renewables was as follows. Hydro energy likely to be exploited amounts to 378 million kilowatts, of which about 11 per cent is already developed. Biomass energies, including stalks, firewood and organic waste, amounts to 260 million tons of standard coal or 70 per cent of domestic energy consumption and 50 per cent of the total energy consumption. China benefits from 9.6 million square kilometers of land area with an annual solar luminosity quantity greater than 600,000 joules per square centimeters. The Chinese Energy State Council also approximates wind energy of 1.6 billion kilowatts, representing about 10 per cent of exploitable energy. Exploitation of geothermal resources is just beginning. In 2002, geothermal resources exploitation was equal to about 462.6 billion tons of standard coal. More than 100,000 tons of this is exploitable energy. Ocean energy also amounts to more than 20 million kilowatts of exploitable resources.

In this chapter we will review Chinese alternative and renewable energies. After exploring these new resources we will provide a summary of China's status with respect to the phenomenon of global warming and pollutant emissions.

10.1 NUCLEAR ENERGY

According to the International Energy Administration, China plans on intensifying its nuclear energy program in order to keep inflationary energy prices down and global markets balanced and efficient. China already owns nine nuclear power stations, primarily along the Eastern Sea. China's goal is to install more than 40 GW of nuclear power capacity by 2020 (Figure 10.1).

Since China began producing nuclear energy in 1992, it has made significant technological advances and uses nuclear as a cleaner alternative to coal. While Europe is the biggest user of nuclear energy, most of its global competitors are also developing nuclear technologies as a substitute for scarcer alternatives. In 2003, nuclear was cheaper, more efficient and cleaner than equivalent energies. Yet, some countries are still reluctant to use nuclear energy because it is associated with increased security risks in times of rising geopolitical uncertainty and tensions. Germany passed a law prohibiting nuclear energy production in 2002, while France relies on nuclear power for 80 per cent of its national energy production. Due to rising energy demand and its inability to control market prices, China is compelled to use nuclear energy to keep global commodities prices down and avoid inflationary pressures on its emerging middle class. China aims to build between six and eight nuclear power plants a year in the next twenty years, and to multiply its nuclear capacity fourfold by 2020.

The China National Nuclear Corporation (CNNC) built eight new reactors in May 2004. The CNNC also created new plants in Lingdong and Sanmen and six in Yangjiang. There are two reactors in Tianwan. The Chinese adhere to international safety standards with European Siemens control systems. Additionally, a nuclear reactor in Lanzhou is used for safe storage of nuclear waste. Japan remains one of the highest consumers of nuclear energy (Figure 10.2) but consumption is expected to level off after 2010 while the Chinese nuclear energy function slope is expected to continue to increase at a growing rate. Chinese dependency on nuclear energy

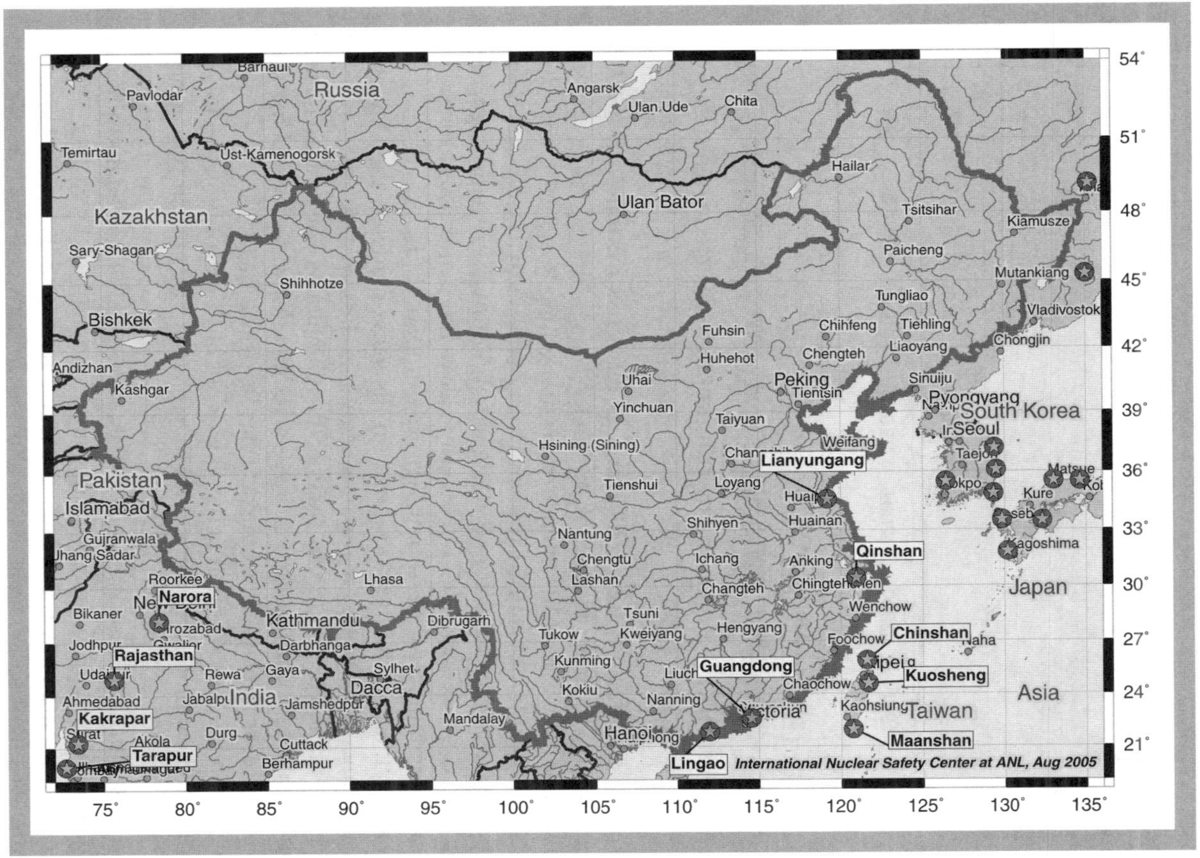

Figure 10.1 Chinese nuclear power stations

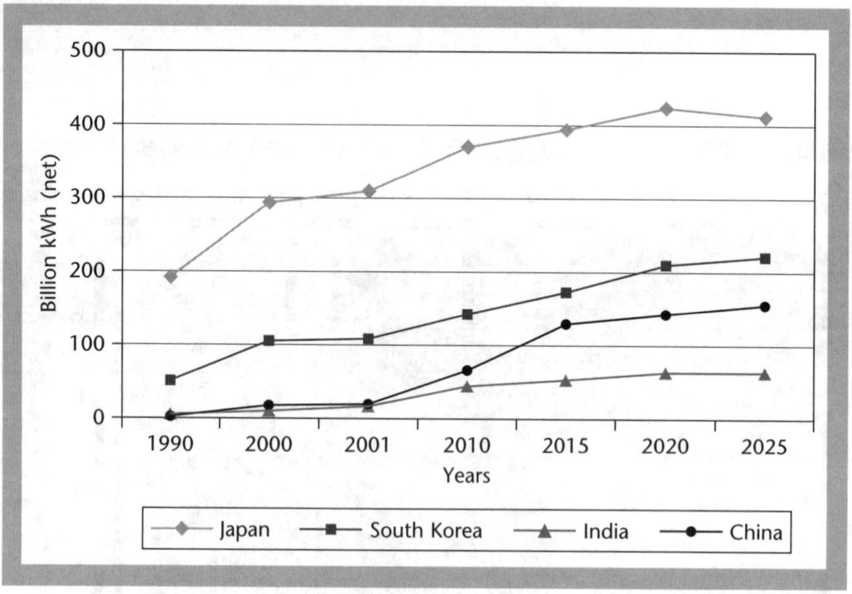

Figure 10.2 Nuclear energy consumption, selected countries

has risen significantly since 2001 and is expected to double from 2010 (70 billion kWh) to 2025 (near 150 billion kWh).

Daya Bay 1 is in the Shenzhen area in Guangdong Province. Guangdong Nuclear Power Joint Venture Company Ltd operates Daya Bay 1. The French National Company, Framatome, designed and built the reactor in partnership with China. Daya Bay 2 is the twin of Daya Bay 1 and is also run by the Guangdong Nuclear Power Joint Venture Company Ltd.

The Ling'ao reactors are designed on the same model as the Daya Bay reactors and are also in Guangdong Province, less than a mile from the Daya Bay plant. They were also designed by Framatome. The Ling'ao nuclear power plant has four reactors. Chinese reactors are always created in pairs: for instance, as at Daya Bay, Ling'ao 1 has its peer Linag'ao 2.

Qinshan 1 is the exception in that it was not built with a twin reactor; the second in the pair did not come until March 2004. The nuclear power plant was constructed in 1981, when the reactor was named Qinshan. The second reactor is Qinshan 2, also known as Qinshan 2-A. China started the Qinshan 2 construction in 1985, two months before the start of Daya Bay 1 (also known as Guangdong 1). Qinshan 1 was connected to the grid 20 months before Daya Bay 1. The reactor was shut down for a year for maintenance in August 1998, and returned to service in September 1999. The Qinshan 2 reactor is the same model as Pakistan's Chasnupp 1. Renamed as 2-A when China decided to construct reactors in pairs, it was the first of the two reactors constructed under phase 2 of the Qinshan project, it is of Chinese design and manufacture. According to the International Atomic Energy Agency, it was

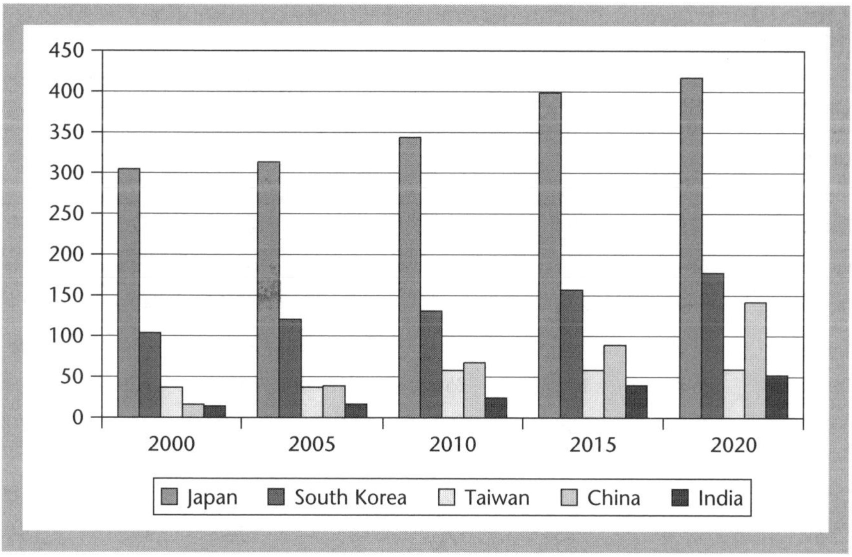

Figure 10.3 Nuclear energy, 2000–20 (thousand MW)
Sources: Energy Information Administration (2003), *International Energy Annual 2003*, Washington DC, 2005, www.eia.doe.gov/iea projections; EIA System for Analysis of Global Energy Markets (2006); Bustelo (2005).

built with imported foreign parts and devices and assembled in China. Unit 2-A came into operation on 2 February 2002. Unit 2-A was also identified as 2-1. Qinshan 2-B came into operation on 4 March 2004, a month ahead of schedule. Both 2-A and 2-B are light water pressurized reactors. About 25 per cent of the world's commercial nuclear reactors hold a 1000 MWe capacity or more. The largest French reactors have a 1200 net MWe capacity which is the largest ever built according to EIA. China owns a pair of Russian-built reactors of this approximate capacity: Tianwan-1 and Tianwan-2, the largest reactors in China. Qinshan 3-A, is the first Canada Deuterium Uranium Reactor (CANDU) constructed in China, and it and its twin, 3-B, are China's first heavy water pressurized reactors (PHWR). Qinshuan 3-A and 3-B are also based on a CANDU unit. Tianwan-1 started operations in 2004. China uses a number of nuclear technologies, including pressurized light water reactors, which are the most common type of commercial reactor, found in nuclear industries throughout the world. Two-thirds of the commercial nuclear reactors in the United States are of this type. China also has a number of pressurized heavy water reactors. The VVER model is a Soviet acronym for a type of light water pressurized reactor that is domestically built or exported. For light water pressurized reactors designed by countries outside the former Soviet Union, the acronym PWR is used instead of VVER.

The International Energy Administration predicts that most Asian countries will increase nuclear energy utilization in the first part of this century (Figure 10.3). Table 10.1 shows predictions for Asia and selected countries.

Table 10.1 Projected nuclear generated electricity consumption, 1990–2025 (billion kWh)

Country	1990	2000	2001	2010	2015	2020	2025
United States	577	754	769	794	812	816	816
France	298	394	401	447	478	520	550
Germany	145	161	163	137	107	15	0
Japan	192	294	309	369	394	426	411
Canada	69	69	73	108	110	118	98
Russia	115	122	125	141	154	129	99
South Korea	50	104	107	141	171	209	220
India	6	14	18	46	55	66	66
China	0	16	17	66	129	142	154

Note: Selected countries, for projections for all countries, see source.
Source: Energy Information Administration, *International Energy Outlook 2004*.

Table 10.2 provides a description of most commercial nuclear reactors in China.

According to a multilateral agreement with Asian partners in February 2007, China has given assurances about the dismantlement of Chinese nuclear reactors and the use of nuclear plants only to enable efficient management of energy policies and to alleviate emission pollution.

Qinshan Nuclear Power Plant

Qinshan Nuclear Power Plant is located in Hangzhou Bay, Zhejiang Province. It is the first nuclear power station to use the pressurized water reactors that were developed, designed and constructed by China. The total installed capacity is 4.1 million kW. It is designed to function for 30 years. The Qinshan nuclear power plant generated its first electricity in December 1991. With the construction of the Qinshan power plant China became the seventh country in the world capable of producing its own nuclear power station, after the United States, Britain, France, Russia, Canada and Sweden.

Daya Bay Nuclear Power Plant

Daya Bay Nuclear Power Plant is in Shenzhen, Guangdong Province. Daya was the second nuclear power plant built in China and the first large-scale one for commercial purposes. Daya Bay required an initial investment of

Table 10.2 China's commercial nuclear reactors

Unit name (alternative names)	Type	Capacity MWe Net	Gross	Status (as of March 31, 2004)
(a) Operational units: reactors fully licensed to operate and that are either in service or are shut down but expected to return to service				
Daya Bay 1 (Guangdong 1)	PWR	944	984	Connected to the grid on 31 August 1993
Daya Bay 2 (Guangdong 2)	PWR	944	984	Connected to the grid on 2 February 1994
Ling'ao 1 (Lingao A)	PWR	938	990	Connected to the grid on 26 February 2002
Ling'ao 2 (Lingao B)	PWR	938	990	Connected to the grid on 15 December 2002
Qinshan 1 (formerly, Qinshan)	PWR	279	300	Connected to the grid on 15 December 1991
Qinshan 2-A (formerly, Qinshan 2)	PWR	610	642	Connected to the grid on 2 February 2002
Qinshan 2-B (formerly, Qinshan 3)	PWR	610	642	Connected to the grid on 4 March 2004
Qinshan 3-A	PHWR	665	728	Connected to the grid in April 2003
Qinshan 3-B	PHWR	665	728	Connected to the grid in July 2003
Total Capacity		**6.593**	**6.988**	
(b) Units in the construction pipeline				
Tianwan-1 (Lianyungang)	VVER	1	1.06	Under construction
Tianwan-2 (Lianyungang)	VVER	1	1.06	Under construction
Total Capacity		**2**	**2.12**	

Source: Energy Information Administration (2003).

US$4 billion. It has two pressurized water reactors imported from France and Britain. Daya holds an installed capacity of 1.8 million kW and generates 12.6 billion kWh annually for Guangdong, Hong Kong and other urban areas in South China. The station was put into operation in 1994 and the government subjected the plant to technical verifications in 1996. The plant is operated as a joint venture by two private entities, the Guangdong Provincial Nuclear Power Investment Co. and Hong Kong Nuclear Power Investment Co. Ltd.

Guangdong Ling'ao Nuclear Power Plant

The Guangdong Ling'ao Nuclear Power Plant is located in the east of Shenzhen city, Guangdong Province, approximately 1.2 kilometers from the Daya Bay power plants. Ling'ao was constructed in May 1997 with a total investment of US$4 billion. It consists of four nuclear generating units, two of which were installed and generating electricity by 2002.

Zhejiang Sanmen Nuclear Power Plant

Zhejiang Sanmen Nuclear Power Plant is in Sanmen County, Zhejiang Province. Zhejiang Sanmen Nuclear Power Plant consists of four 1 million-kW energy generating units. Zhejiang's total investment is worth 10 billion yuan.

Jiangxi Maozidingshan Nuclear Power Plant

Jiangxi Maozidingshan nuclear power plant is located in Jiujiang city, Jiangxi Province. The Jiangxi Maozidingshan nuclear power station consists of six 600,000-kW generating units. This station feeds energy to the Central China and East China urban electricity networks.

Shandong 1 Nuclear Power Plant

Shandong 1 is in Haiyang County, Shandong Province. The nuclear power station consists of four 1 million-kW pressurized water reactors. The total investment in Shandong 1 was 40 billion yuan. The plant started operations at the end of 2003.

Lianyungang Nuclear Power Plant

Lianyungang Nuclear Power Plant is in Lianyungang city, Jiangsu Province. Its nuclear facilities comprise four 1 million-kW generating units. Its implementation was studied and approved in March 1996. The Lianyungang plant is based on two Russian-made wwer-1000 type of reactor engines. They were installed in 1999, and were put into operation in 2004 and 2005.

According to the British Petroleum Statistical Report of 2006, China started to consume nuclear energy heavily in 1993, with a consumption of 1.6 terawatts per hour, thereafter consumption followed an exponential, shape reaching a significant peak at 53 terawatts per hour in 2001. The Chinese nuclear consumption function increases at an accelerating rate in parallel to alternative energy demands to meet the demands of China's new middle class. Table 10.3 (BP, 2006) provides an overview of historical upward consumption (particularly significant since the 1970s) of nuclear energy by various nations. Figures 10.4 and 10.5 enable a comparison of Chinese nuclear consumption with that of other nations, with European and Eurasian countries outconsuming all others since around 1980.

10.2 HYDROPOWER

According to the International Energy Administration, China intends to substantially increase its reliance on hydropower to sustain traditional energy demand. China's strategic plan is to have 200 to 240 GW of hydroelectricity or 7 to 9 GW of new hydropower capacity per year. The Chinese Energy State Council also aims to speed up the development of small hydraulic power energy. By 2010, the installed capacity of small hydraulic power stations ought to reach 27.88 million kilowatts. To reach this policy target, China will need to construct a dam equivalent to the Three Gorges Dam every two years. Hydropower is an environmentally clean way to meet China's expanding energy demand and the country already has 380 GW of exploitable hydropower. However, China's difficulty in relying exclusively on hydropower is water scarcity and the distances from water points and dams to load centers. China is following Japan in substituting reliance on oil by producing more energy from hydropower stations. Thus, Japan built many hydrostations in the 1970s to decrease reliance on oil. Initially, hydropower is very expensive, with the construction of large dams, but these will prove effective in the long run in China, balancing increased demand and potential inflationary prices. The International Energy Administration estimates that the proposed 7 to 9 GW of hydropower capacity per year will require $13 to $23 billion per year. During the 1990s, many hydropower projects were principally funded by the Chinese government, for instance, the Three Gorges and the South-North Water Diversion projects. With the rise of many foreign investment banking institutions and foreign ventures in the 1990s, more dams were built in partnership with international entities, in part to keep international interests with regards to energy prices in balances. The International Energy Administration also reports that the international funding institutions, such as the IMF and the World Bank, impressed upon the Chinese government the need to be more environmentally aware. Huadian, Guodian, Datang and China Power Investment Company initiated a partnership with Hong Kong's China Light and Power company to create a US$4.9 billion thermal and hydropower plant in southern China to feed the southeastern China region. Chinese dams are financed via municipal bonds repaid through taxes and tolls. Additional taxes and usage revenues, such as a tax of 0.0001 RMB on each kWh of

Table 10.3 World consumption of energy from nuclear power, 1965–96 (TWh)

	1965	1966	1967	1968	1969	1970
USA	3.8	5.8	8.1	13.2	14.7	23.0
Canada	–	–	–	–	–	–
Mexico	–	–	–	–	–	–
Total North America	**3.8**	**5.8**	**8.1**	**13.2**	**14.7**	**23.0**
Argentina	–	–	–	–	–	–
Brazil	–	–	–	–	–	–
Chile	–	–	–	–	–	–
Colombia	–	–	–	–	–	–
Ecuador	–	–	–	–	–	–
Peru	–	–	–	–	–	–
Venezuela	–	–	–	–	–	–
Other S. & Cent. America	–	–	–	–	–	–
Total S. & Cent. America	**–**	**–**	**–**	**–**	**–**	**–**
Austria	–	–	–	–	–	–
Azerbaijan	n/a	n/a	n/a	n/a	n/a	n/a
Belarus	n/a	n/a	n/a	n/a	n/a	n/a
Belgium & Luxembourg	–	^	0.1	0.1	^	0.1
Bulgaria	–	–	–	–	–	–
Czech Republic	–	–	–	–	–	–
Denmark	–	–	–	–	–	–
Finland	–	–	–	–	–	–
France	1.1	1.6	2.9	3.5	5.0	5.7
Germany	0.1	0.3	1.5	2.0	5.3	6.5
Greece	–	–	–	–	–	–
Hungary	–	–	–	–	–	–
Iceland	–	–	–	–	–	–
Republic of Ireland	–	–	–	–	–	–
Italy	3.6	3.9	3.2	2.6	1.7	3.2
Kazakhstan	n/a	n/a	n/a	n/a	n/a	n/a
Lithuania	n/a	n/a	n/a	n/a	n/a	n/a
Netherlands	–	–	–	^	0.3	0.4
Norway	–	–	–	–	–	–
Poland	–	–	–	–	–	–
Portugal	–	–	–	–	–	–
Romania	–	–	–	–	–	–
Russian Federation	n/a	n/a	n/a	n/a	n/a	n/a
Slovakia	–	–	–	–	–	–
Spain	–	–	–	0.1	0.8	0.9
Sweden	^	^	0.1	^	0.1	0.1
Switzerland	–	–	–	–	0.5	2.5
Turkey	–	–	–	–	–	–
Turkmenistan	n/a	n/a	n/a	n/a	n/a	n/a

1971	1972	1973	1974	1975	1976	1977	1978	1979
40.1	56.9	87.9	120.0	181.6	201.2	264.1	291.30	268.6
4.0	6.7	14.3	13.9	11.8	15.9	26.8	32.9	33.1
–	–	–	–	–	–	–	–	–
44.1	**63.6**	**102.2**	**133.9**	**193.4**	**217.1**	**290.8**	**323.9**	**301.7**
–	–	–	1.0	2.3	2.3	1.6	2.9	2.7
–	–	–	–	–	–	–	–	–
–	–	–	–	–	–	–	–	–
–	–	–	–	–	–	–	–	–
–	–	–	–	–	–	–	–	–
–	–	–	–	–	–	–	–	–
–	–	–	–	–	–	–	–	–
–	–	–	–	–	–	–	–	–
–	**–**	**–**	**1.0**	**2.3**	**2.3**	**1.6**	**2.9**	**2.7**
–	–	–	–	–	–	–	–	–
n/a	n/a	n/a	n/a	n/a	n/a	n/a	n/a	n/a
n/a	n/a	n/a	n/a	n/a	n/a	n/a	n/a	n/a
–	^	0.1	0.1	6.8	10.0	11.9	12.5	11.4
–	–	–	0.9	2.6	5.0	5.9	5.9	6.2
–	–	–	–	–	–	–	–	–
–	–	–	–	–	–	–	–	–
–	–	–	–	–	–	2.7	3.3	6.7
9.3	14.6	14.7	14.7	18.2	15.8	18.0	30.5	40.0
6.2	9.5	12.1	14.5	24.1	29.6	41.3	43.9	52.1
–	–	–	–	–	–	–	–	–
–	–	–	–	–	–	–	–	–
–	–	–	–	–	–	–	–	–
–	–	–	–	–	–	–	–	–
3.4	3.8	3.0	3.3	3.3	3.8	3.4	4.4	4.7
n/a	n/a	n/a	n/a	n/a	n/a	n/a	n/a	n/a
n/a	n/a	n/a	n/a	n/a	n/a	n/a	n/a	n/a
0.4	0.3	1.1	3.0	3.1	3.6	3.7	3.5	3.2
–	–	–	–	–	–	–	–	–
–	–	–	–	–	–	–	–	–
–	–	–	–	–	–	–	–	–
–	–	–	–	–	–	–	–	–
n/a	n/a	n/a	n/a	n/a	n/a	n/a	n/a	n/a
	^	0.2	0.5	0.2	0.4	0.1	^	2.1
2.6	4.7	6.6	7.2	7.5	7.6	6.5	7.6	6.7
0.1	1.5	2.1	2.1	12.0	16.0	19.9	23.8	21.0
2.8	3.6	6.3	6.2	7.4	7.9	8.1	8.0	9.4
–	–	–	–	–	–	–	–	–
n/a	n/a	n/a	n/a	n/a	n/a	n/a	n/a	n/a

(*Continued*)

Table 10.3 (Continued)

	1980	1981	1982	1983	1984	1985	1986
USA	264.3	287.0	297.7	309.1	344.9	403.9	435.8
Canada	35.7	36.0	38.3	48.6	52.2	60.5	71.3
Mexico	–	–	–	–	–	–	–
Total North America	**300.1**	**323.1**	**336.0**	**357.7**	**397.1**	**464.4**	**507.1**
Argentina	2.3	2.8	1.9	3.4	4.6	5.8	5.7
Brazil	–	–	–	–	1.6	3.4	0.1
Chile	–	–	–	–	–	–	–
Colombia	–	–	–	–	–	–	–
Ecuador	–	–	–	–	–	–	–
Peru	–	–	–	–	–	–	–
Venezuela	–	–	–	–	–	–	–
Other S. & Cent. America	–	–	–	–	–	–	–
Total S. & Cent. America	**2.3**	**2.8**	**1.9**	**3.4**	**6.2**	**9.1**	**5.9**
Austria	–	–	–	–	–	–	–
Azerbaijan	n/a	n/a	n/a	n/a	n/a	–	–
Belarus	n/a	n/a	n/a	n/a	n/a	–	–
Belgium & Luxembourg	12.6	12.9	15.7	24.1	27.7	34.6	39.4
Bulgaria	6.2	9.1	10.7	12.3	12.7	13.1	12.1
Czech Republic	–	–	–	–	–	2.4	6.1
Denmark	–	–	–	–	–	–	–
Finland	7.0	14.7	16.8	17.7	18.9	18.9	18.9
France	61.3	105.3	108.9	144.3	191.2	224.2	254.2
Germany	55.6	65.5	74.4	78.1	104.3	138.6	130.5
Greece	–	–	–	–	–	–	–
Hungary	–	–	^	2.5	3.8	6.5	7.4
Iceland	–	–	–	–	–	–	–
Republic of Ireland	–	–	–	–	–	–	–
Italy	4.9	5.2	9.2	8.2	6.7	7.0	8.8
Kazakhstan	n/a	n/a	n/a	n/a	n/a	–	–
Lithuania	n/a	n/a	n/a	n/a	n/a	9.5	9.9
Netherlands	3.9	3.4	3.6	3.3	3.5	3.9	4.2
Norway	–	–	–	–	–	–	–
Poland	–	–	–	–	–	–	–
Portugal	–	–	–	–	–	–	–
Romania	–	–	–	–	–	–	–
Russian Federation	n/a	n/a	n/a	n/a	n/a	99.3	105.2
Slovakia	4.5	5.1	5.8	6.2	7.2	9.4	11.7
Spain	5.2	9.6	8.8	10.7	23.1	28.0	37.5
Sweden	26.5	37.7	39.0	41.0	50.9	58.6	69.9
Switzerland	13.6	14.5	14.3	14.8	17.4	22.6	22.6
Turkey	–	–	–	–	–	–	–
Turkmenistan	n/a	n/a	n/a	n/a	n/a	–	–

1987	1988	1989	1990	1991	1992	1993	1994	1995	1996
479.2	554.7	557.2	607.2	644.8	651.3	642.4	674.1	708.8	710.2
77.3	82.9	79.9	72.9	84.9	80.6	93.9	107.8	97.8	92.8
–	–	0.4	2.9	4.2	3.9	4.9	4.2	8.4	7.9
556.5	**637.6**	**637.5**	**683.0**	**734.0**	**735.8**	**741.3**	**786.2**	**815.1**	**810.9**
6.5	5.8	5.0	7.3	7.8	7.1	7.8	8.2	7.1	7.5
1.0	0.6	1.8	2.2	1.4	1.8	0.4	0.1	2.5	2.4
–	–	–	–	–	–	–	–	–	–
–	–	–	–	–	–	–	–	–	–
–	–	–	–	–	–	–	–	–	–
–	–	–	–	–	–	–	–	–	–
–	–	–	–	–	–	–	–	–	–
–	–	–	–	–	–	–	–	–	–
7.4	**6.4**	**6.9**	**9.5**	**9.2**	**8.8**	**8.2**	**8.3**	**9.6**	**9.9**
–	–	–	–	–	–	–	–	–	–
–	–	–	–	–	–	–	–	–	–
–	–	–	–	–	–	–	–	–	–
42.0	43.1	41.2	42.7	42.9	43.5	41.9	40.6	41.4	43.3
12.4	16.0	14.6	14.7	13.2	11.6	14.0	15.3	17.3	18.1
10.7	11.8	12.4	12.6	12.1	12.3	12.6	13.0	12.2	12.8
–	–	–	–	–	–	–	–	–	–
19.5	19.4	19.0	19.1	19.4	19.1	19.8	19.3	19.1	19.7
265.5	275.5	303.9	314.1	331.3	338.4	368.2	360.0	377.2	397.3
141.7	156.8	161.7	152.5	147.5	158.8	153.5	151.2	154.1	161.6
–	–	–	–	–	–	–	–	–	–
11.0	13.4	13.9	13.7	13.7	14.0	13.8	14.1	14.0	14.2
–	–	–	–	–	–	–	–	–	–
–	–	–	–	–	–	–	–	–	–
0.2	–	–	–	–	–	–	–	–	–
–	–	0.1	–	0.5	0.5	0.4	0.4	0.1	0.1
9.2	12.8	16.6	17.0	17.0	14.6	12.3	7.7	11.8	13.9
3.6	3.7	4.0	3.5	3.3	3.8	3.9	4.0	4.0	4.2
–	–	–	–	–	–	–	–	–	–
–	–	–	–	–	–	–	–	–	–
–	–	–	–	–	–	–	–	–	–
–	–	–	–	–	–	–	–	–	1.4
124.7	134.8	136.3	118.3	120.0	119.6	119.2	97.8	99.4	109.0
11.5	11.5	12.2	12.0	11.7	11.1	11.9	12.1	11.4	11.3
41.3	50.5	56.1	54.3	55.6	55.8	56.1	55.3	55.4	56.3
67.6	69.4	65.6	68.2	76.8	63.5	61.4	73.2	70.0	73.3
23.0	22.8	22.8	23.6	23.0	23.4	23.4	24.4	24.9	25.1
–	–	–	–	–	–	–	–	–	–
–	–	–	–	–	–	–	–	–	–

(Continued)

Table 10.3 (Continued)

	1965	1966	1967	1968	1969	1970
Ukraine	n/a	n/a	n/a	n/a	n/a	n/a
United Kingdom	15.1	20.2	23.3	26.2	29.1	26.0
Uzbekistan	n/a	n/a	n/a	n/a	n/a	n/a
Other Europe & Eurasia	1.9	2.1	2.3	3.1	3.6	4.4
Total Europe & Eurasia	**21.8**	**28.2**	**33.3**	**37.7**	**46.5**	**49.8**
Iran	–	–	–	–	–	–
Kuwait	–	–	–	–	–	–
Qatar	–	–	–	–	–	–
Saudi Arabia	–	–	–	–	–	–
United Arab Emirates	–	–	–	–	–	–
Other Middle East	–	–	–	–	–	–
Total Middle East	**–**	**–**	**–**	**–**	**–**	**–**
Algeria	–	–	–	–	–	–
Egypt	–	–	–	–	–	–
South Africa	–	–	–	–	–	–
Other Africa	–	–	–	–	–	–
Total Africa	**–**	**–**	**–**	**–**	**–**	**–**
Australia	–	–	–	–	–	–
Bangladesh	–	–	–	–	–	–
China	–	–	–	–	–	–
China Hong Kong SAR	–	–	–	–	–	–
India	–	–	–	–	0.7	1.3
Indonesia	–	–	–	–	–	–
Japan	^	0.4	0.8	0.7	0.9	3.3
Malaysia	–	–	–	–	–	–
New Zealand	–	–	–	–	–	–
Pakistan	–	–	–	–	–	–
Philippines	–	–	–	–	–	–
Singapore	–	–	–	–	–	–
South Korea	–	–	–	–	–	–
Taiwan	–	–	–	–	–	–
Thailand	–	–	–	–	–	–
Other Asia Pacific	–	–	–	–	–	–
Total Asia Pacific	^	**0.4**	**0.8**	**0.7**	**1.7**	**4.6**
TOTAL WORLD	**25.7**	**34.4**	**42.2**	**51.6**	**62.9**	**77.3**
Of which European Union 25#	19.9	26.1	31.1	34.6	42.4	42.9
OECD	23.8	32.3	40.0	48.5	58.5	71.6
Former Soviet Union	1.9	2.1	2.3	3.1	3.6	4.4
Other EMEs	–	–	–	–	0.7	1.3

1971	1972	1973	1974	1975	1976	1977	1978	1979
n/a	n/a	n/a	n/a	n/a	n/a	n/a	n/a	n/a
27.5	29.4	28.0	33.6	30.3	36.2	40.0	37.2	38.3
n/a	n/a	n/a	n/a	n/a	n/a	n/a	n/a	n/a
5.4	9.8	14.3	21.5	28.6	38.1	51.0	52.5	59.6
57.7	**77.3**	**88.6**	**107.5**	**144.2**	**174.0**	**212.5**	**233.1**	**261.4**
–	–	–	–	–	–	–	–	–
–	–	–	–	–	–	–	–	–
–	–	–	–	–	–	–	–	–
–	–	–	–	–	–	–	–	–
–	–	–	–	–	–	–	–	–
–	–	–	–	–	–	–	–	–
–	–	–	–	–	–	–	–	–
–	–	–	–	–	–	–	–	–
–	–	–	–	–	–	–	–	–
–	–	–	–	–	–	–	–	–
–	–	–	–	–	–	–	–	–
–	–	–	–	–	–	–	–	–
–	–	–	–	–	–	–	–	–
–	–	–	–	–	–	–	–	–
–	–	–	–	–	–	–	–	–
–	–	–	–	–	–	–	–	–
1.0	0.9	2.0	1.8	2.1	2.7	1.9	2.3	2.3
–	–	–	–	–	–	–	–	–
7.1	8.8	9.5	18.6	21.7	36.7	28.2	52.0	61.9
–	–	–	–	–	–	–	–	–
–	–	–	–	–	–	–	–	–
^	0.2	0.4	0.5	0.6	0.5	0.3	0.2	0.1
–	–	–	–	–	–	–	–	–
–	–	–	–	–	–	–	–	–
–	–	–	–	–	–	0.1	2.3	3.2
–	–	–	–	–	–	0.1	2.7	6.3
–	–	–	–	–	–	–	–	–
–	–	–	–	–	–	–	–	–
8.1	**9.9**	**11.8**	**20.9**	**24.5**	**39.9**	**30.5**	**59.5**	**73.8**
110.0	**150.8**	**202.6**	**263.3**	**364.3**	**433.3**	**535.4**	**619.3**	**639.6**
49.6	63.8	68.0	79.0	105.6	123.0	147.5	166.8	186.3
103.6	139.9	185.9	237.6	328.1	384.7	474.6	553.0	562.5
5.4	9.8	14.3	21.5	28.6	38.1	51.0	52.5	59.6
1.0	1.1	2.3	4.3	7.6	10.5	9.8	13.9	17.6

(Continued)

Table 10.3 (Continued)

	1980	1981	1982	1983	1984	1985	1986
Ukraine	n/a	n/a	n/a	n/a	n/a	53.3	42.7
United Kingdom	37.0	38.0	44.0	49.9	54.0	61.1	59.1
Uzbekistan	n/a	n/a	n/a	n/a	n/a	–	–
Other Europe & Eurasia	73.9	87.4	102.8	115.3	148.2	9.3	6.8
Total Europe & Eurasia	**312.2**	**408.3**	**454.1**	**528.3**	**669.6**	**800.3**	**846.9**
Iran	–	–	–	–	–	–	–
Kuwait	–	–	–	–	–	–	–
Qatar	–	–	–	–	–	–	–
Saudi Arabia	–	–	–	–	–	–	–
United Arab Emirates	–	–	–	–	–	–	–
Other Middle East	–	–	–	–	–	–	–
Total Middle East	–	–	–	–	–	–	–
Algeria	–	–	–	–	–	–	–
Egypt	–	–	–	–	–	–	–
South Africa	–	–	–	–	4.1	5.6	9.3
Other Africa	–	–	–	–	–	–	–
Total Africa	–	–	–	–	**4.1**	**5.6**	**9.3**
Australia	–	–	–	–	–	–	–
Bangladesh	–	–	–	–	–	–	–
China	–	–	–	–	–	–	–
China Hong Kong SAR	–	–	–	–	–	–	–
India	2.4	2.5	2.2	3.1	4.0	4.5	5.0
Indonesia	–	–	–	–	–	–	–
Japan	82.6	85.6	105.5	108.8	126.7	151.9	166.4
Malaysia	–	–	–	–	–	–	–
New Zealand	–	–	–	–	–	–	–
Pakistan	0.1	0.2	0.2	0.3	0.3	0.4	0.5
Philippines	–	–	–	–	–	–	–
Singapore	–	–	–	–	–	–	–
South Korea	3.5	2.9	3.8	9.0	11.8	16.7	28.3
Taiwan	8.2	10.7	13.1	18.9	24.6	28.7	26.9
Thailand	–	–	–	–	–	–	–
Other Asia Pacific	–	–	–	–	–	–	–
Total Asia Pacific	**96.8**	**101.8**	**124.7**	**140.0**	**167.5**	**202.3**	**227.2**
TOTAL WORLD	**711.3**	**836.0**	**916.7**	**1029.4**	**1244.5**	**1481.8**	**1596.4**
Of which European Union 25#	218.4	297.4	326.3	385.9	491.3	602.7	657.6
OECD	618.2	723.4	785.8	876.2	1044.3	1248.9	1372.2
Former Soviet Union	73.9	87.1	100.2	111.4	143.8	167.3	160.5
Other EMEs	19.2	25.5	30.6	41.9	56.5	65.5	63.7

Notes: ^ Less than 0.05; n/a not available; # Excludes Estonia, Latvia and Lithuania prior to 1985 and Slovenia prior to 1991; growth rates are adjusted for leap years.

Source: British Petroleum (2006).

1987	1988	1989	1990	1991	1992	1993	1994	1995	1996
50.4	72.0	66.5	76.2	75.1	73.8	75.2	68.8	70.5	79.6
55.2	63.5	71.7	65.7	70.5	76.8	89.4	88.3	89.0	94.7
–	–	–	–	–	–	–	–	–	–
8.8	8.9	6.0	4.6	4.0	4.0	4.0	4.6	5.1	6.9
898.3	**986.0**	**1024.7**	**1012.8**	**1037.6**	**1044.5**	**1080.9**	**1050.0**	**1076.9**	**1142.8**
–	–	–	–	–	–	–	–	–	–
–	–	–	–	–	–	–	–	–	–
–	–	–	–	–	–	–	–	–	–
–	–	–	–	–	–	–	–	–	–
–	–	–	–	–	–	–	–	–	–
–	–	–	–	–	–	–	–	–	–
–	–	–	–	–	–	–	–	–	–
–	–	–	–	–	–	–	–	–	–
–	–	–	–	–	–	–	–	–	–
6.5	11.0	11.7	8.9	9.6	9.8	7.6	10.2	11.9	12.4
–	–	–	–	–	–	–	–	–	–
6.5	**11.0**	**11.7**	**8.9**	**9.6**	**9.8**	**7.6**	**10.2**	**11.9**	**12.4**
–	–	–	–	–	–	–	–	–	–
–	–	–	–	–	–	–	–	–	–
–	–	–	–	–	–	1.6	14.8	12.8	14.3
–	–	–	–	–	–	–	–	–	–
5.3	6.1	4.0	6.4	5.4	6.4	6.2	4.9	7.6	8.4
–	–	–	–	–	–	–	–	–	–
189.4	174.9	186.6	195.7	209.3	218.0	248.6	259.5	287.8	297.5
–	–	–	–	–	–	–	–	–	–
–	–	–	–	–	–	–	–	–	–
0.3	0.2	0.1	0.4	0.4	0.6	0.4	0.6	0.5	0.3
–	–	–	–	–	–	–	–	–	–
–	–	–	–	–	–	–	–	–	–
39.3	40.1	47.4	52.9	56.3	56.5	58.1	58.7	67.0	73.9
33.1	30.7	28.3	32.9	35.3	33.8	34.4	34.9	35.3	37.8
–	–	–	–	–	–	–	–	–	–
–	–	–	–	–	–	–	–	–	–
267.4	**252.0**	**266.4**	**288.3**	**306.8**	**315.3**	**349.3**	**373.3**	**411.1**	**432.3**
1736.1	**1892.9**	**1947.1**	**2002.6**	**2097.2**	**2114.3**	**2187.3**	**2228.0**	**2324.6**	**2408.2**
678.9	731.4	778.4	775.4	805.8	815.6	848.7	843.3	864.4	907.2
1477.9	1594.0	1656.0	1713.7	1807.4	1830.8	1903.8	1959.7	2042.7	2096.1
188.6	224.4	220.8	211.5	212.7	208.5	207.1	174.8	182.1	204.9
69.6	74.5	70.2	77.4	77.1	74.9	76.3	93.6	99.8	107.2

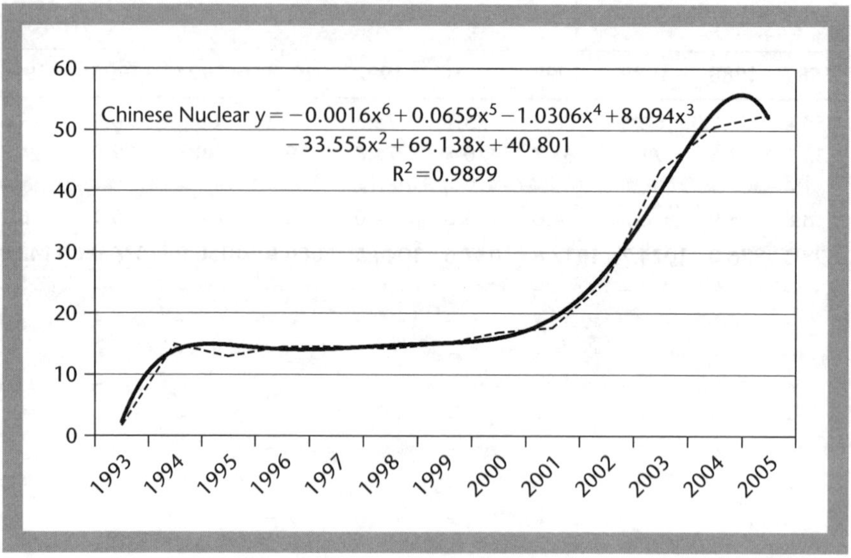

Figure 10.4 Chinese nuclear consumption function, 1993–2005 (TWh)
Note: Growth rates are adjusted for leap years.
Source: British Petroleum (2006).

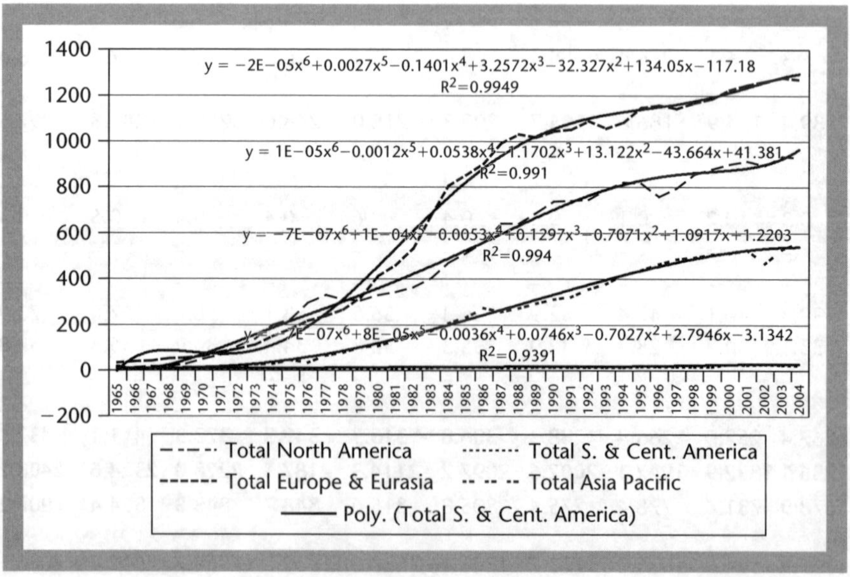

Figure 10.5 Other countries' nuclear consumption functions, 1965–2004 (TWh)
Source: British Petroleum (2006).

electricity generated by a new dam, finance new hydropower energy projects. Heilongjiang Province currently has a hydro capability of 113 MW; Jiangsu Province, 100 MW; Fujian Province, 200 MW; Guangdong Province, 100 MW; and Hebei Province, 200 MW.

China has substantial natural resources for hydropower development, but has not yet reached full capacity. Nonetheless, some of the largest Chinese dams can generate 676,000 MW. According to the International Energy Administration, Chinese exploitable reserves are estimated to be 379,000 MW, and economically feasible exploitable energies approximate about 290,000 MW or 1260 TWh. The International Energy Administration observed in 1998 that China's total installed hydropower capacity was 65,000 MW with an annual power output of 208 TWh which accounts for less than 25 per cent of the total economically exploitable resources. And as noted in the British Petroleum statistical report of 2006, Chinese hydro-electricity energy demand was 22.1 terawatts per hour in the mid-1960s rising exponentially to 400 terawatts per hour in 2005. Compared with international hydro-electricity consumption functions, Chinese demand is rising faster than that of its peers. If we compare the demand to demand in mature markets no longer anticipating drastic intra-social changes and restructuring, we may estimate that China's demand – much of it from the new middle class – will reach beyond 1000 terawatts per hour by 2020. Most mature peers, including the United States and Europe, plateaued at the 900 terawatt line. According to the Chinese government's strategic policies for renewables, installed capacity should rise to 110 GW by 2010, 160–180 GW by 2020 and 260–290 GW by 2050. The power output is expected to reach 350 TWh by 2010 and 500–563 TWh by 2020.

In 1998, China constructed large and medium-sized hydropower projects amounting to 40,000 MW, enabling an annual power output of approximately 128 TW. Since then, various projects have been initiated to increase hydro-electricity capacity to 31,800 MW and beyond. China's Three Gorges project is the largest of its kind in the world and has been operating efficiently since the beginning of 2006.

Medium and large capacity hydropower plants

Ertan hydropower plant is on the Yalong River in Sichuan Province and has a capacity of 3300 MW. It was completed on 4 December 1999 and was the largest Chinese hydropower facility of the twentieth century. The Lijiaxia hydropower station is on the Yellow River in Qinghai Province and has a capacity of 2000 MW. It was completed on 3 December 1999. The Guangzhou pump storage facility benefits from a capacity of 2400 MW and was completed on 14 March 2000. Altogether, hydropower stations have a generating capacity of one million kW and include the following: Gezhouba,

Baishan, Longyangxia, Liujiaxia, Geheyan, Wanjiazhai, Ertan, Wuqiangxi, Tianshengqiao, Yantan, Tianhuangping, Shuikou, Manwan, Dachaoshan, Waterfall Valley on Dadu River, Qingjiang Step, Geheyan and Lijiaxia. Table 10.4 provides a list of hydropower plants in China.

North China Wanjiazhai Key water control project

Wanjiazhai Key hydro-electricity plant stands on the Yellow River in the northwest of Pianguan County, Shanxi Province and started its operations in 1993. The Wanjiazhai Key water control project has an installed generating capacity of 1.08 million kW and has reached a capacity of 2.75 billion kWh. It is the first generating unit with a capacity of 180,000 kW and ramified the domestic Chinese power grid on 28 November 1998.

Northeast China Fengman hydropower station

Fengman hydropower is the first northeastern Chinese station and is located 24 kilometers from Jilin on the upper reaches of the No. 2 Songhua River and 250 kilometers from the Baishan hydropower station. Fengman hydropower station controls a drainage area of 42,500 square kilometers. Fengman reservoir, which is also called the Songhua lake, is 180 kilometers long and has a surface area of 550 square kilometers. The deepest part is 75 meters and the storage capacity is 10.8 billion cubic meters. The dam is 1080 meters long and the highest section is 91 meters. The top of the dam is 266.5 meters above sea level. On top of the dam is a 1.2-meter-high wave prevention wall. The dam has 10 turbo-generator units, one of which has a total generating capacity of 809,000 kW.

Baishan Hydropower Station

Baishan hydropower station is on the Songhua River at Huadian, Jilin Province. It is the largest hydropower station in the northeast grid and has a generating capacity of 1.5 million kW. It started operations in May 1975 and stored water from September 1982. By the beginning of 1985, the dam had three generating units with a total capacity of 900,000 kW. Baishan had an installed generating capacity of 600,000 kW as of June 1992. It has a 149.5-meter-high heavy-pressure dam, with four 12 × 13 meter tunnels above the 404-meter-high spillway dam, three 6 × 7 meter tunnels for discharging water, an underground powerhouse with an installed generating capacity of 900,000 kW and another powerhouse on the surface with an installed generating capacity of 600,000 kW. The reservoir has a storage capacity of 6.812 billion cubic meters. The dam's highest normal storage

Table 10.4 Large hydropower stations completed or under construction (over 1000 MW)

Name	Location	Installed capacity MW	Year of operation and completion
Three Gorges	Hubei	18,200	2003, 2009
Ertan	Sichuan	3300	1998, Completed in 1999
Gezhouba	Hubei	2715	1981, Completed in 1988
1st, 2nd phases of Guangzhou pump storage	Guangdong	2400	1993, Completed in March 2000
Lijiaxia	Qinghai	2000	1997, Completed in 1999
Xiaolangdi	Henan	1800	In progress
Tianhuangping	Zhejiang	1800	1997, completed in 2000
1st, 2nd phases of Baishan	Jilin	1500	1983, Completed in 1994
Shuikou	Fujian	1400	1993, Completed in 1996
Dachaoshan	Yunnan	1350	In progress
Tianshengqiao, 2nd stage	Guizhou, Guangxi	1320	1994, Completed in 1998
Longyangxia	Qinghai	1280	1987, Completed in 1989
Manwan	Yunnan	1250	1993, Completed in 1995
Yantan	Guangxi	1200	1992, Completed in 1994
Geheyan	Hubei	1200	1993, Completed in 1994
Tianshengqiao, 1st stage	Yunnan, Guizhou, Guangxi	1200	In progress
Wuqiangxi	Hunan	1200	1994, Completed in 1996
Liujiaxia	Gansu	1160	1969, Completed in 1974
Wanjiazhai	Shanxi	1080	1998, completed in 2000
Fengman	Jinlin	1004	1943, Completed in 1998

Source: *China Power*, 10, 1999.

water level is 413 meters. The capacity for water control storage is 3.54 billion cubic meters while the flood control storage capacity is 950 million cubic meters.

The step hydropower stations

The step hydropower stations are on the Songjiang River near Changbai Mountain in South Jilin Province and Wusong County near the No. 2 Songhua River. The dam was part of the Chinese government's eight-year energy plan in the 1990s. The three step stations are in Xiaoshan, Shanggou and Shilong and the project also includes the Songshan and Sandiao Songjiang water diversion works on the Songjiang River. These stations have a generating capacity of 510,000 kW and an annual energy production of 831 million kWh. The Songjian station required an investment of 1.67 billion yuan. The Xiaoshan hydropower station is the first at the foot of Changbai mountain and started operating in July 1994. The second generating unit has a capacity of 80,000 kW and started to operate in December 1997.

Lianhua hydropower station

Linhua hydropower station is near Lianhua township, Linkou County, Mudanjiang, Heilongjiang Province. The station is the first large modern water conservancy project in Heilongjiang Province. The station has four 137,500 kW generating units and cost 4.7 billion yuan. It started operating in November 1992 and connected to the national domestic power grid in December 1996. Another three hydropower stations were constructed in 1998.

East China Tianhuangping hydropower station

The East China Tianhuangping station is in Anji County, Zhejiang Province, near Tianhuang mountain. It has a reservoir at 900 meters and a lower one at Xitiao Stream in the Taihu Lake drainage area. The Tianhuangping reservoirs hold 8.85 million cubic meters and 8.77 million cubic meters. The station has a generating power holding resistance of 607.5 meters. The heads of the two reservoirs average 570 meters and the headrace tunnels average 1428 meters in length. Total installed generating capacity is 1.8 million kW. The annual generation capacity is 3.16 billion kWh, and its efficiency is 75 per cent. The East China Tianhuangping station is financed by the Shanghai Municipality, Jiangsu, Zhejiang and Anhui provinces and the East China Power Group Co. Investments amount to 7.118 billion yuan. The project started on 1 March 1994, and ended in June 2000. Construction of the upper reservoir started in December 1993 and was finished in October 1997. The first generating unit holds a capacity of 300,000 kW and started producing energy in September 1998.

Dongjin hydropower station

The station is at Dongjin township, Xiushui County, Jiangxi Province. It has functioned since June 1995 and holds two 30,000-kW generating units.

Shuikou hydropower station

Shuikou hydropower station started in late 1996 and was, at the time, the largest in Eastern China. It is near Xiapu village, Minqing County, Fujian Province and close to the Minjiang river. The dam was created with funding from the World Bank. It consists of seven generating units with installed generating capacity of 1.4 million kW. The capacity of the No. 3 generating unit is 200,000 kW. Construction began in May 1986 and lasted until November 1996. Shuikou hydropower station generates a capacity of 4.95 billion kW.

Shangyuan reservoir power station

Shangyuan reservoir power station is near Panxi town, Fuding County, Fujian Province, west of Tailao mountain. Shangyuan is one of 34 large and medium-sized hydropower stations in the province. The State Planning Commission and the Ministry of Water Resources approved the reservoir on 29 October 1991. The dam came into operation in 1992 and required 168 million yuan in funding. There are four main sections to the station: a dual arch 84.2-meter-high dam; a 6011-meter intake tunnel, 3.8 meters in diameter; a 578-meter pressure steel pipe, 2.2 meters in diameter; the powerhouse, and the 31-kilometers 110-volt transmission line and one supporting project. Shangyuan has a storage capacity of 73.5 million cubic meters and produces electricity from three 12,500 kW generating units giving an annual output of 108 million kWh.

Central and South China Qingjiang step power stations

Qingjiang power station extends for 423 kilometers along the Yangtze River in Hubei Province and generates 3.29 million kW or 11.5 billion kWh a year. The Qingjiang Hydropower Development Company was established in 1987 with three step power stations at Gaobazhou, Geheyan and Shuibuya constructed during the next 15 years. These stations generate 3 million kW for the central China area. Annual production can increase to 10 billion kWh with 15,000 cubic meters of floodwater per second.

Geheyan hydropower station

The Geheyan hydropower station is in Changyang County, Yichang, Hubei Province. Geheyan station consists of a dam, a power reservoir and lifting machines. The dam is a heavy-pressure dam and the storage capacity is 3.4 billion cubic meters. The four generating units have a capacity of 1.211 million kW. The station was the first phase in the Qingjiang step power stations project. Construction at Geheyan started in December 1986 and was completed in December 1994. The power station generated 6 billion kWh of electricity in 1996 and earned 1.8 billion yuan.

Gezhouba hydropower plant

Gezhouba hydropower plant is ramified to the Chinese electric central grid and connects the central China grid and the east China grid. Gezhouba hydropower plant has 21 domestic turbo-generating units with a total generating capacity of 2.715 million kW. Fourteen generating units are installed in the Dajiang power station producing 1.75 million kW and there are seven in the Erjiang power plant producing an output of 965,000 kW. The plant produces 15.7 billion kWh of electrical energy per year. The first generating unit came on line in 1987 and had produced 200 billion kWh of electricity valued at 16 billion yuan by October 1997.

Gaobazhou hydropower station

Gaobazhou hydropower station is the lowest dam of the three step power stations on the Qingjiang River and is in Zhicheng, Hubei Province. It is 50 kilometers from the Geheyan hydropower station on the upper section of the river, 12 kilometers from the Yangtze and Qingjiang rivers and 45 kilometers from the Gezhouba hydropower station. The dam is 439.5 meters long, 57 meters high and stores rainwater from 15,650-square kilometers. Storage capacity is 536 million cubic meters. The station has three 84,000-kW generating units to provide 898 million kWh of electricity annually. The first project phase started in October 1996 and the second-phase project in October 1998. The investment totals 3.076 billion yuan.

Wuqiangxi hydropower station

Wuqiangxi is the largest water conservancy project in Hunan Province. The station is in Yuanling County, Hunan Province, near the Yuanshui River, and is equipped with five 240,000 kW turbo-generator units with an installed generating capacity of 1.2 million kW. Wuqiangxi provides 5.37 billion kWh of electricity annually, one-fifth of the province's total. The storage capacity is 4.27 billion cubic meters. The station required 7 billion yuan investment and a low-interest loan of 25.09 billion Japanese yen from the Overseas Economic Cooperative Fund of Japan. The Wuqiangxi was approved in 1986 and in operation by December 1994.

Zhexi hydropower station

Zhexi station is 10 kilometers west of Anhua County, in Hunan Province and has six turbo-generator units to generate a capacity of 447,500 kW. The station is able to increase its capacity to 500,000 kW.

Dongjiang hydropower station

Dongjiang was built in 11 years. Dongjiang was the first dual-arched thin dam designed and constructed by China, the second of its kind in the world and the first in Asia. The station has stored water since August 1986. The reservoir has a water surface of 160 square kilometers and a storage capacity of 8.12 billion cubic meters, about half that of Dongting Lake.

Yantan hydropower station

Yantan is on the Hongshui River. It is the first large hydropower station in Yantan Province with a generating capacity of more than one million kW. The first phase of Yantan consisted of four generating units with a total capacity of 1.21 million kW to produce 5.3 billion kWh of electricity annually. The dam is 525 meters long and 110 meters high. Construction of the main body started in 1985, and the river was blocked in 1988. The first generating unit began to supply electricity to the local grid in 1992. The other three generating units were installed and put into operation in June 1995, and began to generate electricity for the local grid in June 1996. The project investments totalled 3.532 billion yuan.

Zhaoping hydropower station

Zhaoping station is in Guangxi Zhuang Autonomous Region. The station became operational and was connected to the Wuzhou area grid in April 1994. It is equipped with three generating units with a total generating capacity of 63,000 kW to give 304.5 million kWh of electricity annually and required investments of 125 million yuan.

Bailongtan hydropower station

Bailongtan hydropower is in Guangxi Zhuang Autonomous Region, and is the seventh of the ten step hydropower stations on Hongshui River. It has an installed generating capacity of 192,000 kW and is the first to function automatically. Bailongtan was constructed from February 1993 and the dam started to generate energy in December 1995.

Hemianshi hydropower station

Hemianshi Station is in Xindu town, in Hexian County, Guangxi. It was built in 1970 but did not produce hydro-electricity until September 1974. Hemianshi hydro-station has four generating units able to generate capacity of 68,000 kW. The station provides 360 million kWh of electricity a year. The storage capacity is 296 million cubic meters, and the regulated storage capacity is 112 million cubic meters. Water can irrigate a total of 6867 hectares of farmland.

Yemao hydropower station

Yemao is in Guangxi Zhuang Autonomous Region. It is located in Yizhou and is the sixth hydropower station on Longjiang River. Guangxi Hydropower Engineering Bureau approved the contract for the construction of the station in June 1992. The first generating unit started to supply the local grid in January 1996. Yemao dam is 462.58 meters long and 34.7 meters high. The station is equipped with three 12,500-kW generating units, which provide 187.5 million kWh of electricity annually. The station cost 166 million yuan.

Southwest China Taipingyi hydropower station

The Taipingyi hydropower station is on the Minjiang River, Aba Prefecture, Sichuan Province and is equipped with four 65,000-kW generating units. The project required a total investment of 1.45 billion yuan. Huaneng Group Company and the State Energy Investment Company invested in the project. The construction of the station started in 1991, and the No. 1 generating unit joined the local grid on 7 November 1994.

Ertan hydropower station

Ertan hydropower station is on Yalong River in Panzhihua at the borders of Miyi and Yanbian counties, Sichuan Province. Ertan generates 3.3 million kW capacity and can provide 17.035 billion kWh of electricity annually. Its dam is 240 meters high. It is one of the main power stations in the Sichuan grid and the second largest next to the Three Gorges Project. The project took 10 years to create and began operation in October 1998. The station was completed in 2000. Ertan has a 240-meter-high dam with a storage capacity of 5.8 billion cubic meters and the water surface will cover 10,100 hectares. It has six 550,000-kW turbo generating units. Total investment for the Ertan hydropower station is 33 billion yuan, of which 24.369 billion yuan are financed by the State Development and Investment Company, Sichuan Provincial Investment Company, and Sichuan Provincial Power Company. Foreign funds amount to US$930 million, of which US$780 million represent loans from the World Bank and the remaining US$150 million comes from financial private entities. The 500-kV transmission line is 2097 kilometers long and includes a three-loop line from Ertan to Zigong, a two-loop line on the same tower from Zigong to Chengdu, a second-loop line from Zigong to Chongqing for 162 kilometers a single-loop line from Chongqing to Changshou for 100 kilometers and a single-loop line from Ertan to Panzhihua for 20 kilometers. Substations include Zhaoju controller in Liangshan, Honggou substation in Zhigong, Longwang substation in Chengdu, Chengjiaqiao substation in Chongqing, Changshou substation and Panzhihua substation. Their transforming capacity totals 5.25 million kVa. The 220-kV supporting line is 1340 kilometers long with a transforming capacity of 3.38 million kVa. A 550,000 kW sub-grid generating capacity began operating in August 1998 with connection to the local grid and a second sub-grid started operating in December 1998, some 21 months earlier than scheduled.

Daqiao reservoir hydropower station

Daqiao reservoir is in Mianning County, Liangshan Yi Autonomous Prefecture of Sichuan Province in the Anning river drainage area. The reservoir is 91 meters high. The normal storage level is 2020 meters above sea level and the capacity is 658 million cubic meters with a readjusted storage capacity of 593 million cubic meters. The station has an installed generating capacity of 90 megawatts providing 350 million kWh of electricity annually. Daqiao cost 419.53 million yuan. Operations started in June 1997.

Gongzui hydropower plant

Gongzui plant is situated at Shawan town, Leshan, Sichuan Province, and has Gongzui and Tongjiezi hydropower stations under its administration, with a generating capacity of 1.3 million kW and a designed annual output of electricity of 6.628 billion kWh. Gongzui has seven 100,000 kW generating units in the station designed to produce 3.4178 billion kWh of electricity each year. Its construction started in 1966, the first generating unit started to operate in December 1971 and all of them were operational as of 1978. Tongjiezi hydropower station has four 150,000 kW generating units with an annual output of 3.21 billion kWh. Its construction started in March 1985 and the first generating unit began to supply the domestic grid in October 1992. All the generating units were operating by December 1994. The Gongzui hydropower plant had provided a total of 84.698 billion kWh by October 1997, and its output value was 5.756 billion yuan.

Pobugou hydropower station on the Dadu River

Pobugou hydropower station stands on the upper reaches of the Dadu River in Sichuan Province. It has an installed generating capacity of 3.3 million kW, and the reservoir has a storage capacity of 5.177 billion cubic meters. The station is 186 meters high and the underground powerhouse is 290 meters long, 32 meters wide and 69 meters high.

Shilongba hydropower station

Shilongba was the first hydropower station in China and is located southwest of Dianchi lake, about 45 kilometers from Kunming. Construction at Shilongba started in July 1910, and it began producing electric power in September 1912. The station is installed with two 375-HP machines.

Manwan hydropower station

The Manwan station in Yunnan Province is on the middle reaches of the Lancang River and on the borders of Yunxian and Jingdong counties. It was the first large hydropower station on the Lancang River and is equipped with six 250,000 kW generating units, with a total installed generating capacity of 1.5 million kW. Its annual production of electricity is 7.795 billion kWh. The first two generating units were put into operation in 1993 to add 4 million kWh of electricity to the Yunnan grid. The third and fourth units connected to the grid in 1994. The fifth began to provide power in June 1995.

Dachaoshan hydropower station

Dachaoshan hydropower station is in western Yunnan Province on the borders of Yuxian and Jingdong counties. The station is the second 1 million kW hydropower station on the Lancang River after the Manwan station.

Dachaoshan generates a capacity of 1.35 million kW. It has an annual output of 5.9 billion kWh and can reach 7.021 billion kWh. The station's dam is 118 meters high and the top is 481 meters long. Its storage capacity is 890 million cubic meters. The Dachaoshan station required an estimated investment of 8.87 billion yuan. The project started in 1993 and ended in 1996. The generating unit started to generate power in 2001. The whole station was completed and put into operation in 2003.

Dongfeng hydropower station

Dongfeng station is on the Wujiang River below Dushui hydropower station. It is located in a soluble stone canyon, 8 kilometers downstream from the confluence of the Liuchong and Sancha rivers. Its total installed generating capacity is 510,000 kW and the annual output of electricity can reach 2.42 billion kWh. Wujiang Development Company from Guizhou, Guiyang and the Ministry of the Power Industry constructed the No. 9 Hydropower Bureau. Construction at Dongfeng began in 1984 and 1989 and the entire project was completed and fully operative by December 1995.

Tianshengqiao second-step hydropower station

The Guiyang Hydropower Survey and Design Institute designed the Tianshengqiao hydropower station which is the largest diversion-type station in China. Its total installed generating capacity is 1.32 million kW, and the first generating unit started working on 20 December 1992. The hydropower station has three 9.56 kilometer long intake tunnels with an inner diameter of 8.7 to 9.8 meters, the largest underground projects in China. The side slope of the powerhouse area is 380 meters high, a major achievement by the Ministry of Power Industry, which won the ministry's second prize for scientific and technological progress. The dam was the first to use the new technology for rolling and pressing concrete.

Tianshengqiao first-step hydropower station is able to generate a total capacity of 1.2 million kW and has an annual output of 5.226 billion kWh. Tianshengqiao's station is able to increase the output to 8.839 billion kWh.

Lubuge hydropower station

Lubuge station is on the Huangni River in Yunnan and Guizhou provinces near Guangxi Zhuang region. Lubuge has generated 600,000 kW of electricity with its four turbo-generating units since it came on line in 1989. The Lubuge dam is 103.8 meters high and has a reservoir storage capacity of 111 million cubic meters. Its system consists of a 9387 meter pressure tunnel and two 470 meter pressure steel pipes.

Tanghe hydropower station

Tanghe station is on the Tanghe River. Tanghe is 47 kilometers from Xigaze and generates capacity of 6400 kW, which is distributed to the Xigaze area of Tibet. Work on the Tanghe hydropower station took 10 years and the station began operations in 1980. Shandong Province provided the initial investments of 73.5 million yuan. Tanghe was upgraded in July 1996 with a 415,000-cubic-meter reservoir, and the hydropower output of electricity can reach 34.73 million kWh.

Yanghu hydropower station

Yanghu hydropower is on the northern slope of the Himalayas at 4440.5 meters. It is south of Yarlung Zangbo River and 80 kilometers from Lhasa. Yanghu reservoir covers 620 square kilometers with a storage capacity of 15 billion cubic meters. It was designed in 1973 and constructed in 1985. The Party Central Committee and the State Council remodeled it in August 1989 with an additional pump storage power station. The station has four generating units and a reserve with a total generating capacity of 112,500 kW. Yanghu hydropower station cost 1.5 billion yuan, construction started in July 1991 and the station started working six years later. The station has a head of 840 meters and an intake tunnel 5600 meters long. The Yanghu hydropower station is the largest and most advanced power station in Tibet and supplies electricity to Lhasa, Lhaze and central Tibetan grids.

Woka River first-step hydropower station

Woka River is in Sangri County, Tibet, and the Woka River station is a power station with a low dam and a generating capacity of 20,000 kW. Water comes from two head projects and the intake tunnels consist of main and tributary tunnels of a total length of 6593 meters. Woka holds five water transmission aqueducts, one back siphon and six floodwater aqueducts.

Zhalong hydropower station

Zhalong hydropower station is in Tibet, on the upper part of the Nujiang River in Nagqu County in Tibet. Zhalong hydropower is 30 kilometers from Nagqu town and has a generating capacity of 10,800 kW. The station includes a masonry dam, spillway, tunnels, and a powerhouse as well as transmission and transfer facilities. The Zhalong dam is 39 meters high with a maximum storage capacity of 140 million cubic meters. The adjusted storage capacity is 59 million cubic meters.

Jaggang hydropower station

Jaggang hydropower station is at an elevation of 4751 meters, on Qiangtang Plateau in northern Tibet. In 1997 Jaggang was the highest hydropower station in the world. The first generating unit was in operation in 1997, and the station was tested and approved in September 1998.

Northwest China Qingtongxia hydropower station

The northwest China Qingtongxia station is in the Ningxia Hui region on the Yellow River. Qingtongxia dam is 42.7 meters high and 697 meters long and has an installed generating capacity of 272,000 kW.

Liujiaxia hydropower station

Liujiaxia station is in Yongqing County in Gansu Province. Construction began in 1964. Liujiaxia's five units started operation in 1974 with a total installed generating capacity of 1.22 million kW.

Yanguoxia hydropower station

Yanguoxia's hydropower station is on the Yellow River in Linxia Gansu Province. Construction began in 1958 and the first generating unit began operation in 1961, shortly before it was completed. Yanguoxia's dam is 57.2 meters high, 15.9 meters wide on the top and 321 meters long. Its storage capacity is 220 million cubic meters. Yanguoxia's hydropower has eight axial flow water turbines with a total generating capacity of 352,000 kW.

Daxia hydropower station

Daxia hydropower station is in Gansu Province, at Xixiakou on the Yellow river, 65 kilometers northeast of Lanzhou. The project was approved in May 1991 and construction began in October 1991. Its first generating unit generated capacity of 75,000 kW in 1996 and was connected to the local domestic grid. Daxia's last engine was connected to the domestic grid in June 1998, six months ahead of schedule. Daxia hydropower stations are built on the section from Longyang Gorge to Qingtong Gorge. Daxia is also the first project along the well-known Three Small Gorges, consisting of Small Gorge, Big Gorge and Wujin Gorge. The hydrostation holds four 75,000-kW generating units and one 24,500-kW axial-flow water turbine. Daxia's hydrostation cost 2.5 billion yuan.

Xuelongtan hydropower station

Xuelongtan hydropower is in Menyuan County in Qinghai Province and is the first of the 13 step hydropower stations on the Datong River. Xuelongtan hydropower generates capacity of 20,000 kW, and is able to generate an annual total production of 110 million kWh. Xuelongtan started in May 1994. Xuelongtan holds 250,000 cubic meters with more than 280 meters of open tunnels.

Nina hydropower station

Nina hydropower station is one of the 13 large and medium-sized hydropower stations near the Yellow River and is the third step power station between Longyang Gorge and Qingtong Gorge. Nina generates capacity of 160,000 kW and has an annual electricity production of 736 million kWh. Nina hydropower investments cost 1.562 billion yuan.

Lijiaxia hydropower station

Lijiaxia hydropower station is on the Yellow River, in Qinghai Province on the borders of Janca County and Hualong Autonomous County. The Lijiaxia hydropower station holds five 400,000-kW turbo-generating units that are able to produce 2 million kW with an annual electricity production of 5.9 billion kWh. Lijiaxia cost 5 billion yuan. The dam is 165 meters high and 45 meters wide at the base with a reservoir storage capacity of 1.648 billion cubic meters of water. Building started in July 1987 and the diversion tunnels started operating in 1991. The main section was completed is November 1998. By May 1998, most turbines and units came into operation and were connected to the domestic grid.

Dashankou hydropower station

Dashankou hydropower station is in Hejing County, Xinjiang Uygur Autonomous Region, near the Kaidu River drainage area, and has an installed generating capacity of 80,000 kW. Dashankou's total investment amounts to 206.61 million yuan. The station produces 310 million kWh annually. Construction began in June 1985, the first generating unit came into production on 28 December 1991, and the second on 9 April 1992. Dashankou generated 754 million kWh of electricity by 1994. The hydropower station's industrial output reached 59.53 million yuan.

The British Petroleum statistics on hydropower consumption estimate that China will substantially rely on hydropower production to feed its rising demand. The Chinese consumption curve has risen exponentially since the mid-1960s while hydropower consumption functions leveled or plateaued at a maximum consumption of 900 level terawatts per hour. Given the population size and the rising energy demand, the Chinese consumption of hydroelectricity is expected to increase by three to four times by 2020. Table 10.5 and Figures 10.6 and 10.7 show Chinese hydroelectricity consumption in comparison to other countries from 1965 to 2005.

10.3 BIOMASS

Between 2000 and 2010, biomass energy is estimated to increase from a generated capacity of 2.5 million tons to 17 million tons of standard coal. Chinese biomass production consists of converting methane energy from various wastes from breweries, sugar refineries, livestock and fowl-breeding farms into high quality energies. Figure 10.8 shows the estimated output of traditional biomass and new renewables between 2000 and 2020.

Large and medium-sized methane projects supplied energy to 7.55 million households in 2000 and this is expected to increase to 12.35 million households by 2010. The Chinese Energy State Council aimed to generate a capacity of 2.26 billion cubic meters, equal to 1.8 million tons of standard coal, by 2000 and 4 billion cubic meters, equal to 3.14 million tons of standard coal, by 2010. Modern technologies also encourage biomass energy in China. Fifty per cent of Chinese farmers have high-efficiency woodburning cooking stoves, which save 30 per cent to 50 per cent of annual fuel consumption. Methane applications are used frequently in Chinese households. There are 5.25 million methane tanks in China, with an annual production capacity of over 1.2 billion cubic meters of methane. China has over 600 methane projects with a generating capacity of 100 cubic meters of methane, each capable of supplying up to 84,000 households. According to the International Energy Administration, biomass is the most cost-effective renewable resource in China. However biomass is insufficient to feed the rising domestic demand. It is not that renewable energy is in itself costly, it is the combination of complementary alternative energies that is needed to respond adequately to rising domestic demand that causes costs to rise.

Estimates of China's biomass total 700 million tons of coal equivalent (Mtce), including 120 Mtce of crop stalks, 90 Mtce of firewood, and other residual resources. Biomass renewables, including firewood consumption,

Table 10.5 Global hydropower consumption, 1965–2005 (TWh)

	1965	1966	1967	1968	1969	1970
USA	199.0	199.9	227.2	228.2	256.0	253.5
Canada	117.1	128.8	133.1	136.3	148.2	155.3
Mexico	8.9	10.2	11.1	12.6	13.5	15.0
Total North America	**324.9**	**338.9**	**371.4**	**377.1**	**417.8**	**423.8**
Argentina	1.2	1.2	1.3	1.5	1.3	1.5
Brazil	24.0	26.2	27.4	28.7	30.7	39.8
Chile	4.0	4.2	4.3	3.6	4.0	4.3
Colombia	3.5	3.8	4.3	4.8	5.3	5.9
Ecuador	0.3	0.4	0.4	0.5	0.5	0.6
Peru	2.6	2.8	3.2	3.4	3.7	3.8
Venezuela	1.4	1.4	1.7	2.8	3.3	4.1
Other S. & Cent. America	3.2	4.7	4.9	5.1	5.5	6.2
Total S. & Cent. America	**40.2**	**44.8**	**47.4**	**50.3**	**54.4**	**66.2**
Austria	17.7	17.6	18.0	18.5	17.0	21.5
Azerbaijan	n/a	n/a	n/a	n/a	n/a	n/a
Belarus	n/a	n/a	n/a	n/a	n/a	n/a
Belgium & Luxembourg	1.2	1.3	1.2	1.0	1.1	1.1
Bulgaria	2.0	2.0	2.0	1.3	1.8	2.2
Czech Republic	2.2	2.1	1.9	1.5	1.2	1.7
Denmark	^	^	^	^	^	^
Finland	9.5	10.6	11.8	10.6	8.8	9.5
France	46.9	52.2	45.4	50.4	53.0	56.7
Germany	17.3	18.7	17.9	18.0	15.2	19.0
Greece	0.8	1.9	1.8	1.5	2.3	2.9
Hungary	0.1	0.1	0.1	0.1	0.1	0.1
Iceland	1.7	1.9	1.9	2.1	2.4	3.7
Republic of Ireland	1.0	0.9	0.8	0.8	0.6	0.7
Italy	46.4	47.8	46.4	47.0	45.6	44.8
Kazakhstan	n/a	n/a	n/a	n/a	n/a	n/a
Lithuania	n/a	n/a	n/a	n/a	n/a	n/a
Netherlands	–	–	–	–	–	–
Norway	49.4	48.7	53.3	60.2	57.4	57.8
Poland	0.9	0.9	1.0	1.1	0.9	1.9
Portugal	4.1	5.4	5.6	5.3	6.4	5.9
Romania	1.0	1.0	1.5	1.6	2.2	2.8
Russian Federation	n/a	n/a	n/a	n/a	n/a	n/a
Slovakia	2.2	2.1	1.8	1.7	1.3	2.0
Spain	20.0	27.8	23.1	24.9	31.2	28.4
Sweden	46.4	45.5	49.3	48.8	41.8	41.5
Switzerland	24.4	27.9	30.2	29.9	28.6	29.8
Turkey	2.2	2.4	2.4	3.2	3.5	3.1
Turkmenistan	n/a	n/a	n/a	n/a	n/a	n/a

1971	1972	1973	1974	1975	1976	1977	1978	1979
272.3	278.7	278.2	307.3	306.2	289.8	225.9	286.3	285.9
157.0	171.3	178.8	197.9	195.0	202.8	198.5	217.0	212.6
14.5	15.4	16.3	16.8	15.2	17.2	19.2	16.2	18.0
443.7	**465.4**	**473.3**	**522.0**	**516.4**	**509.9**	**443.6**	**519.5**	**516.5**
1.5	1.5	3.0	5.0	5.1	4.9	5.7	7.7	10.5
43.2	50.7	57.9	65.7	72.3	82.9	93.5	102.7	116.6
4.4	5.2	5.3	6.0	6.1	6.2	6.5	6.8	6.9
6.3	7.0	7.6	8.6	9.5	10.0	10.3	11.9	13.1
0.6	1.1	0.6	0.6	0.6	0.6	0.9	1.1	1.0
4.3	4.4	4.8	5.2	5.5	5.8	6.0	6.2	6.7
5.4	6.0	6.2	7.7	8.9	10.6	12.2	12.2	14.2
7.6	7.1	7.9	7.9	7.9	8.5	9.7	10.3	10.9
73.3	**83.0**	**93.2**	**106.7**	**116.0**	**129.6**	**144.7**	**159.0**	**179.9**
17.0	17.5	19.2	22.3	23.7	20.5	24.9	24.9	28.0
n/a	n/a	n/a	n/a	n/a	n/a	n/a	n/a	n/a
n/a	n/a	n/a	n/a	n/a	n/a	n/a	n/a	n/a
1.2	1.5	1.5	1.6	0.9	0.9	0.7	0.8	0.9
2.2	2.1	2.6	2.1	2.5	3.0	3.5	2.9	3.3
1.2	1.3	1.1	1.7	1.8	1.4	2.0	1.8	2.3
^	^	^	^	^	^	^	^	^
10.7	9.9	10.1	12.1	12.2	9.4	12.1	9.7	10.9
48.9	48.8	47.7	56.2	60.1	48.8	76.6	68.8	67.3
15.3	14.9	16.8	19.2	18.4	15.2	18.8	19.8	19.8
2.9	3.0	2.5	2.2	2.2	2.1	2.1	3.3	3.9
0.1	0.1	0.1	0.1	0.2	0.2	0.2	0.1	0.1
4.0	4.7	5.4	4.8	5.4	5.6	4.6	4.9	5.2
0.8	0.5	0.8	0.8	0.8	0.9	1.0	1.0	1.2
43.4	47.0	39.1	37.8	37.4	43.3	56.2	47.4	45.3
n/a	n/a	n/a	n/a	n/a	n/a	n/a	n/a	n/a
n/a	n/a	n/a	n/a	n/a	n/a	n/a	n/a	n/a
–	–	–	–	–	–	–	–	–
63.9	68.1	73.6	76.6	77.4	82.0	72.2	80.9	89.0
1.9	1.9	1.8	2.5	2.4	2.1	2.4	2.4	2.5
6.3	7.6	8.1	7.8	6.3	5.1	10.2	10.4	10.9
4.5	7.3	7.5	8.5	8.7	8.1	9.3	10.6	11.3
n/a	n/a	n/a	n/a	n/a	n/a	n/a	n/a	n/a
1.5	1.6	1.3	2.3	2.0	2.0	2.4	2.3	1.9
33.3	35.8	29.1	31.7	26.4	22.3	40.7	41.6	47.6
52.0	53.8	59.9	57.3	57.7	54.9	53.5	57.8	61.2
30.0	25.4	27.8	28.9	33.1	26.8	35.8	32.5	30.8
2.5	3.4	2.6	3.4	5.9	8.4	8.6	9.3	10.3
n/a	n/a	n/a	n/a	n/a	n/a	n/a	n/a	n/a

(*Continued*)

Table 10.5 (Continued)

	1980	1981	1982	1983	1984	1985
USA	282.0	266.5	315.5	338.7	327.6	287.2
Canada	222.5	225.0	257.9	266.0	286.2	303.7
Mexico	16.9	24.6	22.9	20.7	23.6	26.2
Total North America	**521.4**	**516.1**	**596.3**	**625.4**	**637.4**	**617.1**
Argentina	15.0	14.5	17.5	18.2	19.6	20.7
Brazil	128.9	130.8	141.1	151.5	166.6	178.4
Chile	7.3	7.6	8.5	8.9	9.3	10.4
Colombia	14.3	14.2	15.0	15.1	16.7	18.3
Ecuador	1.2	1.1	1.2	2.3	4.2	3.3
Peru	7.0	8.0	8.4	8.1	8.6	9.4
Venezuela	14.6	15.1	15.8	18.1	20.2	22.6
Other S. & Cent. America	11.9	12.1	12.5	17.8	18.7	21.0
Total S. & Cent. America	**200.2**	**203.4**	**219.9**	**240.0**	**264.0**	**284.0**
Austria	29.1	30.8	30.9	30.6	29.5	31.6
Azerbaijan	n/a	n/a	n/a	n/a	n/a	1.3
Belarus	n/a	n/a	n/a	n/a	n/a	^
Belgium & Luxembourg	1.1	1.7	1.5	1.6	1.8	1.9
Bulgaria	3.7	3.6	3.0	3.3	3.3	2.2
Czech Republic	2.4	2.1	1.7	1.7	1.3	1.7
Denmark	^	^	^	^	^	^
Finland	10.2	13.7	13.1	13.6	13.2	12.3
France	70.2	73.0	71.3	70.9	67.5	64.1
Germany	20.3	21.7	21.4	20.6	20.2	19.4
Greece	3.8	3.3	3.5	2.3	2.9	2.8
Hungary	0.1	0.2	0.2	0.2	0.2	0.2
Iceland	5.0	5.0	3.8	3.9	3.9	3.8
Republic of Ireland	1.2	1.2	1.2	1.2	1.0	1.2
Italy	43.0	44.3	42.6	42.8	44.0	44.6
Kazakhstan	n/a	n/a	n/a	n/a	n/a	5.2
Lithuania	n/a	n/a	n/a	n/a	n/a	0.4
Netherlands	–	–	–	–	–	–
Norway	84.0	93.3	92.9	106.0	106.3	102.9
Poland	3.3	3.0	2.6	3.3	3.3	3.9
Portugal	9.8	8.9	9.6	8.0	10.2	10.8
Romania	12.6	12.7	11.8	10.0	11.3	11.9
Russian Federation	n/a	n/a	n/a	n/a	n/a	159.7
Slovakia	2.4	2.2	2.1	2.2	2.0	2.7
Spain	30.8	23.2	27.4	28.9	33.4	33.0
Sweden	59.2	60.2	55.6	64.1	68.5	71.6
Switzerland	34.5	36.1	37.0	36.0	30.9	33.0
Turkey	11.3	12.6	14.2	11.3	13.4	12.0
Turkmenistan	n/a	n/a	n/a	n/a	n/a	0.6

1986	1987	1988	1989	1990	1991	1992	1993	1994
297.0	255.4	228.4	274.7	295.8	291.9	255.6	283.3	262.8
310.7	316.3	307.6	291.5	296.9	308.5	316.5	321.7	327.9
20.0	18.3	21.0	24.4	23.5	21.9	26.3	26.5	20.3
627.7	**590.0**	**556.9**	**590.6**	**616.3**	**622.3**	**598.4**	**631.5**	**611.0**
21.0	21.9	15.8	13.3	18.2	16.4	19.5	24.2	27.2
182.4	185.6	199.1	204.7	206.7	217.8	223.3	235.1	242.7
11.3	12.1	11.5	9.6	9.0	13.1	16.8	17.0	17.3
21.4	21.5	24.5	26.3	27.5	27.7	22.4	28.0	32.3
4.0	4.8	4.8	5.0	5.0	5.1	5.1	5.8	6.7
9.9	10.7	10.4	10.4	10.2	11.2	9.7	11.7	12.8
25.1	30.8	34.2	34.7	37.0	44.5	47.3	47.5	51.3
32.4	38.8	39.7	43.5	50.0	51.3	49.9	55.9	59.1
307.5	**326.1**	**340.1**	**347.5**	**363.5**	**387.2**	**394.0**	**425.1**	**449.3**
31.7	36.7	36.5	36.2	32.5	32.7	36.1	38.0	36.9
0.8	0.8	0.7	0.7	0.7	1.8	1.7	2.4	1.8
^	^	^	^	^	^	^	^	^
1.9	2.0	2.0	1.8	1.7	1.8	1.8	1.5	1.9
2.3	2.5	2.6	2.7	1.9	2.4	2.1	1.9	1.5
1.9	2.5	2.1	1.6	1.5	1.3	1.6	1.6	1.8
^	^	^	^	^	^	^	^	^
12.4	13.8	13.4	13.0	10.9	13.2	15.1	13.5	11.8
65.3	72.9	78.8	50.6	57.3	61.5	72.5	67.9	81.0
20.3	22.3	22.5	20.7	19.7	18.5	21.1	21.5	22.5
3.3	3.0	2.6	2.2	2.0	3.2	2.4	2.5	2.9
0.2	0.2	0.2	0.2	0.2	0.2	0.2	0.2	0.2
4.0	4.1	4.4	4.3	4.2	4.2	4.3	4.5	4.5
1.3	1.1	1.2	1.0	1.0	1.0	1.0	1.0	1.2
44.5	42.6	43.5	37.5	35.1	45.6	45.8	44.5	47.7
4.8	5.7	7.0	7.3	7.4	7.2	6.9	7.6	9.2
0.4	0.4	0.4	0.4	0.4	0.3	0.3	0.4	0.7
–	–	–	^	0.1	0.1	0.1	0.1	0.1
96.8	103.8	109.5	118.7	121.4	110.6	117.1	119.6	112.8
3.8	4.1	4.2	3.8	3.3	3.4	3.6	3.5	3.8
8.5	9.2	12.3	6.1	9.3	9.2	5.1	8.7	10.7
10.8	11.2	13.6	12.6	11.0	14.2	11.7	12.8	13.0
164.3	162.5	160.9	159.7	166.8	168.1	172.6	175.2	175.8
2.1	2.4	2.3	2.6	2.5	1.9	2.4	3.9	4.6
27.4	28.2	36.2	20.0	26.2	28.4	20.6	25.7	28.8
61.5	72.8	70.5	71.9	73.0	63.7	74.9	75.4	59.4
33.9	35.8	36.8	30.8	31.0	33.4	34.1	36.6	39.9
11.9	18.6	29.0	17.9	23.2	22.7	26.6	34.0	30.6
0.1	0.8	1.0	0.1	0.7	0.7	0.7	–	–

(Continued)

Table 10.5 (Continued)

	1995	1996	1997	1998	1999	2000
USA	314.0	350.7	360.1	326.6	322.8	278.4
Canada	335.4	354.4	350.9	331.9	343.9	358.4
Mexico	27.5	31.4	26.4	24.6	32.8	33.1
Total North America	**676.9**	**736.5**	**737.4**	**683.1**	**699.4**	**669.9**
Argentina	26.9	23.0	28.1	26.5	21.7	28.8
Brazil	253.9	265.8	279.0	291.5	293.0	304.4
Chile	18.4	16.8	18.8	15.9	13.6	19.1
Colombia	32.2	35.5	31.4	30.5	33.5	30.7
Ecuador	5.5	6.4	6.6	6.5	7.2	7.6
Peru	12.9	13.3	13.2	13.8	14.5	16.2
Venezuela	51.5	53.8	57.3	57.9	60.6	62.9
Other S. & Cent. America	63.2	72.2	76.2	76.5	78.2	81.6
Total S. & Cent. America	**464.6**	**486.9**	**510.6**	**519.1**	**522.2**	**551.3**
Austria	38.5	35.6	37.3	38.7	41.7	43.5
Azerbaijan	1.6	1.5	1.7	2.0	1.5	1.5
Belarus	^	^	^	^	^	^
Belgium & Luxembourg	2.1	2.1	2.2	2.5	2.2	2.6
Bulgaria	2.3	2.9	2.9	3.3	3.0	2.7
Czech Republic	2.3	2.4	2.1	1.9	2.2	2.3
Denmark	^	^	^	^	^	^
Finland	12.9	11.8	12.0	14.7	12.7	14.6
France	75.9	70.2	67.5	66.0	77.6	72.5
Germany	24.2	21.7	20.9	21.2	23.3	26.0
Greece	3.8	4.5	4.1	3.9	4.8	4.1
Hungary	0.2	0.2	0.2	0.2	0.2	0.2
Iceland	4.7	4.8	5.2	5.6	6.0	6.4
Republic of Ireland	1.0	1.0	0.9	1.2	1.1	1.1
Italy	41.9	47.1	46.7	47.4	51.8	50.9
Kazakhstan	8.3	7.3	6.5	6.1	6.1	7.5
Lithuania	0.8	0.9	0.8	0.9	0.9	0.6
Netherlands	0.1	0.1	0.1	0.1	0.1	0.1
Norway	122.5	103.9	110.9	116.3	121.8	142.2
Poland	3.8	3.9	3.8	4.3	4.3	4.1
Portugal	8.5	14.9	13.2	13.1	7.6	11.7
Romania	16.7	15.8	17.5	18.9	18.3	14.8
Russian Federation	177.0	154.0	157.5	158.7	160.9	165.3
Slovakia	5.2	4.5	4.4	4.6	4.8	5.0
Spain	24.5	41.7	37.3	39.0	30.7	36.5
Sweden	67.7	51.5	69.1	73.6	71.7	78.6
Switzerland	36.0	29.7	35.1	34.6	41.0	38.2
Turkey	35.5	40.5	39.0	42.2	34.7	30.9
Turkmenistan	–	–	–	–	–	–

2001	2002	2003	2004	2005	Change 2005 over 2004 (%)	2005 share of total (%)
219.2	267.0	278.6	271.1	**267.8**	−1.0	9.1
333.4	350.7	337.7	337.7	**361.0**	7.2	12.2
28.5	25.0	19.9	25.2	**27.8**	10.5	0.9
581.1	**642.6**	**636.1**	**634.1**	**656.5**	**3.8**	**22.2**
37.0	35.9	33.8	30.3	**34.8**	15.1	1.2
267.9	286.1	305.6	320.8	**340.4**	6.4	11.5
21.7	23.2	22.6	21.3	**25.9**	22.0	0.9
31.5	33.7	36.0	39.9	**39.6**	−0.4	1.3
7.1	7.5	7.2	7.3	**7.5**	3.3	0.3
17.6	18.0	18.6	17.5	**19.0**	8.7	0.6
60.4	59.5	60.5	70.1	**77.9**	11.4	2.6
75.2	79.1	80.5	78.8	**80.9**	2.9	2.7
518.4	**543.1**	**564.9**	**586.0**	**626.0**	**7.1**	**21.2**
43.5	42.0	38.4	39.9	**39.6**	−0.4	1.3
1.3	2.0	2.5	2.8	**3.0**	9.4	0.1
^	^	^	^	^	0.3	♦
2.5	2.5	2.2	2.5	**2.6**	2.9	0.1
1.7	2.7	3.2	3.3	**3.5**	7.7	0.1
2.5	2.8	1.8	2.6	**3.0**	18.4	0.1
^	^	^	^	^	−16.4	♦
13.6	10.7	9.4	14.9	**13.7**	−7.6	0.5
79.4	66.7	64.9	65.1	**56.6**	−12.8	1.9
27.8	28.4	24.1	27.5	**28.0**	2.1	0.9
2.7	3.5	5.3	5.2	**5.6**	9.4	0.2
0.2	0.2	0.2	0.2	**0.2**	−1.9	♦
6.6	7.0	7.1	7.1	**7.0**	−1.4	0.2
0.9	1.3	1.0	1.0	**0.9**	−11.3	♦
53.9	47.3	44.3	49.9	**42.5**	−14.6	1.4
8.1	8.9	8.6	8.1	**8.6**	7.3	0.3
0.7	0.8	1.0	0.9	**0.8**	−12.4	♦
0.1	0.1	0.1	0.1	**0.1**	−8.8	♦
121.0	129.7	106.1	109.3	**136.6**	25.3	4.6
4.2	3.9	3.3	3.7	**3.8**	2.7	0.1
14.4	8.3	16.1	10.1	**5.0**	−51.0	0.2
14.9	16.0	13.3	16.5	**20.2**	22.7	0.7
175.9	164.2	157.2	180.1	**174.9**	−2.6	5.9
5.1	5.5	3.7	4.2	**4.7**	12.6	0.2
43.9	26.5	43.8	34.4	**23.0**	−33.1	0.8
79.2	66.4	53.5	56.1	**68.5**	22.6	2.3
42.7	36.9	36.8	35.5	**33.1**	−6.5	1.1
24.0	33.7	35.3	46.0	**39.7**	−13.6	1.3
–	–	–	–	–		–

(Continued)

Table 10.5 (Continued)

	1965	1966	1967	1968	1969	1970
Ukraine	n/a	n/a	n/a	n/a	n/a	n/a
United Kingdom	4.7	4.5	5.3	4.5	4.4	5.7
Uzbekistan	n/a	n/a	n/a	n/a	n/a	n/a
Other Europe & Eurasia	94.5	106.4	103.8	121.2	133.7	143.3
Total Europe & Eurasia	**396.7**	**429.7**	**426.5**	**455.0**	**460.6**	**486.2**
Iran	1.7	1.8	2.0	2.1	2.3	2.5
Kuwait	–	–	–	–	–	–
Qatar	–	–	–	–	–	–
Saudi Arabia	–	–	–	–	–	–
United Arab Emirates	–	–	–	–	–	–
Other Middle East	0.4	0.5	0.6	0.7	0.8	1.0
Total Middle East	**2.1**	**2.3**	**2.6**	**2.8**	**3.1**	**3.4**
Algeria	0.4	0.4	0.4	0.6	0.4	0.6
Egypt	1.7	1.8	2.0	3.0	4.0	4.7
South Africa	–	–	–	–	–	–
Other Africa	11.4	12.0	13.6	15.6	18.0	19.7
Total Africa	**13.6**	**14.2**	**16.0**	**19.1**	**22.4**	**24.9**
Australia	7.7	7.3	7.6	7.9	8.7	10.4
Bangladesh	–	–	–	–	–	–
China	22.1	22.5	19.4	22.9	23.3	24.0
China Hong Kong SAR	–	–	–	–	–	–
India	19.2	20.0	22.4	25.8	28.7	30.4
Indonesia	1.8	2.4	2.7	3.3	3.4	4.2
Japan	70.4	80.2	70.1	70.4	78.1	78.1
Malaysia	0.6	0.7	0.7	0.8	1.0	1.3
New Zealand	10.0	10.8	11.2	11.8	11.6	12.7
Pakistan	2.2	2.2	2.6	3.1	3.3	2.9
Philippines	0.8	0.8	1.1	0.7	0.8	1.5
Singapore	–	–	–	–	–	–
South Korea	0.7	1.0	1.0	0.9	1.4	1.2
Taiwan	3.7	3.8	3.6	5.2	4.0	3.6
Thailand	1.1	1.3	1.7	1.6	1.2	2.0
Other Asia Pacific	12.6	12.7	12.7	12.9	13.0	13.3
Total Asia Pacific	**152.7**	**165.7**	**156.8**	**167.4**	**178.6**	**185.7**
TOTAL WORLD	**930.3**	**995.7**	**1020.6**	**1071.8**	**1136.8**	**1190.3**
Of which European Union 25#	221.4	239.5	231.3	235.5	230.9	243.5
OECD	712.9	758.6	780.5	799.2	840.5	864.2
Former Soviet Union	85.3	96.2	92.8	109.0	118.5	128.1
Other EMEs	132.1	140.9	147.3	163.6	177.8	198.0

1971	1972	1973	1974	1975	1976	1977	1978	1979
n/a	n/a	n/a	n/a	n/a	n/a	n/a	n/a	n/a
4.3	4.3	4.6	4.8	5.0	5.1	5.2	5.2	5.5
n/a	n/a	n/a	n/a	n/a	n/a	n/a	n/a	n/a
146.0	145.4	143.6	158.0	150.6	162.1	177.8	191.6	208.8
494.0	**506.0**	**506.9**	**542.7**	**541.0**	**530.3**	**621.0**	**630.3**	**668.2**
2.7	3.5	3.0	3.3	3.4	3.8	4.0	6.2	5.4
–	–	–	–	–	–	–	–	–
–	–	–	–	–	–	–	–	–
–	–	–	–	–	–	–	–	–
–	–	–	–	–	–	–	–	–
1.1	1.1	0.8	1.6	2.1	2.6	3.3	3.8	4.4
3.8	**4.6**	**3.8**	**4.9**	**5.5**	**6.4**	**7.2**	**10.0**	**9.8**
0.3	0.5	0.8	0.5	0.3	0.4	0.3	0.3	0.3
5.0	5.2	5.2	6.1	6.8	8.0	9.0	9.9	9.6
0.1	0.8	0.2	1.1	1.1	1.9	1.9	1.9	1.2
21.1	24.3	25.2	28.3	28.9	30.5	32.0	32.6	34.8
26.5	**30.8**	**31.3**	**36.0**	**37.2**	**40.7**	**43.3**	**44.7**	**45.9**
11.8	11.8	12.7	14.3	15.3	14.6	14.1	15.2	14.9
–	0.3	0.3	0.3	0.5	0.5	0.5	0.5	0.6
29.1	32.9	36.8	41.7	43.6	44.2	46.1	43.2	48.5
–	–	–	–	–	–	–	–	–
33.4	32.4	34.5	33.2	39.7	41.5	45.3	56.2	54.2
1.4	1.2	1.6	1.7	1.8	1.1	1.8	1.3	1.4
82.1	86.7	74.9	81.1	83.2	88.3	75.5	72.9	83.4
1.3	1.2	1.3	1.3	1.3	1.2	1.0	1.1	1.3
14.4	15.3	15.1	15.2	16.7	15.5	14.7	15.7	18.4
3.8	4.0	4.2	4.3	4.9	5.3	6.3	7.8	8.5
1.7	1.4	1.8	2.2	2.2	2.6	1.9	2.8	2.8
–	–	–	–	–	–	–	–	–
1.3	1.4	1.3	1.9	1.7	1.8	1.4	1.8	2.3
3.1	3.6	3.4	4.7	5.3	4.3	4.0	5.0	4.6
2.3	2.1	2.3	3.0	3.0	3.2	2.9	1.9	3.1
13.5	13.7	13.6	14.0	14.1	14.3	14.6	14.8	14.9
199.2	**208.2**	**203.8**	**218.9**	**233.1**	**238.3**	**230.0**	**240.1**	**258.8**
1240.4	**1298.0**	**1312.3**	**1431.2**	**1449.2**	**1455.2**	**1489.8**	**1603.5**	1679.1
240.8	249.5	243.7	260.4	257.4	234.3	309.2	297.5	309.5
894.5	931.8	930.3	1008.6	1012.5	987.2	979.6	1050.2	1080.3
129.7	126.5	126.1	136.1	129.7	139.7	151.2	164.0	179.6
216.2	239.8	255.8	286.5	307.1	328.4	359.0	389.3	419.2

(Continued)

Table 10.5 (Continued)

	1980	1981	1982	1983	1984	1985
Ukraine	n/a	n/a	n/a	n/a	n/a	10.7
United Kingdom	5.1	5.4	5.6	6.5	6.1	6.9
Uzbekistan	n/a	n/a	n/a	n/a	n/a	5.4
Other Europe & Eurasia	214.7	215.0	201.3	204.6	231.7	58.8
Total Europe & Eurasia	**657.9**	**673.2**	**654.5**	**673.7**	**706.0**	**716.7**
Iran	5.6	6.1	6.4	6.3	5.8	5.6
Kuwait	–	–	–	–	–	–
Qatar	–	–	–	–	–	–
Saudi Arabia	–	–	–	–	–	–
United Arab Emirates	–	–	–	–	–	–
Other Middle East	4.3	4.4	4.2	3.8	3.5	3.2
Total Middle East	**9.8**	**10.5**	**10.6**	**10.0**	**9.3**	**8.8**
Algeria	0.3	0.4	0.5	0.2	0.5	0.7
Egypt	9.8	10.2	10.5	9.8	9.6	8.7
South Africa	1.0	2.1	2.6	2.6	2.6	2.8
Other Africa	36.4	37.7	37.9	34.9	34.7	38.6
Total Africa	**47.5**	**50.3**	**51.4**	**47.6**	**47.3**	**50.7**
Australia	14.3	14.7	13.7	12.9	13.9	15.2
Bangladesh	0.6	0.6	0.6	0.8	0.8	0.6
China	58.2	65.5	74.4	86.4	86.8	92.4
China Hong Kong SAR	–	–	–	–	–	–
India	55.4	59.1	49.8	48.0	53.7	51.8
Indonesia	1.3	1.3	1.4	2.1	2.3	2.8
Japan	90.1	89.8	81.1	90.6	72.2	87.5
Malaysia	1.7	1.8	1.8	2.1	4.1	3.8
New Zealand	19.4	19.7	18.3	19.8	20.4	19.7
Pakistan	8.9	9.3	10.4	12.1	12.5	13.0
Philippines	3.4	3.7	3.8	3.0	5.2	5.6
Singapore	–	–	–	–	–	–
South Korea	2.0	2.7	2.0	2.7	2.4	3.7
Taiwan	2.9	4.8	4.8	5.0	4.4	6.9
Thailand	1.3	2.8	3.5	3.4	4.1	3.7
Other Asia Pacific	15.4	16.0	16.9	17.1	18.5	19.6
Total Asia Pacific	**274.9**	**291.7**	**282.6**	**305.8**	**301.3**	**326.2**
TOTAL WORLD	**1711.7**	**1745.2**	**1815.2**	**1902.6**	**1965.3**	**2003.6**
Of which European Union 25#	292.0	294.8	290.4	298.4	305.1	311.9
OECD	1074.0	1084.9	1149.8	1207.1	1206.0	1203.7
Former Soviet Union	183.6	186.6	174.6	179.6	202.5	215.0
Other EMEs	454.1	473.8	490.9	515.9	556.8	585.0

1986	1987	1988	1989	1990	1991	1992	1993	1994
10.7	9.6	11.9	10.1	10.7	11.9	8.1	11.2	12.3
7.0	6.3	7.1	6.7	7.2	6.2	7.1	5.7	6.6
4.6	6.8	7.2	5.5	6.7	6.0	6.3	7.4	7.2
60.3	63.3	71.3	65.5	65.0	65.5	64.5	64.5	67.7
699.0	**745.9**	**791.6**	**712.2**	**734.5**	**740.8**	**768.2**	**793.3**	**798.8**
7.1	8.2	7.5	7.5	6.4	6.8	8.8	9.7	8.0
–	–	–	–	–	–	–	–	–
–	–	–	–	–	–	–	–	–
–	–	–	–	–	–	–	–	–
–	–	–	–	–	–	–	–	–
2.7	3.8	3.8	2.8	2.1	2.3	3.0	2.9	3.7
9.8	**12.0**	**11.4**	**10.3**	**8.5**	**9.1**	**11.8**	**12.6**	**11.6**
0.3	0.5	0.2	0.3	0.1	0.3	0.2	0.4	0.2
9.3	8.7	8.8	9.7	9.8	10.0	10.3	10.7	11.2
3.4	3.4	4.6	3.8	2.9	3.8	2.1	1.5	2.6
41.0	39.9	41.3	44.0	46.0	48.5	48.2	45.6	46.2
54.0	**52.5**	**54.8**	**57.7**	**58.8**	**62.5**	**60.9**	**58.2**	**60.2**
15.0	14.8	15.0	15.0	15.5	15.9	16.4	16.8	16.4
0.5	0.6	0.8	0.9	0.9	0.8	0.8	0.6	0.8
94.5	100.0	109.1	118.4	126.7	124.7	130.7	151.8	167.4
–	–	–	–	–	–	–	–	–
52.6	49.0	54.6	63.4	66.4	73.9	70.0	70.7	80.2
4.5	4.6	5.0	6.3	5.7	5.7	8.6	7.8	6.9
80.8	82.1	87.8	97.0	93.9	103.4	88.4	105.0	76.7
4.1	4.9	5.6	5.3	4.0	4.4	4.4	4.9	6.5
22.1	21.9	23.0	22.6	23.2	22.9	21.1	23.5	25.8
12.5	16.4	17.2	17.4	17.1	18.3	19.9	21.9	21.5
6.0	5.2	6.3	6.5	6.1	5.1	4.3	5.0	5.8
–	–	–	–	–	–	–	–	–
4.0	5.3	3.6	4.6	6.4	5.0	4.9	6.0	4.1
7.4	7.1	6.2	6.7	8.2	5.5	8.3	6.7	8.9
5.6	4.1	3.8	5.6	5.0	4.6	4.2	3.7	4.5
20.6	20.5	22.6	25.3	28.2	28.6	28.3	29.4	31.4
330.1	**336.4**	**360.6**	**394.8**	**407.0**	**418.9**	**410.2**	**454.0**	**457.1**
2028.1	**2063.0**	**2115.3**	**2113.1**	**2188.7**	**2240.9**	2243.5	**2374.6**	**2388.0**
296.6	324.1	338.7	279.9	288.5	298.6	317.6	321.6	329.2
1189.5	1196.4	1201.3	1177.2	1218.5	1232.1	1222.4	1292.8	1243.7
215.1	220.3	230.9	222.2	235.7	234.7	234.6	245.0	247.2
623.4	646.3	683.1	713.6	734.5	774.1	786.6	836.9	897.1

(Continued)

Table 10.5 (Continued)

	1995	1996	1997	1998	1999	2000
Ukraine	10.2	8.8	10.0	15.9	14.5	11.4
United Kingdom	6.4	4.9	5.7	6.7	8.2	7.8
Uzbekistan	6.2	6.5	5.8	5.8	5.7	5.9
Other Europe & Eurasia	66.6	74.2	67.2	68.9	74.3	70.3
Total Europe & Eurasia	**807.3**	**769.0**	**787.7**	**818.3**	**834.0**	**859.4**
Iran	8.3	7.8	5.9	7.5	5.1	3.8
Kuwait	–	–	–	–	–	–
Qatar	–	–	–	–	–	–
Saudi Arabia	–	–	–	–	–	–
United Arab Emirates	–	–	–	–	–	–
Other Middle East	4.2	5.0	4.8	4.8	3.1	4.3
Total Middle East	**12.5**	**12.8**	**10.8**	**12.3**	**8.3**	**8.1**
Algeria	0.2	0.1	0.1	0.2	0.2	0.1
Egypt	11.5	11.8	12.1	13.8	15.0	14.2
South Africa	1.8	3.6	4.7	4.1	3.3	4.0
Other Africa	49.1	49.6	51.4	53.7	60.5	59.4
Total Africa	**62.6**	**65.1**	**68.3**	**71.8**	**79.1**	**77.6**
Australia	16.1	16.6	16.6	16.2	16.1	16.1
Bangladesh	0.4	0.7	0.7	0.9	0.8	0.9
China	190.6	188.0	196.0	208.0	203.8	222.4
China Hong Kong SAR	–	–	–	–	–	–
India	75.9	69.0	70.2	83.6	82.2	77.0
Indonesia	7.5	8.1	5.2	9.7	9.4	10.0
Japan	87.8	86.9	93.7	104.2	92.7	91.3
Malaysia	6.2	5.2	3.9	4.9	7.5	7.4
New Zealand	27.5	26.0	23.2	25.3	23.5	24.6
Pakistan	22.7	24.8	18.4	24.1	21.5	17.6
Philippines	6.2	7.1	6.1	5.1	7.8	7.8
Singapore	–	–	–	–	–	–
South Korea	5.5	5.2	5.4	6.1	6.1	5.6
Taiwan	8.9	9.0	9.6	10.6	8.9	8.9
Thailand	6.7	7.3	7.2	5.2	3.5	6.0
Other Asia Pacific	33.8	32.7	30.9	30.5	34.0	35.6
Total Asia Pacific	**495.9**	**486.8**	**487.1**	**534.1**	**517.9**	**531.3**
TOTAL WORLD	**2519.7**	**2557.1**	**2601.8**	**2638.7**	**2660.9**	**2697.5**
Of which European Union 25#	325.9	324.6	334.2	347.8	352.6	368.9
OECD	1331.5	1368.2	1394.0	1372.7	1386.6	1386.8
Former Soviet Union	241.2	216.0	217.7	225.8	227.7	230.0
Other EMEs	947.1	972.8	990.1	1040.3	1046.7	1080.6

Notes: ^ Less than 0.05; ♦ Less than 0.05%; + More than 100%; n/a not available; # Excludes Estonia, Latvia and Lithuania prior to 1985 and Slovenia prior to 1991; growth rates are adjusted for leap years.

Source: British Petroleum (2006).

2001	2002	2003	2004	2005	Change 2005 over 2004 (%)	2005 share of total (%)
12.2	9.8	9.2	11.9	**12.4**	4.5	0.4
6.5	7.4	6.0	7.6	**7.6**	0.9	0.3
5.4	7.3	7.6	6.9	**7.3**	5.2	0.2
68.1	66.7	70.8	74.3	**74.5**	0.5	2.5
863.0	**809.1**	**776.6**	**827.7**	**827.0**	**0.2**	**28.0**
4.1	8.0	9.6	11.9	**12.5**	5.0	0.4
–	–	–	–	–	–	–
–	–	–	–	–	–	–
–	–	–	–	–	–	–
–	–	–	–	–	–	–
4.4	4.9	4.6	4.9	**4.9**	0.4	0.2
8.5	**12.9**	**14.2**	**16.8**	**17.4**	**3.6**	**0.6**
0.1	0.1	0.3	0.3	**0.6**	+	♦
14.4	14.0	12.9	13.7	**13.7**	0.3	0.5
3.7	4.1	3.5	3.7	**3.5**	−5.3	0.1
62.3	67.2	67.8	68.1	**70.1**	3.3	2.4
80.5	**85.4**	**84.5**	**85.7**	**87.9**	**2.8**	**3.0**
16.1	16.0	16.2	16.1	**16.3**	1.9	0.6
1.0	1.1	1.1	1.1	**1.2**	2.2	♦
277.4	288.0	283.7	353.6	**401.0**	13.7	13.6
–	–	–	–	–	–	–
72.0	68.5	69.3	83.8	**96.0**	14.9	3.3
11.7	9.9	9.1	9.4	**9.4**	0.3	0.3
91.8	93.1	103.0	102.2	**87.7**	−14.0	3.0
6.4	5.3	5.7	6.2	**6.6**	7.4	0.2
22.6	25.2	23.7	27.2	**24.2**	−10.9	0.8
18.3	20.4	25.6	24.2	**30.7**	27.2	1.0
7.1	7.0	7.9	8.5	**8.4**	−1.5	0.3
–	–	–	–	–	–	–
4.1	5.3	6.9	5.9	**5.2**	−11.2	0.2
9.2	6.4	6.9	6.6	**7.9**	21.0	0.3
6.3	7.5	7.3	6.0	**5.8**	−3.7	0.2
39.4	38.8	42.0	40.9	**39.4**	−3.4	1.3
583.5	**592.5**	**608.5**	**691.7**	**739.8**	**7.2**	**25.0**
2635.0	**2685.6**	**2684.9**	**2842.0**	**2954.6**	**4.2**	**100.0**
387.7	330.1	324.3	333.0	**312.8**	−5.8	10.6
1290.5	1313.0	1289.2	1308.2	**1311.6**	0.5	44.4
239.5	229.6	226.1	252.0	**249.7**	−0.6	8.5
1105.1	1143.0	1169.5	1281.9	**1393.3**	9.0	47.2

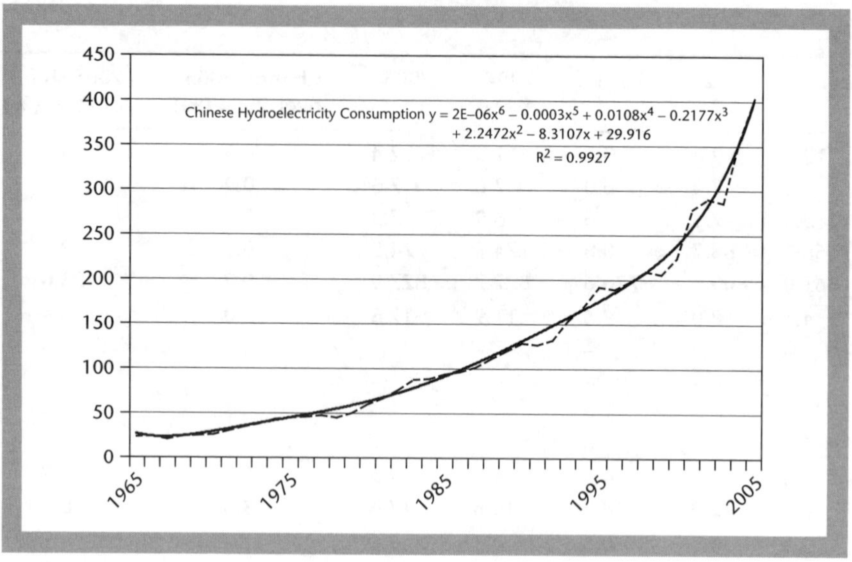

Figure 10.6 Chinese hydroelectricity consumption function, 1965–2005 (TWh)
Source: British Petroleum (2006).

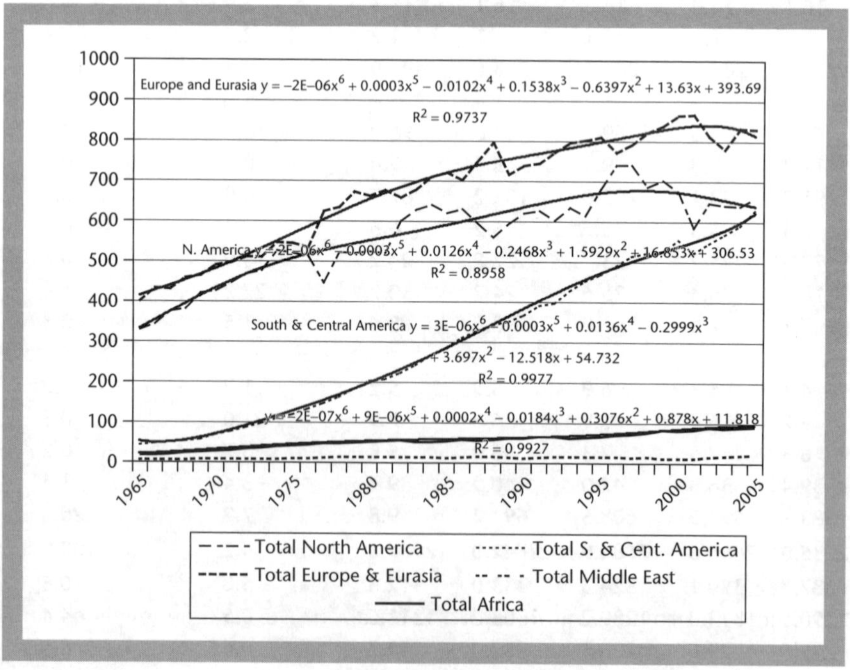

Figure 10.7 Other countries' hydroelectricity consumption functions, 1965–2005 (TWh)
Source: British Petroleum (2006).

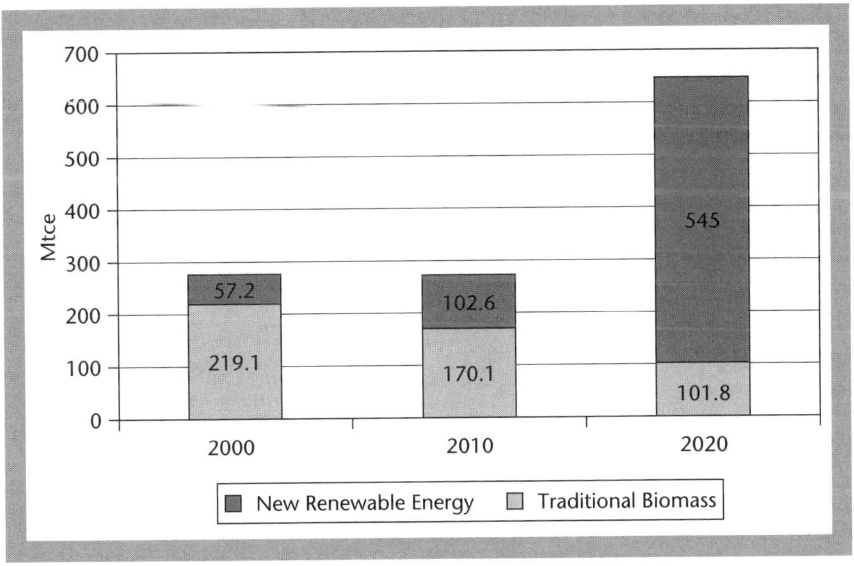

Figure 10.8 Estimated output of traditional biomass and new renewables, 2000–2020
Source: Chinese Energy Ministry, Chinese Energy and Technology Symposium 2003.

was 100 Mtce, an estimated 11 per cent beyond sustainable use in 1998. Furthermore, the Chinese government expects that in the future, all kinds of biomass resources are likely to increase dramatically; with crop stalk use doubling between 1995 and 2010 and municipal solid waste use rising by 160 per cent. According to the International Energy Administration, the area of China is relatively small to produce energy sufficient for 1.3 billion individuals. Chinese irrigated farmland areas decrease as the Chinese authorities prefer to plant forests rather than maintaining farmland. Chinese provinces are also modernizing and building more urban centers at the cost of less farmland, thus biomass production decreased. Various alternative products can be used to transform into energy. For instance, sugarcane slugs or bagass are utilized to generate 800 MW of electricity. About 164 centralized gas demonstration projects use crop stalks from villages and rural areas to produce a total gas production of 45 million cubic meters. However, this methodology is not appropriate for substantial energy and electricity production.

Table 10.6 provides more information about the income levels of rural residents and consumption of energy in daily life.

The Chinese authorities and the NESP also promote the production of biomass technologies and the increased potential for such alternative energy usages. The Chinese authorities estimate that potential for biomass

Table 10.6 Income levels of rural residents and consumption of energy in daily life, 1997

Per household income, yuan/yr	<1500	4500–6500	6500–9000	9000–14,000	>14,000
Energy for daily life kgce/yr	807.0	1033.2	1350.9	1384.4	1645.4
Of which,					
coal, kg	876.1	999.0	1318.7	1556.3	1042.5
LPG, kg	21.6	28.7	68.5	71.2	103.3
Oil products, kg	12.1	18.4	35.8	180.0	229.5
Electricity, kWh	204.8	273.3	335.4	504.5	641.3

Sources: Energy Ministry of China/US Department of Energy, Design for Market-oriented Development Strategy of Bioenergy Technologies in China, Project Expert Team, sample survey.

power in China could reach 3 GW by 2010 and 10–15 GW by 2020. Biomass energy production may be used in three main ways: (1) combust biomass directly to heat; (2) biomass gasification to supply biogas directly or for generation; (3) product liquid fuel. Biomass energy production is performed through direct combustion. According to the International Energy Administration, about 2.5 billion people in the world use such alternatives and the annual consumption amounts to 1.3 billion tons. Biomass development trends optimize utilizations through biomass gasification, biomass generation and product liquid fuel. Biomass gasification and biogas supply is principally used in Chinese rural areas for cooking but is not used much elsewhere. China counts about 500 examples of biomass gasification and biogas supply, and China's annual biogas production totals about 0.15 billion cubic meters. Chinese biogas technologies are modern, highly developed and applied widely throughout the country. According to NESP, approximately 11.1 million farmers in China are involved in biogas projects.

According to the Association of China Energy Research of June 2003, Chinese farmers and biomass producers generated an annual output of about 4.7 billion cubic meters. Biomass is an added value energy resource in the rural and farming areas where traditional energy networks (oil and natural gas pipelines) are still difficult to access.

Biomass technology is well-known and applied to some degree in most countries. Biomass energy may be generated in two ways: by direct combustion or with gas turbines to create a biomass gasfication mechanism. These two technologies are widely used in China with combustion techniques mostly used for commercial purposes. The United States counts the most biomass installations, with more than 350 biomass plants which are principally used to produce pulp, paper and forestry.

10.4 ELECTRICITY GENERATION

According to the Chinese authorities (NESP), the Chinese electricity industry had an overcapacity in the late 1990s. The oversupply that unbalanced market efficiencies was principally due to reduced demand as a result of the closures of inefficient state-owned industrial units. The Chinese market does not yet have adequate infrastructures to enable the trading of electricity. Once domestic markets share enhanced grids with neighbors, and given the globalization of energy markets, the Chinese market will offer more opportunities. The short-term oversupply was partially resolved by temporarily shutting down a few power plants to establish an equilibrium. This power market readjustment lasted until January 2002, until the surging demand of the new middle class created a further disequilibrium around 2003.

Prior to reviewing Chinese energy market policies, let us first consider the Chinese power industry reform in 2002. The State Power Corporation of China was created in 1997, and the power industry reform was initiated by State Council in March 2002. Due to the Chinese entry to the WTO in 2001 and the globalization of international energy trading corporations, new generation companies and new grid companies were established on 29 December 2002. The State Council approved the formation of the State Electricity Regulation Commission (SERC) in February 2003. Power energy reforms are still ongoing and have produced the retail pricing framework shown in Table 10.7.

The main objectives of China's power reform are to break up the monopoly with respect to generation and introduce competition into the power generation (wholesale) market; to reduce costs, improve efficiency, and improve tariff-setting mechanisms in the power sector; to optimize resource allocation and promote domestic and international grid interconnection; and to create an open free power market with balanced monitored competition under government regulation.

The legal structure of the market in China is currently being re-engineered to fit with international market integration. In July 2003, Chinese energy policies regulating electricity were engineering the pricing structure for electricity in the east and northeast of the country. According to Huaneng Power International, one of the few Chinese-owned independent operators, the State Council set a number of short-term objectives: establish a pricing grid mechanism to efficiently maximize resource utilizations and minimize transportation costs; accommodate a competitive bidding power pricing with bid and ask points; focus on more production; establish a preliminary pricing mechanism for transmission and distribution to integrate regional with national power grids and smooth national with international power grids; link retail pricing systemic grid with the on-grid prices; optimize retail pricing structure; and test high voltage users to directly purchase from the power generation companies based on a reasonable price for transmission and distribution. The State Council decided on 4 June 2004 to take the following steps to maintain reliability of supply in periods of peak demand for electricity and avoid congestion of pricing concentrations

Table 10.7 Retail electricity prices by power station for Huaneng International average rate (inclusive of VAT) (RMB/MWh)

	2004	2003	Variation (%)
Dalian	283.62	272.69	4.01
Fuzhou	365.00	331.82	10.00
Nantong	325.18	312.52	4.05
Shang'an	303.25	307.94	−1.52
Shantou Oil-Fired	604.08	672.14	−10.13
Shantou	446.86	435.17	2.69
Dandong	289.05	276.95	4.37
Shidongkou II	342.56	332.85	2.92
Nanjing	321.67	307.31	4.67
Dezhou	332.58	333.34	−0.23
Weihai	394.06	386.50	1.96
Jining	299.89	274.66	9.19
Shidongkou I	285.43	256.64	11.22
Taicang	341.10	312.80	9.05
Changxing	351.94	320.57	9.79
Huaiyin	330.88	317.21	4.31
Xindian 1	320.83	342.41	−6.30
Yushe1	282.10	200.63	40.61
Yingkou2	315.48	n/a	n/a
Jinggangshan2	325.67	n/a	n/a
Luohuang2	286.74	n/a	n/a
Yueyang2	316.52	n/a	n/a
Qinbei3	273.11	n/a	n/a
Average	327.88	318.68	2.89

Source: Austin et al. (2005).

resulting in an illiquid pricing grid. Regulatory measures included: demand-side management policies to control demand for electricity and to ensure supply – this includes restricting power consumption by power intensive industries; differentiated pricing for off-peak and peak consumption; improving the reliability of power distribution and transmission networks; maintaining coal supply and improving coal distribution channels with an emphasis on decreasing transportation costs; accelerating construction of energy and transport projects to ensure supply; and an increased emphasis on energy efficient consumption.

Chinese power reforms aim at re-engineering state-owned power assets, monitoring tariff structures, and creating a transparent and rational power market with efficient balanced competition under government regulation. The state power grid (see Figure 10.9) is decomposed into two grid enterprises: the State Grid Corporation of China (SGCC), and the China Southern Power Grid Co. Ltd. There are five national generation companies (China Huaneng Group, China Datang Group, China Huadian Group, China Guodian Group, and China Power Investment Group). The new power market is regulated by the State Council-approved State Electricity Regulation Commission (SERC). SERC formulates regulatory reforms for power industry management, proposes tariff adjustments, writes licenses for power plants and other facilities, and mediates disputes between trading entities in this industry. The responsibilities of SGCC in the new

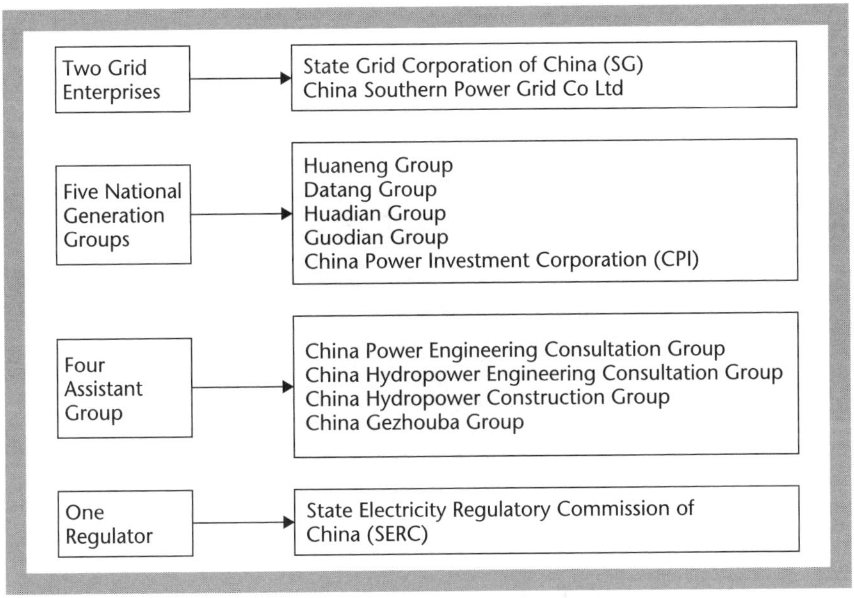

Figure 10.9 Chinese electricity market structure with main national participants
Source: Austin et al. (2005).

reformed power market include: trading and dispatching electricity between regional power grids, coordination with regional power grid companies with respect to daily production issues, participating in investment, construction and operation of relevant trans-regional transmission and electricity transformation and interconnection projects, and monitoring of construction and management of the Three Gorges Dam transmission and transformation projects. In 2002, China's electricity supply–demand balance was overall in equilibrium, but started to exhibit a slim supply shortage. By the end of 2002, the total Chinese power capacity for production had risen to 356 GW (gigawatts). In 2002 power generation equaled 1654 billion kWh. The unit utilization average during 2002 for all thermal power plants in China was 5270 hours and the average capacity utilization coefficient amounted to 60 per cent. To attract more trading competitors in the Chinese markets, the Chinese regulatory entities have provided a series of maps of China's power markets (figures 10.10–10.13).

The Chinese Energy Department also provides information about power energy (see tables 10.8 and 10.9).

According to NESP, the Chinese authorities have built dozens of major new electric power projects since 2003 and expect to create more power plants to meet rising energy demand in the future. Mainly due to the rising economy in most industry sectors, by 2010, installed capacity will need to rise to 600 GW, and total power demand will be 2700 billion kWh. By 2020,

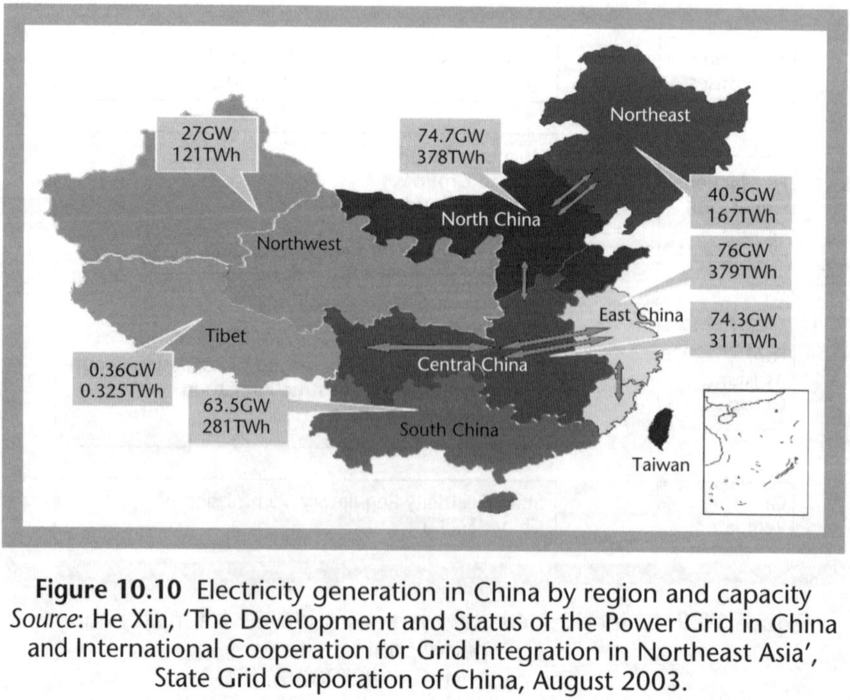

Figure 10.10 Electricity generation in China by region and capacity
Source: He Xin, 'The Development and Status of the Power Grid in China and International Cooperation for Grid Integration in Northeast Asia', State Grid Corporation of China, August 2003.

the installed capacity in China will reach 900 GW, and the total power consumption that must be met by generation will be about 4300 billion kWh. From 2003 to 2010, total power generation (and consumption) in China is expected to rise exponentially at a rate of 6.6 to 7 per cent per year, and from 2010 to 2020, at a rate of 4.5 to 5.5 per cent per year. There are six regional power networks and three isolated provincial power grids. The six regional power networks are the Northeast China Power Network (NEPN), the North China Power Network (NCPN), the East China Power Network (ECPN), the Central China Power Network (CCPN) (this includes the Chuanyu (Sichuan and Chongqing) Power Network (CYPN)), the Northwest China Power Network (NWPN), and the South China Electric

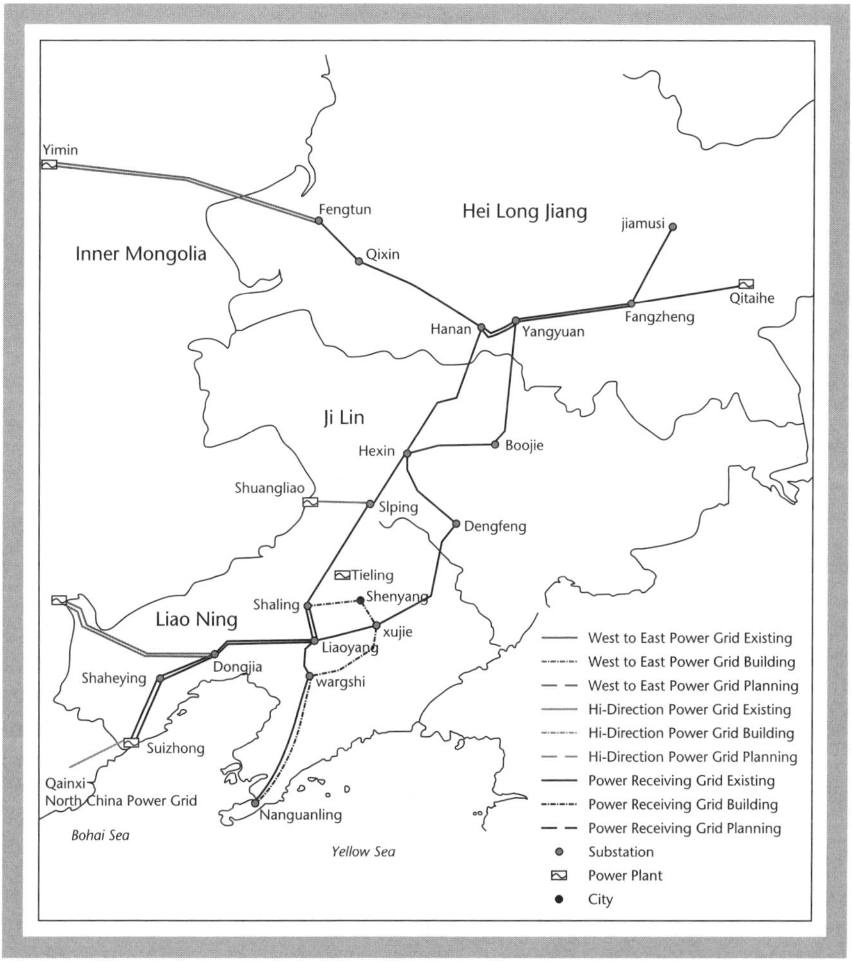

Figure 10.11 500 kV northeast China power grid, 2002
Source: As Figure 10.10.

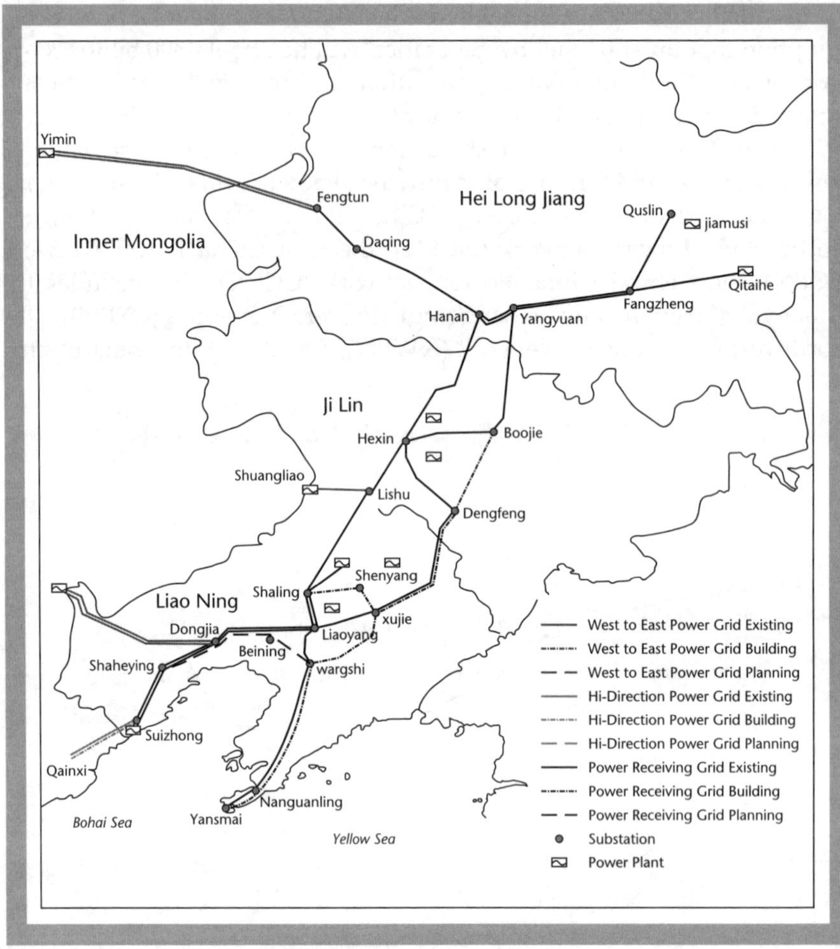

Figure 10.12 500 kV northeast China power grid, 2005
Source: As Figure 10.10.

Joint Venture Power Network (SCPN). The three isolated provincial power grids are the Hainan Power Grid (HNPN), the Xinjiang Autonomous Region Power Grid (XJAR), and the Lhasa Power Grid (Tibet). The nationwide power grid interconnection, west-to-east power transmission, and north-south power transaction project are being connected.

The first regional transmission interconnection was constructed in 1989. It was 1050 km long, with a capacity of 500 kV (kilovolt) direct current (DC) line from the Central China Power Network (at the Gezhouba hydropower station) to the East China Power Network (at Shanghai) with a capacity of 1200 MW. The second was the Yunnan-Guizhou-Guangxi-Guangdong transmission interconnection system. The third was completed in May 2001 with a line of 500 kV AC capacity that connects the Northeast China Power

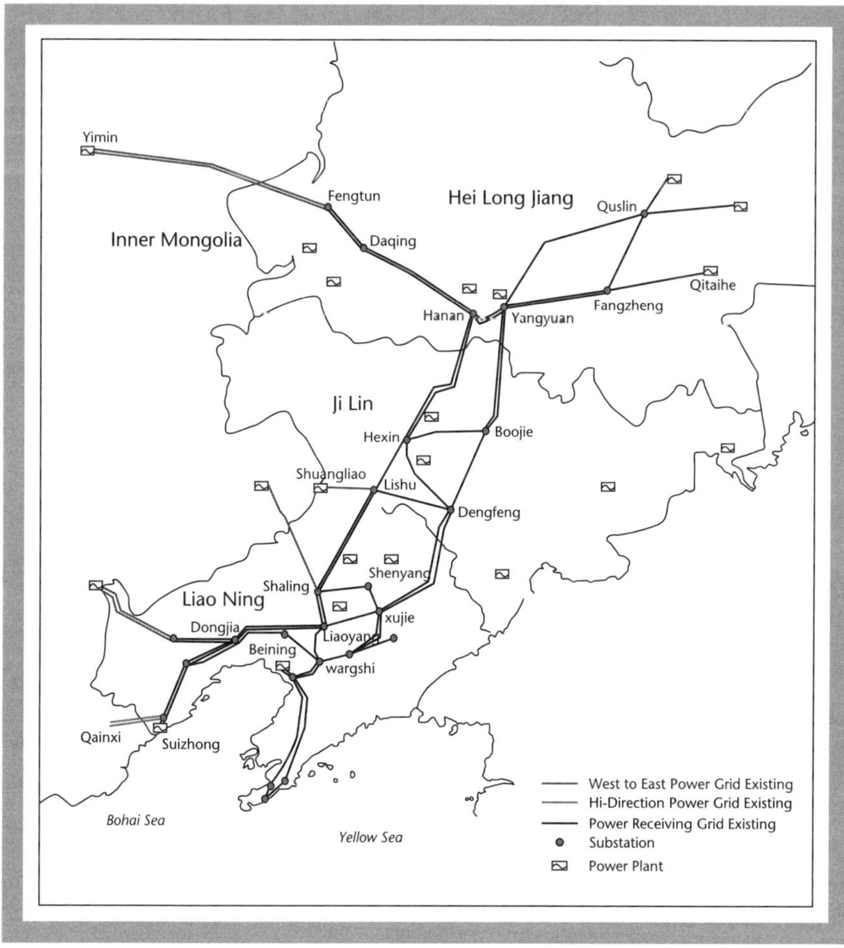

Figure 10.13 500 kV northeast China power grid, 2010
Source: As Figure 10.10.

Table 10.8 Energy consumption and expenditures of herdsmen

	Item	Annual consumption	Unit price	Total expenditures (yuan)
Illumination	Kerosene	18 kg	Rmb 1.60/kg	28.8
Illumination	Candle	730 pieces	Rmb 0.275/piece	200.75
Radio	Dry battery	18 pairs	Rmb 1.05/pair	18.9
Illumination for lamb birth	Dry battery	12 pairs	Rmb1.05/pair	12.6
Total				261.05

Network and the North China Power Network. The fourth was a 500 kV AC transmission line connecting the independent Fujian Power Grid and the East China Power Network to facilitate power transfer between two power grids; a second AC line on this routing was finished in January 2003. In 2002, SGCC connected the Central China Power Network and the Chuanyu Power Network through a single 500 kV AC line; a second line was planned for June 2004. In June 2003, SGCC completed the second DC line from Three Gorges to East China, and in August 2003, SGCC built up the first AC line between North China and Central China. By the end of 2002, the total installed electricity generation capacity in Northeast China had reached 40.5 GW, of which about 14 per cent came from hydropower, with the remaining 86 per cent from thermal power. Power generation totaled 167 TWh (terawatt hours) during 2002, overall generation unit utilization reached an average of 4279 hours in North China over the year, and the average thermal power unit utilization reached 4749 hours per year. By the end of 2002, there were 31 segments of 500 kV AC transmission line in North China, with a total length of 5030 km. In addition, there were 473 segments of 220 kV line, with a total length of 22,845 km. The total capacity of 500 kV substations in North China as of 2002 was 13,806 MVA (mega volt-amps), consisting of 16,500 kV substations. As of 2002, the electric power delivery direction is north to south and west to east inside the Northeast China Power Network. After a feasibility study, SP (State Power) decided to build a 500 kV transmission line between the Northeast China Power Network and the North China Power Network to transfer the surplus power from the former to the major load center in North China Power Network. The line as constructed starts from the Suizhong plant and ends at the Jiangjiaying substation in Tangshan city in the Jing-Jing-Tang power grid. Construction started on the transmission line in 1999, and the line was

Table 10.9 Power supply costs of mini power generator

	100 W wind power generator	Battery	Maintenance	Subtotal
Number (unit)	1	2	–	–
Price (yuan/unit)	700	360	–	–
Life span (year)	10	3	–	–
Average annual costs (yuan)	70	120	30	222
Power output (kWh/year)	–	–	–	260
Power generation costs (Yuan/kWh)	–	–	–	0.85

Source: Chinese Energy Ministry/Department.

put into operation in May 2001. This transmission interconnection provides a bi-directional power supply, and the North China Network and Northeast China Network thus form a synchronized power network. The bi-directional power supply capacity of the line was initially limited to 600 MW. With the second AC line in operation, the bi-directional supply capacity of the line is 1000 to 1200 MW. The Chinese estimated its power production at around 30 gigawatts at the end of 2004.

Electricity energy was backed by the usage of oil supplies and primarily by diesel. The Chinese government 2006 policy reforms promote the installation of more electricity power plants to meet the rising demand of service industries which consume more electricity. Part of the reforms is to design a better connected national power electric grid to ramify to new urban centers from the northeastern coast to more rural central and western areas. In 2003, the Chinese markets opened to neighboring markets such as Russia to smooth trading grids, supply more energy and enhance corporate trading opportunities. The State Grid Corporation of China (SGCC) promoted business initiatives to develop a strategy for 'west to east electricity delivery, bi-directional power supply between the south and the north, interconnection nationwide'.

Due to the insufficient coverage of the grid, China experiences a national shortage of generating capacity with imbalances between supply and demand throughout the country. Consequently some areas exhibit surplus capacity while others experience shortages. The Chinese electric power grid also experiences higher costs at some points of the grid due to insufficient transmission capacity to feed power shortages. Increased construction of power plants since 2002, with added capacity to modernize and meet the new era's market demand which will be geared to more service industries,

Table 10.10 China's electricity generation (billion kWh)

1989	518.2
1990	550.9
1991	600.9
1992	670.6
1993	744.1
1994	816.2
1995	880.9
1996	926.0
1997	972.7
1998	1,014.0

Source: Energy Information Agency.

aims to answer these problems. New plants built in China since 2002 have at least 120 GW of generating capacity. Table 10.10 shows power energy generation through the 1990s, when manufacturing was expanding and new factories were being built.

According to the International Energy Administration, with rising service and manufacturing industries, power energy demand in China will rise by an average of 15 per cent per year, at an increasing rate.

The International Energy Administration also reports that China's largest power plant project ever, the Three Gorges Dam, is estimated to be completed by 2009. The Three Gorges Dam includes 26 separate 700-MW generators and it is able to generate a total of 18.2 GW. In March 2002 the Three Gorges Dam became the China Yangtze Three Gorges Electric Power Corporation. The dam was renovated and improved in June 2003 and began operating with new engines and turbines in July 2003. Part of the project is complementary to a series of hydropower dams on the Yellow River. Shaanxi, Qinghai, and Gansu provinces support the creation of additional reservoirs by the Yellow River Hydroelectric Development Corporation, with 25 new generating stations providing a combined capacity of 15.8 GW. The Chinese government are exploring alternative ways to generate energy as depicted in Table 7.3 above.

Various projections anticipate that Chinese electricity consumption will increase at an accelerating rate, with an average minimum of 4.3 per cent per year until 2025. The author forecasts an exponential rise from 2006 to 2025 with given historical data. The British Petroleum Statistical report of 2006 shows that electricity generation doubled from 1990 to 2000, from 621 to 1368 terawatt hours, respectively. Electricity generation is already rising at an increasing rate, with an annual rise of 12.6 per cent from 2004 to 2005. The rising consumption almost matches the increase in generation, but it is hard to assess whether the Chinese government will be able to maintain the balance between supply and demand and keep electricity market pricing efficient and inflationary trends under control.

According to the International Energy Administration, the Chinese government is reformulating its policies in response to rising demand. In 2002 the Chinese government re-engineered its long term electric power sector under the National Power Industry Framework Reform Plan by the State Council. According to this program, the State Power Corporation (SPC) separated assets to restructure them into 11 regional transmission and distribution companies in December 2002 to incorporate transportation of electricity in the distribution function to energy to various points of the Chinese grid. The pricing of electricity is regulated in relation to the function of available supply for demand within the grid, with applicable tariffs and caps or ceilings in case demand outpaces supply. The Chinese government also created a new electricity law, updating the law of 1995. As part of the restructuring of the Chinese energy program, China also plans to upgrade a unified power grid network by the year 2020. The Chinese authorities also aim to construct three significant regional power grid plants by 2010. These will be installed in northern, southern and central China to generate electricity and ensure a balanced demand throughout the country. According to the new re-engineering plans, the central power grid will be connected to the Three Gorges power grid to feed additional energy to the Central China Power Grid, East China Power Grid, and Chuanyu in Sichuan Province in Chongqing Municipality Power Grid. The re-engineered northern power plant will be connected to the actual North China Power Grid, Northeast China Power Grid, and Northwest China Power Grid. And the newly restructured southern plant will provide additional capacity in Yunnan, Guizhou and Guangdong provinces and the Guangxi Zhuang Autonomous Region. The Chinese authorities anticipate that the re-engineering program will take around 10 years to connect all the various local and regional power grid centers to the main framework and to unify the national power grid network. To realize this goal, the Chinese State Council built a series of hydropower stations along the Honghe, Lancang, Yellow and Yangtze rivers. The authorities are also progressing in connecting each of these new power plants to facilitate the transmission of thermal power from north to south and of hydropower from the west to the east. These additional branches on the grid also contribute to reducing the transportation costs of energy within China. According to the SDPC, the northern power grid center will also be connected to the Russian power grid network by 2020 to alleviate energy costs between both countries and to reduce inflationary pressures on global energy prices. In addition to this international link, the Chinese southern power grid center is also intended to connect with the Thai power grid network. This international ramification is also aimed to alleviate rising energy prices near urban centers such as Hong Kong and Shanghai. Furthermore, the Chinese authorities also plan on building additional thermal power plants in Western Inner Mongolia, Shanxi and Shaanxi provinces, Ningxia Hui Autonomous Region, and Western Henan Province to facilitate the west-to-east power transmission program and to keep Asian zone prices down and in balance with neighboring markets. The Chinese authorities revamped the renewables pricing framework to reduce up-front development costs that create high up-front prices. Thus, the Shanghai Energy Conservation Supervision Center (SECSC) and the Shanghai Economic Commission with the international assistance of the CRS implemented a green pricing program to deliver lower energy prices in Shanghai where greater demand creates inflationary pressures on energy prices.

The Chinese Energy State Council produced further reforms in February 2006. The State Council established some regulatory structures to ensure lower-cost electricity development and energy efficiency. CSEP also consulted a dozen research institutes to establish regulation of the Chinese electricity framework and to monitor its market pricing structures from point to point of the grid, and the government created the State Electricity Regulatory Commission (SERC). International grantees are providing training to new commissioners and SERC staff in best international practices to assure successful reforms, including regulatory incentives for energy efficiency and renewable energy. Table 10.11 and Figures 10.14 and 10.15, represent the Chinese power production function compared to those of international competitors according to British Petroleum data from 1965 to 2005. To learn more about the regulatory and legal structure of Chinese power markets and market re-engineering see http://www.worldbank.org.cn/English/content/roadmap.pdf.

10.5 GEOTHERMAL ENERGY

The International Energy Administration reports that there are 21 countries generating 8000 megawatts of electricity from geothermal energy sources. Global statistics also report that there are 11,300 thermal megawatts in more than 27 nations for direct use applications such as aquaculture, agricultural tools, greenhouse operations, manufacturing machinery, and industrial processing. China uses geothermal energy as a complementary alternative to meet its rising demand. Geothermal energy sources are increasingly being developing in China which uses it principally for heating or cooling devices and machinery in various industries. China principally uses

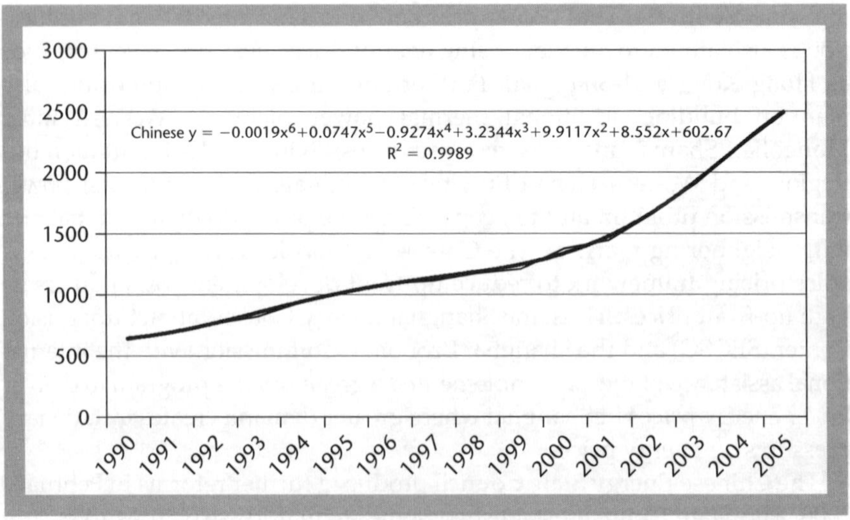

Figure 10.14 Chinese electricity generation function, 1990–2005 (TWh)
Source: British Petroleum (2006).

medium and low temperature geothermal energy to enhance the functioning of industrial machinery.

According to the State Council and the Chinese Department of Energy, China's total exploitable high temperature is around 150°C and China's geothermal resources approximate about 6744 MW. The main geothermal centers are in Tibet and west Yunnan where natural resources are most abundant. It is not expensive to exploit geothermal energy but given its limitations in feeding demand, it is comparatively expensive to use as an add-on to other similar alternative resources or as a supplement to the finite capacity of traditional energies such as oil and natural gas. The Chinese government aims to further develop geothermal resources to meet market demand and to remain environmentally sound. The Chinese energy reform program aims to implement geothermal capacities up to 100 MW by 2010, and targets a generating capacity of between 500–1000 MW by 2050. In 2005, China had only five operational geothermal facilities with a capacity of less than 28 MW: JangBaJin geothermal power plant (Tibet); 25.18 MW; Naqu power plant (Tibet), 1.0 MW; Langjiu power plant (Tibet), 1.0 MW; Fengshun power plant (Guangdong), 0.3 MW; and Huitang power plant (Hunan), 0.3 MW. China's largest geothermal power plant, JangBaJin holds a capacity of 25.18 MW and generates 97 million kWh per year. It is located

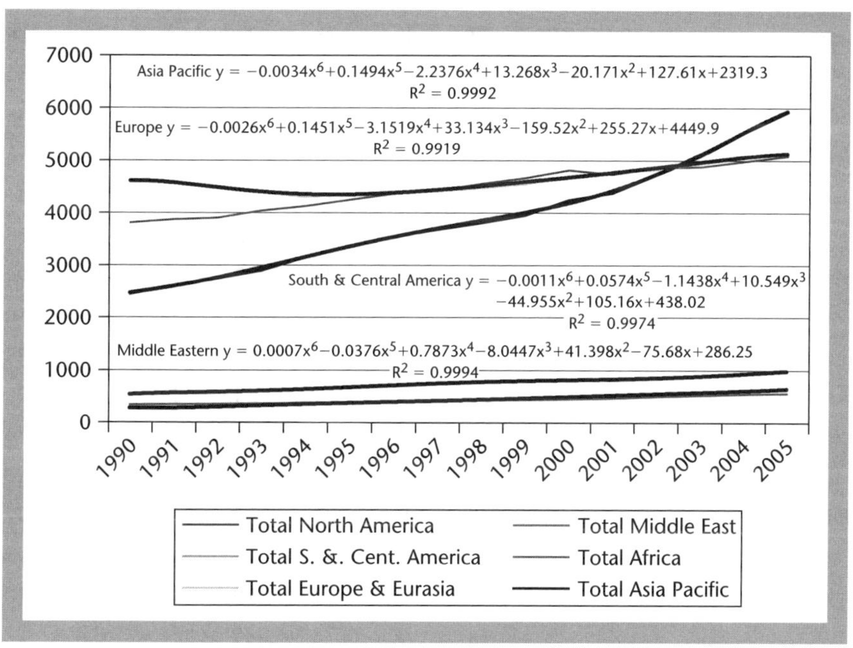

Figure 10.15 Other countries' electricity generating functions, 1990–2005 (TWh)
Source: British Petroleum (2006).

Table 10.11 Electricity generation, 1990–2005 (TWh)

	1990	1991	1992	1993	1994	1995	1996	1997
USA	3185	3223	3235	3354	3407	3517	3611	3661
Canada	482	508	521	532	556	560	573	574
Mexico	122	127	130	135	146	153	163	175
Total North America	**3790**	**3858**	**3886**	**4021**	**4109**	**4229**	**4346**	**4410**
Argentina	51	54	56	63	66	67	70	72
Brazil	223	234	242	252	260	276	291	308
Chile	18	20	22	23	25	27	30	33
Colombia	35	37	36	40	43	45	45	46
Venezuela	61	65	69	71	74	77	79	83
Other S. & Cent. America	121	124	125	132	143	151	156	173
Total S. &. Cent. America	**509**	**534**	**550**	**582**	**612**	**644**	**671**	**715**
Austria	50	51	51	53	53	57	55	57
Azerbaijan	23	23	20	19	18	17	17	17
Belarus	40	39	38	33	31	25	24	26
Belgium & Luxembourg	72	73	73	72	73	76	77	80
Bulgaria	42	39	36	38	38	42	43	43
Czech Republic	63	61	59	59	59	61	64	65
Denmark	26	36	31	34	41	37	54	44
Finland	54	57	57	61	65	63	69	69
France	420	455	463	473	477	494	513	504
Germany	550	539	537	526	528	535	550	550
Greece	35	36	37	38	41	42	43	44
Hungary	28	30	32	33	34	34	35	35
Iceland	5	4	5	5	5	5	5	6
Republic of Ireland	15	15	16	16	17	18	19	20
Italy	217	222	226	223	232	241	244	251
Kazakhstan	87	86	83	78	67	67	59	52
Lithuania	28	29	19	14	10	14	17	15
Netherlands	72	74	77	77	80	81	85	87
Norway	122	111	118	120	113	123	105	112
Poland	136	135	133	134	135	139	143	143

1998	1999	2000	2001	2002	2003	2004	2005	Change 2005 over 2004 (%)	2005 share of total (%)
3797	3876	3990	3924	4050	4076	4168	**4239**	2.0	23.3
562	577	605	582	581	568	576	**594**	3.3	3.3
182	192	204	210	214	217	222	**233**	5.3	1.3
4541	**4645**	**4800**	**4716**	**4845**	**4861**	**4967**	**5066**	**2.3**	**27.9**
74	81	89	90	85	92	100	**106**	5.6	0.6
322	335	349	329	346	364	387	**405**	4.9	2.2
36	38	40	43	44	47	51	**52**	3.5	0.3
46	44	42	43	46	47	49	**49**	1.5	0.3
87	86	89	95	97	98	104	**112**	7.6	0.6
182	185	191	190	200	209	218	**227**	4.4	1.2
746	**769**	**801**	**789**	**816**	**857**	**909**	951	**4.9**	**5.2**
57	60	62	62	63	60	65	**61**	−5.4	0.3
18	18	19	19	19	21	22	**23**	4.4	0.1
24	27	26	25	26	27	27	**27**	1.2	0.1
81	82	85	81	86	88	90	**90**	1.0	0.5
42	38	41	44	43	43	42	**43**	3.6	0.2
65	65	73	75	76	83	84	**83**	−1.8	0.5
41	39	36	38	39	46	40	**36**	−10.3	0.2
70	69	70	74	75	84	86	**70**	−17.8	0.4
511	524	541	550	559	567	574	**575**	0.6	3.2
553	555	564	586	587	608	616	**619**	0.8	3.4
46	50	54	54	55	58	60	**61**	2.3	0.3
37	37	35	36	36	34	34	**36**	6.3	0.2
6	7	8	8	8	8	9	**9**	1.0	♦
21	22	24	25	25	25	26	**25**	−1.6	0.1
260	266	277	279	284	294	303	**302**	♦	1.7
50	47	52	56	59	64	67	**68**	1.6	0.4
18	14	11	15	18	19	19	**15**	−23.1	0.1
91	87	89	94	96	97	102	**102**	−0.2	0.6
117	123	143	122	131	107	111	**138**	25.2	0.8
143	142	145	146	144	152	154	**157**	2.1	0.9

(Continued)

Table 10.11 (Continued)

	1990	**1991**	**1992**	**1993**	**1994**	**1995**	**1996**	**1997**
Portugal	29	30	30	31	31	33	35	34
Romania	64	57	54	55	55	59	61	57
Russian Federation	1082	1068	1008	956	876	862	847	834
Slovakia	24	23	22	24	25	26	25	24
Spain	152	159	161	161	164	169	177	184
Sweden	147	147	146	146	143	148	140	149
Switzerland	56	58	59	61	66	63	57	63
Turkey	58	60	67	74	78	86	95	102
Turkmenistan	15	15	13	13	10	10	10	9
Ukraine	299	279	253	230	203	194	182	176
United Kingdom	320	323	321	323	327	337	351	351
Uzbekistan	56	54	51	49	47	47	45	46
Other Europe & Eurasia	185	168	152	141	140	137	147	147
Total Europe & Eurasia	**4569**	**4558**	**4449**	**4369**	**4283**	**4341**	**4393**	**4394**
Iran	58	63	67	75	81	84	89	97
Kuwait	18	11	17	20	23	24	25	27
Qatar	5	5	5	6	6	6	7	7
Saudi Arabia	70	75	80	88	97	100	103	107
United Arab Emirates	17	17	19	22	24	25	27	28
Other Middle East	78	73	84	89	97	102	111	119
Total Middle East	**246**	**243**	**272**	**299**	**327**	**341**	**362**	**386**
Algeria	16	17	18	19	20	20	21	21
Egypt	43	45	47	48	50	53	57	61
South Africa	165	168	168	175	182	188	200	210
Other Africa	101	103	105	104	108	112	113	115
Total Africa	**325**	**333**	**338**	**346**	**361**	**373**	**390**	**408**
Australia	156	158	162	166	170	175	180	189
Bangladesh	8	9	10	10	11	12	13	13
China	621	678	754	812	928	1007	1081	1105
China Hong Kong SAR	29	32	35	36	27	28	28	29
India	285	309	327	350	379	413	431	455

1998	1999	2000	2001	2002	2003	2004	2005	Change 2005 over 2004 (%)	2005 share of total (%)
39	43	44	47	46	47	45	**46**	2.2	0.3
53	51	52	54	55	57	56	**59**	5.4	0.3
826	846	878	891	891	912	932	**952**	2.5	5.2
25	28	31	32	32	31	31	**32**	3.5	0.2
196	209	225	237	246	264	280	**292**	4.5	1.6
154	155	146	162	147	135	148	**154**	4.7	0.8
63	70	67	72	67	67	66	**60**	−8.6	0.3
111	117	125	123	132	141	150	**162**	8.6	0.9
9	9	10	11	11	11	11	**12**	8.3	0.1
172	171	169	171	174	180	182	**185**	1.8	1.0
363	368	377	385	387	398	396	**399**	1.2	2.2
46	45	47	47	49	49	50	**48**	−3.9	0.3
151	146	145	145	143	152	158	**161**	1.8	0.9
4458	**4528**	**4670**	**4764**	**4810**	**4930**	**5033**	**5102**	**1.6**	**28.1**
102	110	119	128	138	146	156	**169**	8.5	0.9
30	32	33	35	37	40	43	**45**	7.0	0.2
8	9	9	10	11	12	13	**14**	8.0	0.1
113	120	126	134	142	153	165	**176**	7.5	1.0
33	37	40	43	47	50	53	**56**	6.0	0.3
126	132	134	139	146	147	150	**153**	2.3	0.8
413	**440**	**461**	**489**	**520**	**547**	**579**	**614**	**6.3**	**3.4**
24	25	25	26	27	29	31	**34**	9.0	0.2
65	71	73	80	86	92	97	**104**	7.2	0.6
205	203	211	210	218	234	245	**245**	0.4	1.3
118	121	127	138	153	155	157	**163**	4.4	0.9
412	**420**	**436**	**455**	**483**	**509**	**529**	**546**	**3.3**	**3.0**
200	207	213	217	226	228	239	**243**	1.8	1.3
14	15	16	17	19	20	22	**23**	6.9	0.1
1164	1197	1368	1435	1654	1905	2204	**2475**	12.6	13.6
31	29	31	32	34	36	37	**38**	3.8	0.2
492	526	555	571	593	612	655	**679**	4.0	3.7

(Continued)

Table 10.11 (Continued)

	1990	1991	1992	1993	1994	1995	1996	1997
Indonesia	33	38	41	44	52	59	67	76
Japan	841	871	876	887	944	969	989	1015
Malaysia	25	28	32	36	40	47	53	59
New Zealand	32	33	33	34	35	36	36	37
Pakistan	46	50	54	57	58	64	68	62
Philippines	26	26	26	27	30	34	37	40
Singapore	16	17	18	19	21	22	23	26
South Korea	119	132	148	163	185	205	228	249
Taiwan	90	99	106	115	125	133	142	150
Thailand	46	52	60	66	74	84	91	98
Other Asia Pacific	56	57	55	57	59	60	63	66
Total Asia Pacific	**2430**	**2590**	**2735**	**2879**	**3137**	**3346**	**3530**	**3668**
TOTAL WORLD	**11,870**	**12,116**	**12,230**	**12,495**	**12,828**	**13,274**	**13,693**	**13,980**
Of which: European Union 25 #	2464	2533	2525	2526	2565	2633	2725	2736
OECD	7586	7754	7828	8014	8230	8481	8720	8871
Former Soviet Union	1726	1681	1560	1460	1327	1294	1261	1234
Other EMEs	2558	2681	2841	3021	3271	3499	3711	3875

near Lhasa, Tibet's capital, for which it generates almost half the electricity. Geothermal electricity from this plant averages 0.64 yuan/kWh which is higher than other means and not enough to cover generation costs. In addition to this cost, the local administration subsidizes the plant's operation. According to the Chinese Energy State Council, JangBaJin and Naqu stations are the only Chinese geothermal plants with high temperatures between 140 to 160°C. The temperature of JangBaJin's natural resources can reach a peak of 329.8°C in water 2007 meters deep. However, the vast majority of geothermal plants in China have a much lower water temperature.

Geothermal energy production is highly dependent on environmental and weather factors creating volatility in production which can make it difficult for energy production to be stable and consistent with demand. This unreliability is why geothermal energy is primarily used as a substitute alternative energy.

1998	1999	2000	2001	2002	2003	2004	2005	Change 2005 over 2004 (%)	2005 share of total (%)
77	84	93	102	108	113	118	**123**	4.5	0.7
1020	1037	1058	1040	1058	1083	1108	**1134**	2.6	6.2
60	63	67	72	75	84	91	**96**	6.4	0.5
38	38	39	40	41	41	43	**42**	−0.2	0.2
66	65	65	75	82	84	89	**96**	8.5	0.5
42	41	45	47	48	53	56	**57**	1.4	0.3
28	30	32	33	35	35	37	**38**	4.1	0.2
241	267	295	314	336	352	372	**395**	6.4	2.2
163	169	185	189	199	209	218	**227**	4.4	1.3
94	94	96	102	109	117	126	**131**	4.1	0.7
66	69	74	78	85	94	102	**110**	8.1	0.6
3795	**3933**	**4232**	**4364**	**4703**	**5066**	**5515**	**5906**	**7.4**	**32.5**
14,364	**14,734**	**15,399**	**15,576**	**16,179**	**16,770**	**17,532**	**18,184**	**4.0**	**100.0**
2804	2845	2921	3009	3035	3126	3189	**3193**	0.4	17.6
9089	9310	9626	9613	9829	9961	10195	**10389**	2.2	57.1
1221	1236	1270	1292	1303	1343	1372	**1393**	1.8	7.7
4055	4188	4503	4671	5047	5466	5965	**6403**	7.6	35.2

Notes: * Based on gross output; ♦ Less than 0.05%; # Excludes Slovenia prior to 1991.
Source: British Petroleum (2006).

10.6 WIND ENERGY

Chinese windpower generation started in the 1890s. Nowadays, the Chinese Energy State Council has developed modern technologies to intensify windpower energy, which is cleaner and one of the more productive of alternative energy sources. Figure 10.16 and Table 10.12 show that China, like the UK, does not utilize wind energy to the same extent as international counterparts. For most countries, the consumption of wind energy declined significantly in the 1990s as generation costs rose drastically.

The International Energy Administration estimates that the total capacity of global windpower increased from 2 million kW to 32 million kW from

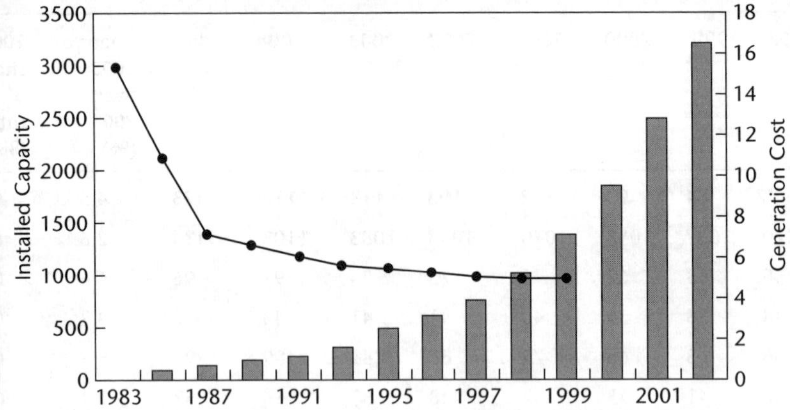

Figure 10.16 Wind power development worldwide, 1983–2002
Source: Chinese Energy Ministry, Chinese Energy and Technology Symposium, 2003.

Table 10.12 The top ten countries in wind power installed capacity, 2002 (10 thousand kW)

	In 2002	Accumulated amount	Ratio (%)
Germany	325	1197	37.3
Spain	149	504	15.7
USA	43	467	14.5
Denmark	53	288	9.0
India	22	170	5.3
Italy	11	81	2.5
Holland	22	73	2.3
UK	5	57	1.8
Japan	13	49	1.5
China	6	47	1.5

Source: Chinese Energy Ministry, Chinese Energy and Technology Symposium 2003.

1990 to 2005. The approximate global annual average rate of increase is 15 per cent. With modern technology, wind energy production costs fell to US$0.05/kWh. China has been investing in modern wind technology since the early 1990s. According to the Chinese Ministry of Energy, this alternative energy source raised production from 20 megawatts in 1990 to 224 MW

Table 10.13 Installed capacity distribution of China's wind farms

Facility	Capacity (MW)
1. Hebei Zhangbei Wind Farm	9.8
2. Inner Mongolia Zhurihe Wind Farm	4.2
3. Inner Mongolia Xilinhaote Wind Farm	30.7
4. Inner Mongolia Huitengxile Wind Farm	1.0
5. Inner Mongolia Shangdu Wind Farm	3.875
6. Liaoning Donggang/Hengshan Wind Farm	17.2
7. Zhejiang Hedingshan Wind Farm	10.2
8. Zhejiang Sijiao Wind Farm	0.3
9. Shandong Rongcheng Wind Farm	0.165
10. Hainan Dongfang Wind Farm	8.755
11. Shandong Changdao Wind Farm	0.11
12. Guangdong Nan'ao Wind Farm	42.43
13. Fujian Pingtan Wind Farm	1.055
14. Xinjiang Dabanheng Wind Farm	64.1
15. Jilin Tongyu Wind Farm	7.2
16. Shanghai Zongming Wind Farm	*
17. Jiangsu Qidong Wind Farm	*
18. Zhejiang Linhaiguocangshan Wind Farm	19.8
19. Jiangxi Panyanghu Wind Farm	*
20. Guangdong Shantou Huilai Wind Farm	12.0
21. Gansu Yumen Wind Farm	1.2

Note: *Not yet in operation.

Source: From information presented at the Workshop on Wind Energy Development for China in the 21st Century.

in 1998. Twenty-four wind farms with a total capacity of 268 MW are listed in Table 10.13.

Chinese wind energy production nonetheless remains far behind its initial 2000 target of 1000 MW. Wind energy in China has not flourished to the same extent as in other countries as a result of slow growth, inadequate government policies and too much bureaucracy, slow and inefficient reforms in the power sector and pricing inefficiencies due to imbalances between supply and demand within China. The economic efficiency of wind energy is illustrated in Tables 10.14 and 10.15.

According to the Chinese Energy State Council, smaller wind generators with a capacity of 100W–5 kW are found throughout China. Additionally, in 1998 the Chinese government installed 156,000 units able to reach capacity

Table 10.14 Prices of China's wind and coal power (yuan/kWh of scale to network)

	(1) Wind power	(2) Coal power	(1)/(2) (per cent)
Xinjiang	0.70	0.32	218
Inner Mongolia	0.71	0.35	202
Liaoning	0.95	0.45	211
Shandong	0.80	0.45	178
Zhejiang	0.79	0.50	156
Fujian	0.79	0.55	143
Guangdong	0.77	0.60	132

Source: Zhao Jiarong (1998).

Table 10.15 The higher cost of wind power in perspective

As shown by this example, the current cost of grid-connected wind power including value added tax could be as high as 0.951 yuan/kWh (11.45 c/kWh) in the period of repayment of both principal and interest. The costs would decrease at 0.37 yuan/kWh (4.45 c/kWh) after the loan payment period. *In this context, the average life-cycle cost would be 0.472 yuan/kWh (5.68 c/kWh)*. The basic assumptions for cost calculation are:

Full load power generation hours: 2500 h/year

Capacity Factor: ~30 per cent

Initial investment: 8300 yuan/kW, of which

  Equity capital investment 20 per cent

  Domestic commercial loans: 80 per cent

  Domestic commercial loan interest: 7.56 per cent

  Debt service period: 7 years

Annual operational costs: 5 per cent of the total investment

Depreciation of wind power generator: 12.5 years

Value added tax: 17 per cent

Source: Chinese Energy Ministry/Department, 2003.

efficiency up to 18.1 MW. The number of users of such energy approximate 159,000, mostly representing local populations in Inner Mongolia. A further seven provinces also have over 1000 small wind generator units: Inner Mongolia owns 140,000, Gansu 3120, Henan 2710, Ningxia 1720, Xinjiang 1680, Shandong 1330 and Anhui 1080.

Smaller units have been found to be uncompetitive with medium to larger scale wind generation projects. Smaller units were also found to be economically inefficient in regard to keeping up with China's substantial energy demand. In 2003, China used 11 sizes of small windpower generators – 50, 100, 150, 200, 300, and 500-watt, and 1, 2, 3, 6, 10, and 20 kW machines. Small windpower generators comprise blades, turbine, gyrorotor, tail wing, supporting pillar, batteries and a pad. The mechanism functions with blades turning the turbine to produce electricity which is stored in batteries. Stored energy is then converted with DC to AC converters and transmitted to Chinese consumers. Wind energy systems generate power of various voltages, such as 12 V, 24 V, 36 V, 110 V and 220 V. In 2002, about 26 provinces, municipalities and regions used wind power.

About 25 manufacturers in China create small household windpower units and China's assembly capacity for small wind power generators is 30,000 units per year. The main manufacturers of windpower engines are Shangdu Husbandry Machinery, the Power Factory in Inner Mongolia and Shangxi Taiyuan 884 Factory. Table 10.16 lists a number of major manufacturers of large turbines.

Thirty domestic manufacturing entities create small-scale wind turbines in the 50 to 300 W range. Village users can assemble about 130,000 portable units to obtain a cumulative capacity of 17 MW. Chinese manufacturers can produce turbines up to 200 kW in size and can also acquire technology for 500 kW size turbines. In addition, groups of wind turbines between 50 to 200 kW size are installed in arrays to produce around 13.4 MW in various regions of China. China also exports wind turbines to the United States, Belgian, German, Swedish, and Danish manufacturers. The Chinese market has also benefited from exports of small windpower generators to neighboring countries in Southeast Asia, as well as to Japan, Germany and Cuba. In addition, China trades new technology and creates commercial joint ventures. Research and development are also part of the Chinese wind energy technology program. The Ministry of Electric Power reported that the national energy renewal program has promoted Chinese wind energy creation since 2003. According to the ministry, the Chinese national wind resource capacity exceeds 250 gigawatts (GW). Most of the wind resource is found in the northern and western regions of China. Additionally, there is some wind along the eastern sea coast. The Ministry of Electricity Energy also reports that wind resources may be used for smaller power developments in rural villages. Wind resources may also be used to supplement inadequate supply within the Chinese national grid in order to provide a more uniform supply and demand balance where there are regional areas suitable for wind energy. The new Chinese energy reforms include the connection to the national grid of all wind farms in the 10 to 30 MW range as well as medium-sized wind farms with capacities of 100 MW or more.

In addition to the Ministry of Electricity Energy, or State Council, the National Meteorological Bureau and the Meteorology Institute formulate policies regulating wind energy allocation and monitor the balance in

Table 10.16 Major large wind power generator manufacturers

Enterprises	Basic status
Danish NEG MICON – Inner Mongolia Wind Power Co.	Founded in 1996, joint venture, mainly assembling of 600 kW generators and tower.
Danish MICON-Zhejiang Power Equipment Manufacturer	Founded in 1996, mainly assembling of large wind power generators. Up to date 17 have been produced.
German NORDEX-Xian Aviation Generator General Co.	Founded in 1997, joint venture designing and manufacturing large wind generators – 600 kW and 250 kW. Designed capacity for 140 units per year, unrealised for technical and market reasons.
German HSM-Luoyang No.1 Tractor Co.	Founded in 1994, joint venture designing and manufacturing large wind generators such as 250 kW wind generators. Ten units produced in 1995 with full German technologies. The planned 60 units output unrealized as the result of fund shortage.
Spanish Lude-Luoyang No.1 Tractor Co.	Founded in 1998, joint venture designing and manufacturing 600 kW wind turbines; plan to produce 2 × 600 kW generators in 1999 with a localization rate reaching 70 per cent.
Beijing Wandian Co. Ltd.	Founded in 1996 with Austria PEHR, designing and manufacturing large wind generators. Purchased one FLODA636 sample generator. Prototype unit produced in 1998 with Chinese blades and tower and remained in operation in Huitengxile Wind Farm. Planned to produce 10 units in 1999 and 30 units in 2000.
Xinjiang Wind Energy Co.	Its subordinated wind farm was located in Xinjiang Dabancheng. At the end of 1998, its installed capacity reached 40 MW. Through more than a decade's operation, rich operational and maintenance experiences were accumulated with strong technical strength. In 1997 it purchased designing technology of Jacobs 600 kW. At present, Chinese-made Jacobs unit is working smoothly in Dabancheng. Its generator, gear box and yawing system are all Chinese made with the components such as blades, high speed brake, hydraulic system and safety mechanism imported.

market pricing efficiencies between supply and demand in China. These meteorological centers also determine wind energy potential based on a simple set of parameters, such as average wind energy density (W/m^2), and annual hours of wind speed above given limit levels in m/s.

The Chinese Center for Renewable Energy Developments (CRED) has created a 'wind concession' approach. The concept is that of an auction involving the buying and selling of wind rights to private developers with the lowest bid per kilowatt-hour gaining a geographic share of the wind market. CRED also implemented a power tariff structure with variables to evaluate the supply and demand of windpower prices. Four batches of wind concession projects, each capable of generating 1100 megawatts (MW), required investments of approximately US $1.1 billion. The wind grid pricing between bid and ask prices produces a market with tariffs of US 4.8–6.8 cents/kWh. This system helps generate a new market in the pricing of wind energy in China to encourage demand. Yet the system is estimated to be inadequate and pricing tariffs are considered too low to be cost efficient with initial investment. Demand would have to be greater and wind energy better utilized throughout China. CRED also completed a guidebook for wind concessions.

However, installation of wind concession programs is expensive. For instance, according to the China Sustainable Energy Program, foreign investors and private developers helped provide capital to design and implement wind concession programs in Guangdong and Jiangsu regions. Those investments cost approximately $200 million in new wind installations over several years. China will also install an additional 700 megawatts of new wind facilities which will cost at least US$700 million. Chinese policies are reinforcing the production of advanced large scale turbine technologies. The Chinese Energy State Council and NDRC intend to install at least 20,000 MW of wind generation by 2020.

CRED, the China Renewable Energy Industry Association (CREIA), and the China Wind Energy Association are producing a series of measures to promote the production of wind energy in China. They have tested the productivity and efficiency of local wind generating projects to encourage further investment in wind turbine manufacturing and have proposed ideas for expanding the wind energy market in China. They are in contact with international partners to benefit from competition with peers and align domestic standards with those of international market participants. They have also developed a series of recommendations to enhance wind energy usage. The Center for Resource Solutions (CRS) also has industry contacts and investments to achieve these goals. Moreover, Clean Development Mechanism (CDM) projects will account for 5 per cent to 7 per cent of total energy investment in developing countries between 2001 and 2010. In addition, the Chinese Energy State Council also participates, with 35 to 45 per cent of CDM carbon credits in 2010. The return on investments from the sales of those credit units can bring up to US$1 billion per year for clean energy projects. Yet, this is considered relatively small compared to the total energy infrastructure investment which is approximately US$100 billion per year.

The Chinese Energy State Council introduced an energy law in February 2006, and the Chinese government enhanced its renewable energy programs at the Beijing renewable energy conference in November 2005. It has increased its investments and resources toward a projected national consumption of 10 to 15 per cent of total consumption by 2020. The Renewable Energy and Energy Efficiency Partnership (REEEP) and the Chinese Centre for Renewable Development (CRED) contributed to research which evaluated 15 per cent of wind energy demand at 764 MW. CRED also reports that mainland wind developments are able to produce as much as 250 GW and Chinese offshore resources 750 GW. The Chinese State Council also targets a total wind generation capacity of 5 GW by 2010 and 30 GW by 2020.

The Chinese Energy State Council is focusing on developing more 200-kilowatt or larger wind-driven generators and windpower field control and management system. The whole government strategic plan is to develop policies and conditions to bring the national windpower capacity to 0.3 to 0.4 million kilowatts, equal to 0.35 to 0.46 million tons of standard coal, by 2000, and 1–1.1 million kilowatts, equal to 1.08 million tons of standard coal, by 2010.

10.7 SOLAR ENERGY

The Chinese Energy State Council promulgates the usage of energy-saving solar buildings, solar energy water heaters and photovoltaic generating systems. Solar energy buildings and solar energy water heaters are most popular with the general population of China. The strategic development aim is to generate a total capacity of solar energy equivalent to 1.23 million tons of standard coal by 2000 and 4.67 million tons of standard coal by 2010. World solar production is also exponential (see Figure 10.17 and Table 10.17).

The Chinese government includes solar energy in its portfolio of alternatives or renewables to alleviate inflationary global energy prices and adequately to meet rising demand. The Chinese government began to develop solar power technologies in the 1970s. The new policy programs plan to further develop solar technological facilities such as those already installed in Tibet which have a total capacity of 100 MW.

China produced 3 MWp of photovoltaic cells in 1999, and its total cumulative amount in operation totals 15 MWp. The solar market does not represent a vast capacity compared to other alternative energies. Solar energy can only ramify 1 per cent of 15 MW to the national grid. Half of the solar energy produced is used for telecommunication projects. Half of Tibet's 889 townships had PV-based satellite stations in 1997. Thirty per cent of the projects are for individual usage and 10 per cent of the telecommunication tools based on solar energy are for industrial use. Pricing regulations vary according to the available supply. Sales of photovoltaic cells on the domestic

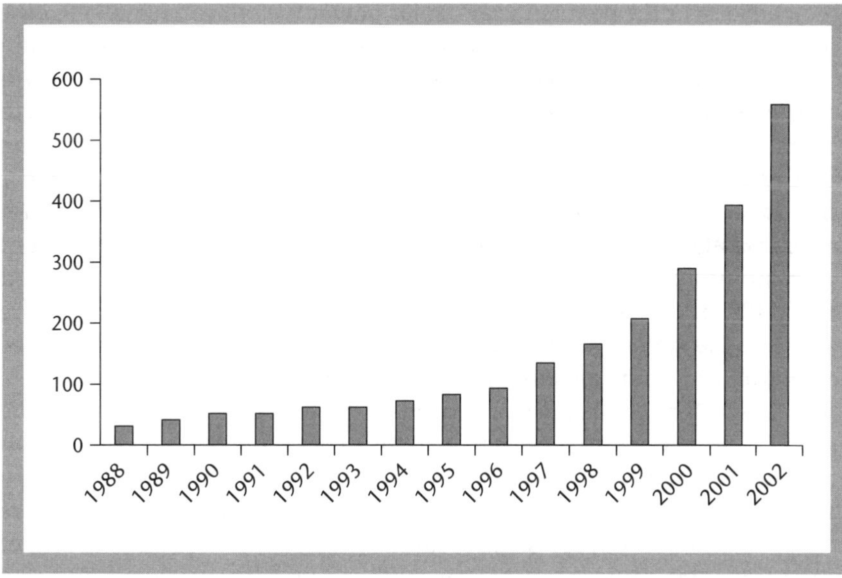

Figure 10.17 World production of solar battery components, 1988–2002 (MW)
Source: China/Spanish Renewable Energy Symposium (2003).

Table 10.17 Number and capacity of minimum large-scale Chinese power stations, 2003

Large-scale centralized power station (>100 kW)	5	5	5
Total (MWp)	288	395	525

Source: Wang Sicheng, *Solar Photovoltaic Generation*, China/Spanish Renewable Energy Technology Symposium, Dec. 2003.

Chinese market rose at an increasing rate by an average of 20 per cent per year since the late 1990s. Pichel (2006) provides more information about the costs and market pricing of solar energy in China. Solar energy is the simplest and most widely used renewable resource principally due to the cheap cost of heating water and using solar energy. Among its manufacturing performances, China ranks first in the world in the production of solar water heaters.

According to Jiménez Viviana (2005) from the Earth Policy Institute, rises in world sales of solar devices will alleviate Chinese domestic pricing. Statistical information reveals that from 1990 to 1998, the total heat collection area used for solar water heaters rose tenfold, that is from 1.5 million square meters to 15 million m^2. Statistics also indicate that 1999 sales reached 5 million m^2, representing a market worth 3 billion yuan. Chinese authorities also predict that over the next ten years, solar power water heaters will increase at an annual rate of 15 per cent to reach 100 million m^2 by 2010, 200–250 million m^2 by 2020 and 400–600 million m^2 by 2050. This exponential growth rate is matched by an increasing industrial and manufacturing rate of 20 to 30 per cent per year providing total capacity of 6 million m^2 each year. The Chinese Energy State Council also reports that about 1000 Chinese firms create, manufacture, sell, and install solar water heaters. Solar water heaters are also being exported to countries in Southeast Asia and Europe.

According to the Chinese Ministry of Energy and a survey conducted among 188 enterprises involved in the solar heating industry, only 44 are owned by the state. The survey, conducted by the Specialized Committee on Solar Energy, indicates that the rest are primarily collectives or private companies. More enterprises are based in the east than in the west of the country. The survey also reveals that about 80 per cent of users of solar heating systems are small to medium-sized enterprises looking to use energy efficiently. Solar heating systems are not adequate or efficient enough for use in large manufacturing units and factories which is why only 20 per cent of large companies have an annual capacity greater than 10,000 m^2. The survey also reports that only two companies exceeded 100,000 m^2 per year.

The Chinese Energy Ministry has also reported rapid development in photovoltaic technology. A new solar heating system uses a single-crystal silicon battery, a multi-crystal silicon battery, a non-crystal silicon battery, a hull cell and a compound battery. This new technology improves the conversion efficiency of batteries. The conversion efficiency of this solar heating system generating energy with batteries is able to reach 15 to 17 per cent in a single crystal silicon battery and 14 to 15 per cent in a multi-crystal silicon battery. The Chinese Specialized Committee on Solar Energy also observes that with this new technology, the highest conversion efficiency is over 25 per cent, which also contributes to decreasing the production cost. The price of a crystal silicon battery component was about US$6.5/Wp in 1995, decreasing to US$5.0/Wp in 2000. It decreased further to US$3.0–3.5/Wp in 2002. The cost of the solar alternative system has decreased by 47 to 54 per cent in 7 years. Given other energy alternatives and renewables, the cost of a crystal silicon solar battery will decrease to US$2/Wp by 2010 and system price to US$3.12/Wp, while the estimated electricity price will increase to US$0.11/kWh. The China National Energy Strategy and Policy 2020 program 'Renewable Energy Strategy and Policy' also reveals that by 2020, the price of a crystal silicon solar battery will decrease to US$1.0/Wp, system price will reduce to US$1.52/Wp, and the electricity price will reach US$0.052/kWh. The report also indicates that production is increasing exponentially. Since 1988, the production of solar batteries rose fifteen-fold increasing from 33.6 MWp to 561 MWp in 2002. Production of solar energy in China is third in the world, behind the United States (second) and Japan (first). Researchers in the Chinese energy department indicate that research

Table 10.18 Major regional distribution of solar water heater manufacturers in China (based on survey data collected in 1997)

Area	Number of enterprises	Annual capacity >10,000 m^2	Annual capacity >5000 m^2
China as a whole	188	35	26
Beijing	15	5	2
Titanjin	2	1	1
Hebei	42	5	4
Jiangsu	31	10	4
Anhui	13	3	2
Shandong	25	5	3
Guangdong	7	2	1
Sichuan	3	1	1
Yunnan	16	2	7
Gansu	4	1	1
Shanxi	3	0	0
Liaoning	6	0	0
Zhejiang	4	0	0
Henan	9	0	0
Guangxi	1	0	0
Hainan	1	0	0
Xinjiang	2	0	0
Heilongjiang	1	0	0

Source: Survey Report on Solar Water Heater Manufacturers, 1997, by Chinese Specialized Committee on Solar Thermal Utilization.

and development technologies in solar energy are lowering costs and are concentrated on high-efficiency and low-cost batteries using special materials such as non-crystal silicon and micro crystal silicon layers thin film battery. According to the statistics, global solar water heating reserves were approximately 100 million m^2 in 2002. This represents approximately 14 million tons of standard coal equivalent. According to the 2020 China National Energy Strategy and Policy half the global reserves are in China. Tables 10.18 and 10.19 provide information on solar water heating in China.

However, solar systems are still not completely reliable as they depend on climate, which is changeable. There are three types of solar energy systems used to create heat. One of them uses a line focusing system, another applies a tower system and a third type uses a system resembling a satellite

Table 10.19 Technical and economic features of Chinese-made solar water heaters

	Integrated	Flat plate	Vacuum
Heat collecting area (m^2/unit)	0.3–1.0	1.0–2.0	1.0–1.2
Average system efficiency (per cent)	45–50	About 45	40–45
Hot water output (kg/day)	60–70	70–100	70–100
Water temperature (C)	40–60	40–60	40–70
Average price (yuan/m^2)	<500	700–1000	1500–2500

Sources: 'Draft of Rural Energy Development Study in the P.R. China', Asian Development Bank, 1996.

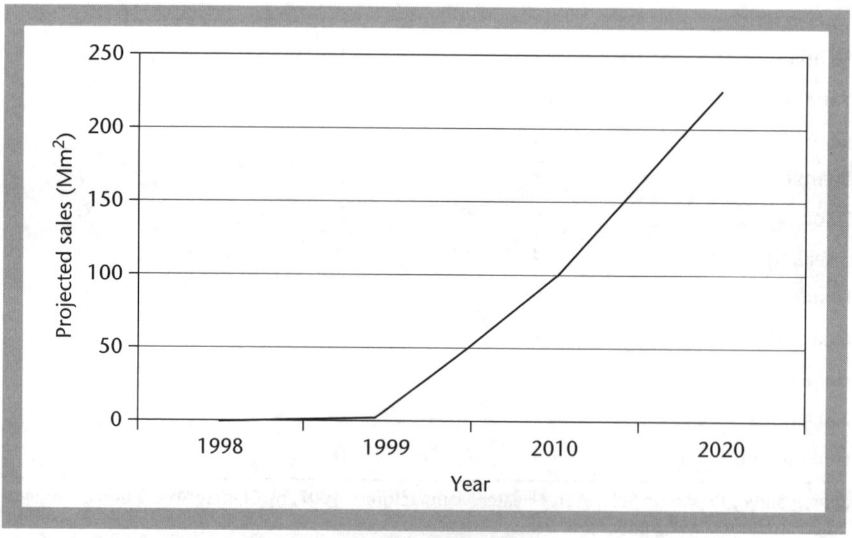

Figure 10.18 Solar water heating sales, 1998–2020
Source: China Strategy (2020: 213).

dish. The solar heating system installed capacity is over 350,000 kW. Researchers Zhang Zhenmin and Wang Gehua forecast in the China National Energy Strategy and Policy Plan 2020 that solar energy heat generation will be applied in mature and developed countries in 2020 as an alternative energy to lower the cost of oil and natural gas and to leverage global market prices (see Figure 10.18).

Solar energy applications are now more advanced and benefit from new technologies and modernization. Appliances such as stoves can utilize solar energy, as well as passive solar houses and solar energy dryers. China currently produces solar energy water heaters of 2.3 million square meters,

Table 10.20 Basic technical and economic features of a large household solar water heater and a coal burning boiler

Solar water heater	Coal burning boiler
Local solar radiation total: 585 × 10^4 KJ/m^2/year	Boiler capacity: 1.0 t/h
Solar energy utilization coefficient: 0.7	Number of boiler: 1 unit
Water heater average conversion rate: 0.65	Boiler efficiency: 0.4
Initial investment: 135,200 yuan	Coal thermal value: 16720 kJ/kg
Of which heat collector: 68,000 yuan	Coal price: 200 yuan/t
	Annual coal consumption: 50 t
	Initial investment: 86,850 yuan of which boiler:45,000 yuan
Annual system operation costs: 50 yuan in first 5 years 150 yuan in the second five years	**Annual operational costs: 11,600 yuan**

Source: 'Economic Performance Intercomparison of Large Solar Water Heating System', Solar Energy, Henan Academy of Sciences, 1998.

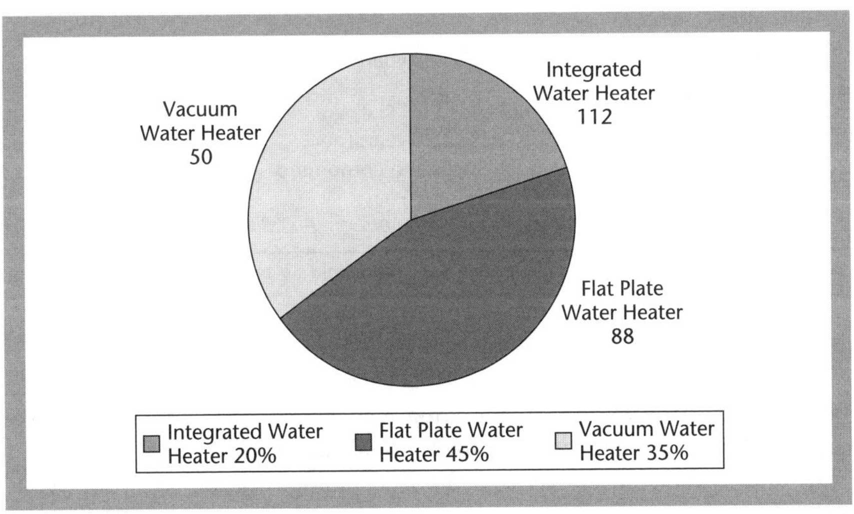

Figure 10.19 China's solar water heater output and market composition, 1997 (10,000 m^2)
Source: Chinese Energy Ministry/Department.

passive-type solar houses of 1.8 million square meters, solar energy crop warm houses of 342,000 hectares, 140,000 solar stoves and solar energy dryers of 13,200 square meters.

An average of one square meter of solar water heater can save 100 to 150 kilograms of standard coal per year (Table 10.20; see also Figure 10.19 and

Table 10.21 Economic performance of solar and gas water heaters

	Solar water heater plus electric heating system used for backup	Gas water heater
Scale	1.5 m^2	10 liters
Initial investment	3200 yuan (including electric heater)	1500 yuan
Life span	10–15 years	5–10 years
Heating efficiency	0.52	0.8
Power tariff	0.6 yuan/kWh	–
Gas tariff	–	1.2 yuan/m^3
Annual power consumption	73 kWh	–
Annual gas consumption	–	364 m^3
Cash interest	2.25 per cent	2.25 per cent
Equipment depreciation	6.7 per cent	6.7 per cent
A) Annual power expenses	43.8 yuan	–
B) Annual gas expenses	–	436.8 yuan
C) Annual depreciation	214.40 yuan	100.5 yuan
D) Annual interest	80.00 yuan	37.5 yuan
Annual total operation expenses (A + B + C + D)	338.20 yuan (not including compound interest)	574.80 yuan (not including compound interest)

Sources: 'Assessment of Economic Performance of Housebold Solar Water Heater', Solar Energy, Lin Jianmin, 1999.

Table 10.21). An average of one square meter of floor space in a solar heated Chinese home saves about 20 to 40 kilograms of standard coal a year, and a solar stove saves 500 to 700 kilograms of straw or wood per year.

As noted above, solar energy photovoltaic power started to develop in China in the 1970s. From 1983 to 1987, China ran seven solar-cell production lines with imported technologies. In 1988 China created a solar-cell production capacity of 4.5 megawatts, against 200 kilowatts before 1984. Solar cells are used not only for telecommunications but also in rural areas where there are no other power sources, accounting for 1.1 megawatts per year. China utilizes solar energy in twenty-eight counties, in over a thousand villages and a thousand islands, many of them hard to access. There are two photovoltaic power stations in Tibet with capacities of 10 kW and 20 kW, providing energy to nine counties without hydraulic power or electricity. Solar cell research also provided an efficient and practical model of the single-silicon cell.

10.8 OCEAN AND SEA ENERGY

According to IEA (2007) modern technologies have been producing energy from the ocean, waves and tides since 1956. China's first tidal stations and engines are in Fujian and Guangdong provinces and are used for pumping water for irrigation. Tidal energy development is mainly focused on the areas of Zhejiang and Fujian. Before 2000, experiments and research in 10,000 grade generating sets and sea engineering developed capacity close to 50,000 kilowatts. The Chinese Energy State Council aims to install power stations producing 30,000 kW by 2010. Technological progress allows the conversion of tidal energy to electric energy.

China's ocean and sea energy evolution progressed in three stages, with the first phase dating back to 1958. From the early 1960s, China established about 40 small tidal power stations with a capacity of a few dozen kilowatts. The second phase started in 1970 with a small number of tidal power stations with larger capacities, among them Jiangxia tidal power plant and Baishakou power plant holding a capacity of 3000 kW each. Tidal power stations constructed during those two early stages were decommissioned principally as a result of technological inadequacies, capacity limitations and substantial conflict between irrigation and navigation purposes. During the 1970s, power from waves and tides was further developed in urban centers such as Shanghai, Guangzhou, Dalian, Qingdao, Beijing and Tianjin. For instance, a Shanghai boiler works invented a prototype for floating wave energy power near the sea borders around 1975. Zhou Shan sea station began operations in 1978. The late 1970s were typified by propeller engines using tidal power, which were able to reach capacities of 5.7 kW with a waving flow rate of 3 m/s. Similar developments were also constructed in Qingdao and Harbin. According to the Chinese Sustainable Energy Program, the Guangzhou Institute of Energy Conversion and the Chinese Academy of Sciences tested new innovations using weather related variables with the 'Lifting Cycle by Fog Drops' and 'Open Cycle of Oceanic Thermal Energy Conversion'. The Chinese ocean and marine energy section ran seven tidal power engines and one tide flood power station with a total capacity of 11 MW.

The third phase in the evolution of sea and ocean energy production in China is marked by modern scientific and technological progress enabling an increase in capacity to 10 MW tidal power stations. This phase also saw testing and surveys on Pingtan Island in Fujian and Dachen Island in Zhejiang. The Xingfuyang tidal power plant on Pingtan Island was constructed in

1985 and has been able to hold a capacity of 1.28 MW since May 1989. In addition, as modern technology advanced, China added robust capacity and stronger engines to increase energy production and holding capacities, such as a 10 MW intermediate testing tidal power station in Jiantiao Port in Zhejiang and Daguanban Port in Fujian and the implementation of the Maluanwan tidal power plant. Also during the later phases, the China Sustainable Energy Program reveals that Shanghai Research Institute of Hydrographic Bureau increased performances and capacities with an additional four-valve type wave energy power conversion engine for transforming tidal waves into energy. About 10 sets were constructed in 1985. Additionally, the Guangzhou Research Institute of Energy Conversion of the Chinese Academy of Sciences invented the BD-102 type wave energy conversion engine. It utilizes a symmetrical wing air turbine that is patented in the National Patent Bureau. Many city centers along the eastern coast used these engines to leverage the usage of sea and marine waves into energy to feed urban centers such as Guangzhou, Shanghai and Tianjin. Similar to hydropower energy, Daiwanshan tidal waves station can hold a capacity of 8 kW with generators holding as much as 3 kW each. According to the China Sustainable Energy Program, more private companies were investing in renewable energy, such as Taiwan Electric Power Company in 1989. During the mid-1980s, the Xian manufacturing hub also produced marine and sea tidal wave generators. Sea and hydro energy tides produced additional stations with a capacity of 70 kW. However, these have limited capacities and are expensive to maintain.

Advanced technologies allow the usage of alternative energies such as tidal energy, wave energy, ocean thermal energy and weather drops, as well as the creation of significant capacities of 0.3 million kW. More testing is ongoing with advanced technologies to utilize ocean energy. The Chinese Energy State Council, the NESP and the China Sustainable Energy Program provide ample information about advanced sea technology developments. China's coasts benefit from substantial multiple natural resources along 32,000 kilometers of coastline and on more than 6500 offshore islands. With these geographical advantages, China is able to leverage from large natural capacities and exploit the potential of tidal and wave energy generation. The International Energy Administration reports that China has exploitable tidal resources of 21.8 GW with annual electricity capacities of 62.4 TWh. Fujian station produces as much as 40 per cent and Zhejiang owns 47 per cent of this total. The administration also notes that Zhejiang benefits from wave energy with as much as 14 GW. Wave energy resources are 12.9 GW, representing a third of total energy resources in Taiwan. Furthermore, the International Energy Administration reports that the Chinese ocean thermal energy produces 1320 to 1480 GW near the Xisha Islands and the offshore areas east of Taiwan.

10.9 COAL LIQUEFACTION AND TRANSFORMATION PROCESSES

China has intensified the search for alternative means to secure energy as consumption continues exponentially to outpace production. To alleviate the inflationary pressures of energy prices, Chinese government policies also focus on investing in alternative energies. Coal is a substitute for oil and natural gas, easy to use energies that are depleting and are partly responsible for economic pricing dislocations. In this section, we consider alternative means to solve the forecast energy shortage.

The market for liquefied petroleum gas has risen exponentially since the 1990s, particularly in rural areas (see Figure 10.20). Table 10.22 illustrates China's reliance on coal mines to generate energy and its labor productivity in comparison to the USA (to which it is second), Australia and South Africa.

According to Credit Suisse Hong Kong research, coal prices in 2010 are estimated to average about US$45 a ton. Crude oil futures on the New York Mercantile Exchange rose 25 per cent through 2005 to reach US$75.78 a barrel on 7 July 2006, which represents the highest peak since New York trading began in 1983. Credit Suisse calculates that coal liquefaction plants are exploitable as long as oil continues to trade at US$30 a barrel. To ameliorate and intensify energy products and maintain prices at affordable levels, other measures have been taken. Shenhua Ningxia is building a separate coal-to-chemicals plant at Ningdong, in the east of the province. The plant, designed to turn 250,000 tons a year of coal into methanol, was intended to be completed at the end of 2006 and to start producing methanol early in 2007. The plant cost 1 billion yuan (US$125 million). The company also intends to spend 10 billion yuan to build a plant to transform coal into raw chemical material, polypropylene, at Ningdong. This is scheduled to start operation in the first half of 2009. According to the International Energy Administration (EIA) (2005; 2006), coal represents 65 per cent of China's primary energy consumption. China is the world's biggest global consumer and producer. According to the British Petroleum Statistical Report (June 2006), China's coal consumption rose from 978 to 1082 million tonnes equivalent from 2004 to 2005. Statistics show that coal consumption and production rose exponentially, in parallel, over the last 20 years. Coal is a traditional energy source used primarily because it is cheap and abundant in China. The sole disadvantage for Chinese coal production is that policymakers have neither reformed its infrastructures nor re-engineered its drilling and extraction machinery. More recent Chinese statistics demonstrate that coal policies remain part of the 2006 renewable energy policy reforms and that coal mines in China are being renovated with new technologies and more efficient advanced machinery.

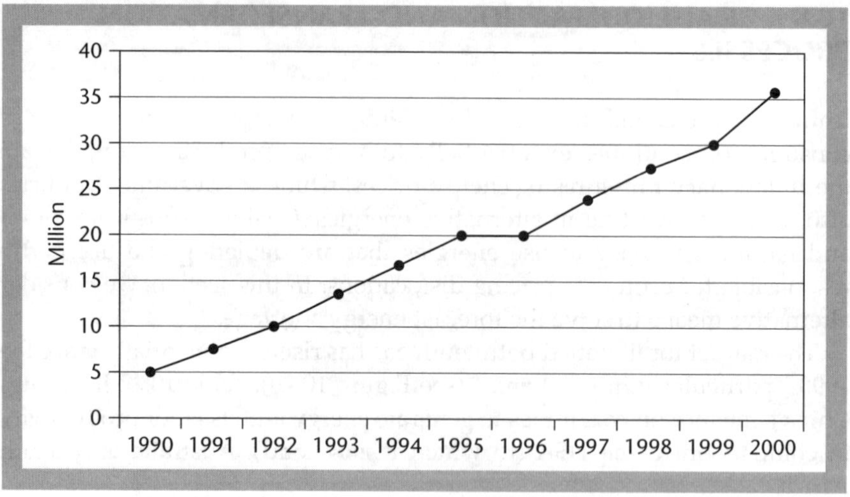

Figure 10.20 Rural families using liquefied petroleum gas
Source: Chinese Energy Ministry, Chinese Energy and Technology Symposium, 2003.

Table 10.22 International comparison in labor productivity and labor cost of coal industry, 1998

	China state-owned key mines	USA	Australia	South Africa
Coal output, Mt	503.5	927	219.4	219.3
Workers, thousand	2629	78	20.3	58.5
Productivity t/p/yr	192	11,846	10,788	3749
Labor cost, US$/t	4.2	3.8	10.5	2.9
Annual wage, US$1000/p	0.88	44.6	76.2	11.1

Note: One US dollar = 8.3 yuan renminbi; Mt = million tons, t = tons and p = persons.
Source: State Bureau of Coal Industry; IEA, Coal Information, 1999 edition.

The International Energy Administration annual energy outlook report (2006) observes that China's coal consumption in 2003 was 1.53 billion short tons and that Chinese consumption represented 28 per cent of the world total. Recent statistical data also shows that coal consumption rose significantly from 2001 to 2003, compensating for the decreasing pace from 1997 to 2000. The Chinese economy has been expanding significantly, boosted by the foreign investments poured into the country during the 1990s, primarily profiting from cheaper forex, limited human rights and labor law reforms. At the time, the Chinese middle class did not have the capacity to consume as much as that of the matured markets, neither did it have the incentives to reform its energy policies with the rise of industrial and manufacturing production. It was not until the late 1990s that China began to experience problems with its domestic rising demand and potentially unsustainable energy needs over the short term. In consequence, China reformed its energy policies significantly in 2006, having experienced massive failures in coal mines and other energy sectors as a result of inadequate capacities, infrastructural inefficiencies and lack of adequate international standards.

However, developments were delayed until the commitment of foreign investment needed to enable reform and renovation. At the same time, China started to use more efficient means of energy such as natural gas, oil, and electricity generation from a variety of sources, such as hydro or nuclear power. Consequently, China's rising dependence on easier but more expensive alternative resources such as oil led to a coal surplus. The Chinese government attempted to export the coal surplus to neighboring markets in Asia, and as part of the energy reforms, coal mines that were too incapacitated or too expensive to renovate were closed. According to statistical data from the International Energy Administration (2006), tens of thousands of small coal mines were closed because inadequate security, employees' safety hazards and overproduction resulting in surplus capacity made them too expensive to maintain. By 2004, according to the World Energy Council and the International Energy Administration the Chinese population has been facing grave challenges in meeting the exponentially rising demand for energy, including coal. Some years saw depressed prices from low demand, while later years exhibit substantial variations. This market volatility is principally due to the opening of markets and adoption of globalization standards and policies, enabling significant populations to increase their lifestyles and energy needs at the cost of others. For instance, Chinese surpluses are eroded through rising exports to try to decrease the pricing trend of coal which was the case in 2002 and especially 2003. But fundamentally, those eroded surpluses, accumulated at the beginning of the century and dissipated cheaply to exports to purchase more expensive means of energy, represented the substantial resources extracted from generations of significant work at very low prices. The author's study of various global oil curving points demonstrates that China's coal will be on the rise again, perhaps around 2010 to 2015, primarily because during the ensuing period, oil prices will vary in disparate geopolitical strategic points of the globe creating more speculation and discrepencies in oil market pricing efficiencies. And in the meantime, it will be more expensive for the Chinese middle class to rely exclusively on oil and natural gas prices as they create an inflationary pressure on their living standards.

So, from the beginning of its reforming agenda, Chinese policy-makers smartly diversified into other alternatives in order to leverage fully from its main capabilities. Thus, coal pricing curves are estimated to continue to rise as a reliable contingency substitute for oil and natural gas, and China's coal demand is projected to rise significantly. Economically speaking, while coal's market share as part of total Chinese energy consumption is projected to decrease at a growing pace and oil and natural gas demands will rise, coal consumption as a whole is forecast to rise in absolute terms. Technological innovations are being installed along with advanced methodologies such as coal gasification, gas liquefactions and a coal pipeline. These allow for the creation, transformation and transport of coal in its alternative forms from point to point within China. For instance, a pipeline transmits gasified coal to Qingdao port. Another new alternative is to produce coal-bed methane, and to this end British Petroleum, Chevron Texaco, and Virgin Oil drilled into Ningxia province in January 2001. According to Chevron Texaco's information for investors, regulators and press officials, Chevron Texaco is the largest foreign investor in coal-bed methane. Other energy firms such as ConocoPhillips performed similar experiments in coal-bed methane in April 2004 in Shaanxi province which is near the west-to-east pipeline zone.

Since its entry into the WTO more advanced technologies have become available to Chinese energy corporations and China has allowed more foreign investment into improvements in its coal industry, to modernize existing coal mines and create more efficient and safer large-scale mines. The China National Coal Import and Export Corporation is the main Chinese corporate entity to benefit from foreign intelligence, expertise and investments to improve capacities and renovate machinery infrastructures. Since around 2002, the Chinese Energy State Council has rapidly reformed and re-engineered coal mining sectors to maintain pricing efficiencies in domestic and international markets. Moreover, China aggregated a number of large old state coal mines into seven private Chinese corporations at the end of 2005. This corporate governance overhaul is part of the CNPC and Sinopec's re-engineering to integrate new technologies and financial structures in its private entities to compete at an international level and to be able to maintain global financial and energy pricing efficiencies. Chinese energy firms aim to move towards foreign venture capital and private equity businesses as part of modernization and internationalization of corporate governance programs and to meet domestic as well as international market needs.

Part of this overhaul consists in integrating new technology such as coal liquefaction technology to create liquid fuels based on coal. The Shenhua Group in Inner Mongolia started coal liquefaction projects in 2005. The disadvantage of this new technology so far is its cost and economic infeasibility. According to the Energy Information Administration report (2006), Chinese energy policies plan on decreasing reliance on coal to less than 60 per cent of total energy use by 2020. China currently relies on coal to meet 66 per cent of its energy needs. In addition, the government is attempting to increase use of clean coal technologies for power generation, which includes coal liquefaction for transport fuels.

The 2006 Energy Information Administration report also asserts that policies over the next ten years suggest integrating the NDRC's Energy Bureau within the State Energy Office, which is independent and an integral part of the Ministry of Energy. This government body also promotes research and development in coal technologies. As part of this program, Chinese government policies are promoting investments in the coal sector of 279 billion RMB or US$34 billion between 2000 and 2010 and another 200 billion RMB or US$24 billion between 2010 and 2020. The targeted investment goals also approximate around US$500 billion and US$700 billion for power generation, principally from coal-fired generation. Since 2002, the Chinese government has developed liquefied natural gas programs, primarily due to rising demand and supply shortages. Thus, according to the ChinaBidding reports of July and August 2006, private Chinese corporations are moving aggressively into more advanced technological research and development. For instance, the International Energy Administration (2006) reported on peak-load capacity and growing concern about pollution from coal firing. Private firms in China have started to use cleaner burning fuel alternatives such as LNG to generate electricity. Reports say that electric power demand rose by about 15 per cent in 2004 and that China experienced a 30-gigawatt nationwide power deficit that year within its 24 provincial grids. Demand on the domestic grids initiated a rising fuel oil demand estimated at 170,000 barrels per day to generate electricity.

The International Energy Administration (2006) forecasts a power shortfall which will become increasingly significant, principally due to the emerging middle class's demand for fuel oil in the power sector. The Chinese government is trying to keep up with this rise by creating new LNG projects while also accelerating imports, from 1 million metric tons in 2006 to between 20.9 and 25.9 million metric tons in 2015. The Chinese government started construction of LNG regasification stations in Guangdong Province that began operations in June 2006, importing initially from Northern West Shelf Australian LNG. The city of Xiuyu in Fujian Province is also part of the 2008 enlargement plans to facilitate growing demands and imports. In parallel to these value added infrastructures constructed from 2005 to 2010, the Chinese government is reviewing more than 10 additional LNG infrastructures along the coastal zones of China to balance import delivery throughout the whole country, to reduce transportation costs and avoid concentrated market inefficiencies surrounding urban centers. According to the Energy Information Administration (2004–2006), the Chinese government plans to build additional LNG terminals on China's northeastern coast, close to Russia, to facilitate imports of natural gas via pipelines from Irkutsk. These gargantuan infrastructural engineering projects aim to regulate international energy pricing more efficiently and to alleviate global market pricing deficiencies and significant economic gaps between regional market participants. The intensification of new infrastructural pipelines, ports and liquefication technologies are part of the globalization smoothing framework to equalize pricing within regional zones and to facilitate the raising of new standards.

Some of the additional infrastructure projects came about as a result of the lack of progress in negotiations between China and Russia with regard to Russian natural gas pricing being indexed to domestic Chinese coal prices. The way to smooth such markets is to introduce derivatives and to introduce an intermediary market to convert one into the other; thus providing a bridge between two very different markets in pricing and infrastructural engineering abilities. This facilitating idea started to appear in late 2005 and is estimated to take some time to develop over a large scale to bring smoothing capacities within those regional zones. LNG pricing in Asian contracts traditionally has been linked to the price of oil, not coal. Recently, China has indexed LNG and natural gas pipeline imports to the price of domestic coal.

Additional LNG projects enhance the industrial and manufacturing developments in surrounding areas. Guangdong and Fujiang projects in Shanghai, Ningbao Zhejiang Province and Qingdao Province participate in the economic improvements of the neighboring regions. These regions of economic growth and rising industrial output are the most substantial in capacities. For instance, Shanghai is both China's largest city and port. As the most future-oriented city in the world, Shanghai provides an example of leadership within its region. Around 2005, Shanghai consumed more than 46 million short tons of coal per year, fulfilling 70 per cent of energy demand. Shandong's soil proves to have substantial natural resources, including coal.

The smoothing and regulation of Chinese energy prices within its own grid has proved a challenge since 2002. As a possible solution to regulate this market between energy alternatives or renewables, the Chinese Energy State Council for LNG developments has large coal-fired power plants, and the government is carefully considering the issue of natural gas and coal prices charged to the power plants. According to the International Energy Administration (2004) and the World Energy Council's Survey of Energy Sources (2004), the cost of coal from Northern China was approximately \$1.92 per million Btu. Statistical reports confirm that the average cost of natural gas from China's west-to-east pipeline is around \$4.22 per million Btu. Chinese coal-fired electricity generation costs on average \$34 per MWh, and Chinese natural gas fired generation \$44 per MWh. Potential users of natural gas in the electric power sector estimate that prices for natural gas need to be in the range between \$3.30 to \$3.60 per million Btu to be able to be feasibly and economically competitive with coal as a similar substitute. As an example, the expected price of natural gas from the Guangdong LNG terminal is \$2.80 per million Btu, with transportation costs, and an additional \$0.40 per million Btu for liquefaction transformation to gas. The economics and pricing of the coal market in China have been the subject of numerous studies since 2004. SSY Consultancy and Research Limited

(2005) released a survey of future trade pricings in 'SSY's Coal Trade Forecast' and Meadows (2005) provides additional information about coal markets in 'Coal Plants Toast' in the *Chronicle Journal*.

The pricing of the Chinese coal market is relatively smooth domestically, especially throughout the eastern and southeastern regions, which have greater coal resources. These areas are economically productive as they account for half China's total GDP and their coal resources contribute to about 17 per cent of total energy resources. Consequently, coal transportation from those regions to the central, western and northern outreach may average 340 miles. The Chinese transportation network is being remodeled and archaic rail infrastructures, potentially hazardous to employees, are being upgraded. Transportation pricing is embedded in the cost of coal energy depending on point to point transmission grids. As early as 1992, the Chinese pricing system was opened to similar international competing markets with bid and ask prices and negotiable spreads to define market efficiencies and liquidity. Chinese pricing markets are composed of variables such as state-run producers, medium-sized wholesalers' intermediaries, large manufacturing and industrial entities and foreign investors and importers. The 2005 statistical data reports that domestic coal prices were around $5 to $7 per short ton higher than the international market price of high quality steam coal (Figure 10.21 represents a coal pricing function by region). This added embedded high transportation and maintenance costs in Chinese domestic prices creates a new opportunistic market of interest to foreign investors and Chinese importers.

Chinese prices are also subject to an added premium from pollution emission units imposed by international standards. Companies running on archaic infrastructures and not modernized according to international standards are subject to penalties if considered too polluting. Some companies may purchase pollution emission units to hedge their coal energy production and not be subject to such regulations. According to the State Environmental Protection Administration (SEPA), new international pollution regulation standards require companies to pay a tax or a fee if they pollute and generate sulfur dioxide emissions. SEPA multiplied the fee tenfold in October 2003 for emissions of sulfur dioxide and nitrogen oxide. Additionally, Chinese coal-fired plants that use inadequate or inefficient archaic polluting infrastructures are subject to a governmental due diligence program to evaluate the adequacy of continuing performances and market efficiencies. Depending on the age and condition of the infrastructural machinery, the government audits approve continued usage.

Since 2003, Chinese financial and commodities markets have been more closely involved with the reforms of market pricing efficiencies, particularly with direct neighbors who have not shared the Chinese experience. Chinese market officials are rapidly reforming internal infrastructures in order to smooth domestic pricing with that of linked and correlated regional

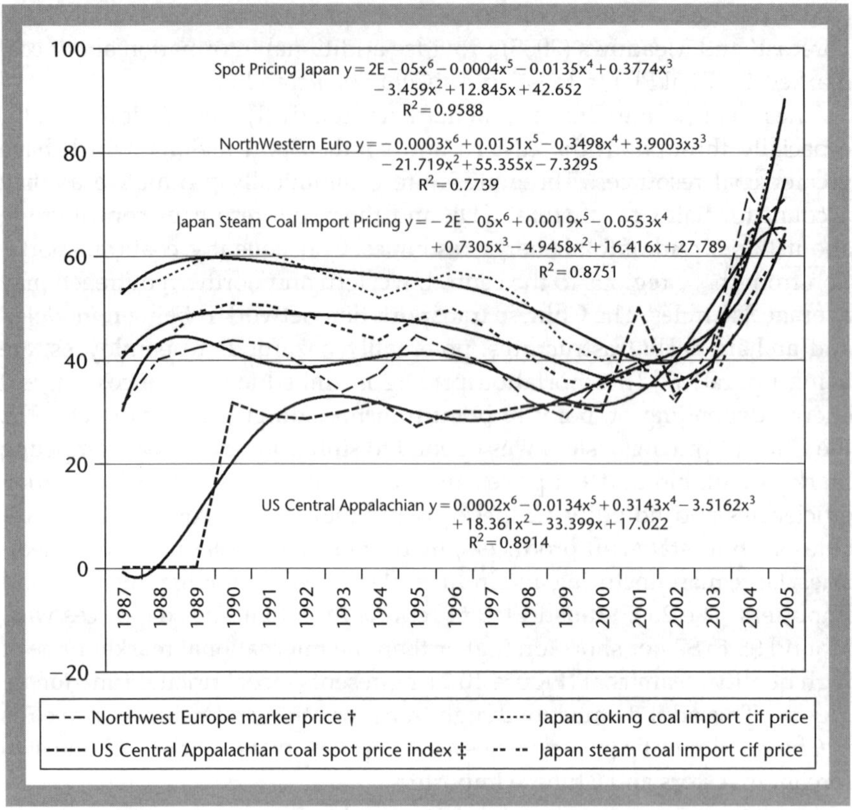

Figure 10.21 Coal prices' pricing functions by regions, 1978–2005 ($/tonne)
Notes: †: From McCloskey Coal Information Services; ‡: price is for CAPP 12,500 Btu, 1.2 SO2 coal, source Platts; cif : cost + insurance + freight (average prices).
Sources: BP (2006); McCloskey Coal Information Services.

competitors. Thus, the Chinese Energy State Council aims to index its import contracts for LNG and natural gas pipelines to bring them in line with domestic coal prices. This is also aimed to smooth the import trading grid on a global scale. Very few commodity-rich countries use transformations of one energy into another and they remain primarily reliant on unique original energies, not least because they are easily exploitable. However, with the rise of global competition for inevitably limited and scarce natural resources, the transformation from gas to liquid (GTL) becomes a more attractive alternative and a means by which to leverage limited natural gas reserves. It is worth noting that as yet only South Africa and Malaysia have commercial GTL operations, although since 2003 similar energy options and derivatives have been planned in other countries rich in energy resources, including Algeria, Australia, Egypt, Iran, Nigeria, and Qatar. According to the International Energy Administration (2006), plant sizes range from 20,000 to 160,000 barrels per day of liquid output. Russia also has significant potential for GTL production as it is highly rich in natural gas resources.

The Chinese government implemented a framework for increasing production of alternative energies to compensate and respond to the rising demands of the new middle class. In response to the exponential rising demand for crude oil, substitute and alternative energy products are also being developed. A professor at Xiamen University in East China's Fujian Province reported to ChinaBidding news that substitute products, such as coal and renewable energies offer ways to avoid full dependence on crude oil and vulnerability to rising global prices. China started to engineer new technologies, such as coal-to-liquid fuel processes, twenty year ago. Zhang Yuzhuo, manager at Shenhua Corporation, China's largest coal producer, revealed that coal liquefaction became more important and was heavily utilized when China became a net oil importer in 1993. The Chinese government created its first coal-to-liquid operation in Pingdingshan, Central China's Henan Province, in 1999. The 500,000-ton annual capacity project closed because the coal grade was not appropriate for the liquefaction process. The 863 program coal-to-liquid fuel project started in 2001. Shenhua Group was the first enterprise to initiate its production and in August 2004, the group started the implementation of a direct coal liquefaction program in Ordos in North China's Inner Mongolia Autonomous Region. The annual capacity is approximately five million tons and costs 24.5 billion yuan or US$3 billion. The first part of the project aims to produce 3.2 million tons of oil derivatives and is targeted to be on production by 2007 while the second phase aims for production by 2010 with an annual capacity of 2.8 million tons.

To help compensate for the rising demand for crude oil, all energy producers are investing into more coal and gas productions. In February 2006, the Lu'an Group initiated another coal liquefaction project with annual capacity of 160,000 tons in Tunliu, Shanxi Province. In April 2006, Yankuang Group also started a huge two-phase coal liquefaction project, costing 100 billion yuan, in Yulin in Northwest China's Shaanxi Province. This investment is expected to produce a total annual output of 10 million tons of oil products by 2020. All projects require NDRC approval, and many other smaller groups and companies have started similar enterprises scattered throughout the country to help meet rising demand. A total of approximately 30 coal liquefaction projects were being considered in the first quarter of 2006 to help meet this rising oil demand and level off global prices. The total capacity of those new seeded investments is estimated to surpass 16 million tons for a total of 120 billion yuan or US$15 billion. China's forecast annual oil output liquefied from coal is to reach 50 million tons by 2020.

Foreign firms are also invested in China. Shell Gas and Power Developments and the Shenhua Ningxia Coal Industry Company Shenhua-Ningmei signed a partnership agreement to study coal liquefaction technology on 11 July 2006 in Yinchuan, capital of Northwest China's Ningxia Hui Autonomous Region. This partnership aims to have a daily production capacity of 70,000 barrels of oil and chemical production. Chinese oil companies also proliferated global partnerships especially in yet unexploited areas such as South America and Africa. In June 2006, South Africa-based Sasol, the global leader in production of fuel using coal, partnered with Shenhua Group to set up two coal-to-liquid plants using a unique Fischer-Tropsch technology in Northwest China. Part of the partnership agreement is to initiate a production center of 80,000 barrels a day in Shaanxi Province, and the other part is to produce 80,000 barrels a day coal-to-liquid project in the Ningxia Hui Autonomous Region. Both plants require an estimated US$5 billion and are targeted to operate by 2012. According to a June 2006 Japanese newspaper *Nihon Keizai Shimbun*, Japan assists China and neighboring Asian nations with the technology to liquefy coal to alleviate rising regional demand for oil and consequent rising prices.

The transformation of coal and oil to produce oil derivatives through liquefaction also creates environmental and ecological challenges. It takes about 3 to 5 years to produce a liquefaction project to meet oil demand, and the financing is linked to yearly international fluctuating oil prices. It is an expensive procedure and inadequate for substantial demand. The different qualitative grades also vary: for instance exploitable coal reserves were 188.6 billion tons at the end of 2002, but the average resource recovery rate is only 30 per cent. The reserves are estimated to last only 30 years if annual coal output is around 1.9 billion tons. According to the NDRC, the coal liquefaction industry will be developed in the five-year period from 2006 to 2010, or the 11th Five-Year Development Program period for China. China's largest corporations are also extending their capabilities to secure future demand and to process and transmute natural gas into alternative energies. According to Chinese media reports, China's three oil companies, China National Petroleum Corporation (CNPC), China Petrochemical Corporation (Sinopec) and CNOOC, have all developed an LNG development plan to import at least 60 million tons by 2020 (Table 10.23 shows proposed imports from 2005–15).

In addition to this increased production, Hong Kong's sole gas supplier, the Hong Kong and China Gas Company, invested HK$780 million or US$100 million in a coal gasification joint venture in North China. In May 2006, the company planned to double the mainland investment to HK$12 billion (US$1.56 billion) until 2009. With a total investment of 6 billion yuan (US$732 million), Towngas currently operates 33 urban projects for piped-gas joint ventures and four mid-stream projects. Gas distribution on the mainland contributed HK$240 million (US$31million) to 2005 profits, compared with

Table 10.23 China's proposed LNG Imports, 2005–15 (million metric tons per year)

Year	Haikou LNG (Hainan)	Shenzhen LNG (Guangdong)	Xiuyu LNG (Fujian)	Ningbao LNG (Zhejian)	Shanghai LNG	Rudong LNG (Jiangsu)	Qingdao LNG (Shandong)	Hebei/ Tianjin LNG	Dalian LNG (Liaoning)	Beihai LNG (Guangxi)
2005	–	–	–	–	–	–	–	–	–	–
2006	–	3.9	–	–	–	–	–	–	–	–
2008	–	3.9	2.6	–	–	–	3.0	–	–	–
2010	1.0	9.9	2.6	4.0	4.0	3.0	5.0	–	–	–
2012	2.0	12.9	2.6	4.0	4.0	3.0	5.0	2.0	3.0	–
2015	3.0	12.9	5.0	10.0	10.0	5.0	5.0	3.0	6.0	3.0

Sources: *International Gas Report*, No. 522 (22 April 2005); Wu and Fesharaki (2005); FACTS Inc., *China Oil & Gas Monthly* (May 2005).

HK$132 million (US$17 million) in 2004. Towngas is adopting the same business model as electricity supplier and piped gas distributor XinAo Gas. Those companies trade shares in Hong Kong and hold majority interests in three coal-mining companies in Inner Mongolia, Yunnan Province and Heilongjiang Province to enhance coal gasification projects. These projects are to improve air cleaning, to reduce the impact of fuels on the environment, and to protect against pollution.

10.10 ADVANCED TECHNOLOGIES

This section is intended to provide a summary of new advanced technologies that enable environmental protection and economic energy consumption efficiencies. China is currently focusing on engineering, technologies and sciences to sell to international partners (Middle East) in exchange for natural resources such as oil or other forms of energy. The following information was provided by the Chinese Department of Energy and is also publicly available. While many manufacturing corporations rapidly depreciated as a result of archaic architectures, which is the case mostly in the mining and coal industry, the Chinese government and corporations have invested in new technologies to enhance economic efficiencies, production, and global competitivity. Table 10.24 shows the various types of technologies under development, and the United States Energy Information Administration (2006) provides cost comparisons for coal, nuclear, natural gas and wind power (Table 10.25).

Various advanced scientific methods are being used to improve economic efficiencies in energy production. For instance, the ten-year Chinese government strategic guidelines encourage the use of innovative technologies such as microbe oil technology. According to the Xinhua News Agency of 29 June 2004, a new microbe oil displacement technology was discovered in the Shengli Oilfield in East China's Shandong Province. This new technology increased production in the 40-year-old oilfield. The new technology consists of pumping air into oil reserves to mix microbes. The water concentration falls in the oil reserves and enables oil recovery. The air facilitates the microbe oil displacement technology. This creates daily output savings of 25 per cent. The oil water extraction does not pollute the atmosphere and it is an economically efficient solution. Shengli oilfield is 40 years old and has a daily average output of 520,000 barrels. This new technology will become more widely used and is intended to improve and increase oil recovery efficiency. China has not yet reached its economic apotheosis. A net importer of oil since 1993, in 2003, China imported almost 100 million tons of crude oil and consumed about 34 per cent of the world availability.

Alongside new advanced technologies, petroleum conservation and substitution technologies are also being developed. They focus on clean coal,

Table 10.24 Phased assessment of China's renewable energy technology development

Types of Technologies	Maturity and development phase			
	R&D	Demonstration	Early commercialization	Commercialized
Small hydropower				*
Solar water heater				*
Passive solar house				*
Solar stove				*
Solar dryer		*		
Solar cell			*	
Grid-connected wind turbines			*	
Small and mini wind turbines			*	
Geothermal power generation				*
Geothermal heating				*
Traditional bioenergy technology				*
Small methane tank				*
Large & medium methane tech			*	
Municipal organic waste power generation		*		
Biomass gasification		*		
Other modern bio energy techniques	*			
Wave power generation	*			
Tidal power generation			*	
Ocean thermal energy conversion	*			
New hydrogen manufacturing tech	*			
Hydrogen storage techniques	*			

Source: Chinese Energy Ministry/Department.

Table 10.25 Levelized cost comparison for new generating capacity in the United States (2004 dollars per MWh)

Cost Element	Technology			
	Coal	Natural gas	Wind	Nuclear
Capital	30.4	11.4	40.7	42.7
O&M	4.7	1.4	8.3	7.8
Fuel	14.5	36.9	0.0	6.6
Total[a]	53.1	52.5	55.8	59.3

Notes: a. Includes transmission hookup costs; O&M = operations and maintenance.
Source: Energy Information Administration (2006).

natural gas substitution, fuel oil technologies, methanol and ethanol substitution technologies, and oil saving technologies such as process energy optimization, plasma oil-free ignition, fuel emulsification, and fuel additives.

Clean coal technologies focus on the development of large-scale and advanced coal washing, dressing and processing technologies, coal liquefying technologies, large-scale coal gasification technologies, integrated water gas pulp preparation and application technologies, more than 410 ton/hour large-scale circulating fluidized bed technologies, integrated gasification combined circulating (IGCC) power generation technologies, and high-efficiency/low-pollution coal-fired power generation technologies.

Electricity saving technologies favor the development of high efficiency motors, high-voltage and high-power frequency conversion speed regulation technologies, high-efficiency electro optical source and ballast technologies, S9-and-above transformer and amorphous alloy iron-core transformer technologies, cold and thermal storage technologies, and energy saving via efficient domestic appliances, electrolytic and electroplated power supply, transmission and transformer substation network systems and industrial electrical furnaces.

Multi-gang supply technologies deal with the development of cogeneration, central heating and thermal energy ladder-type utilization technologies.

Residue-heat and top pressure recovery technologies also participate in the development and popularization of 75-ton/hour and above grade dry coke quenching technologies, high-capacity full blast furnace gas power generation technologies, 1000 cubic meters and above blast furnace top differential pressure power generation technologies, blast furnace hot-blast stove residue-heat recovery technologies, converter gas recovery technologies, oilfield blow-off concentration and recovery technologies, and large-scale residue-heat boiler and advanced heat exchange technologies.

Building energy conservation technologies also focus on the development of sealing strips for doors and windows, multi-layer airing windows, exterior thermal insulating composite wall construction, heat calculation against the household, technologies such as temperature control, heat supply pipelines and network regulation and control, heat reflecting thermal insulation and heat protection, solar-energy buildings, high-efficiency lighting systems, and computer simulation.

'Three wastes' comprehensive utilization technologies also focus on the development of high-capacity gangue power generation technologies, hollow brick production technologies through full gangue semi-stiff extrusion process, technologies to produce cement raw material with gangue instead of clay, road construction and backfill technologies, high value-added pulverized coal ash comprehensive utilization technologies, organic waste water comprehensive utilization technologies, smelting waste liquor recycling technologies, ground-level crude oil, sump oil and mud recycling technologies, and petrochemical waste gas recovery technologies.

Paragenous and associated mineral resources comprehensive utilization technologies also focus on the development of high value-added utilization and fine and in-depth processing technologies for coal-bed gas, coal series paragenous and associated minerals, schreyerite magnetite, and metals.

Renewable resources recovery and utilization technologies also deal with the development of practical recovery, sorting and processing technologies for renewable resources such as waste rubber, waste plastics, waste domestic appliances, waste computers and waste batteries.

Oil conservation and substitution projects consist of replacing fuel oil consumption industries with clean coal and natural gas as key contents such as water gas pulp, coal gasification, fuel oil conservation projects with energy optimization by using substitutes such as plasma oil-free, methanol and ethanol.

Clean coal projects also include thermal coal optimization and processing technologies such as washing, dressing, shaping, blending, screening and grinding. The process uses small and medium-sized coal-fired industrial boilers with fine fire coal, screened lump coal, fixed sulfur shaped coal, and advanced technologies such as circulating fluidized bed, and pulverized coal firing as key contents with large-sized circulating fluidized bed boilers. The full-process consists of liquid sulfur dioxide extraction with pollution control technologies such as desulfurization while washing and dressing. The Chinese energy department promotes the industrialization of clean coal technologies.

Electric motor speed regulation projects include high-voltage and high-power frequency conversion with the manufacturing of components and electro-technical frequency converter. This includes low-voltage, small and medium-sized power frequency conversion speed regulation, with structure-simplified special frequency converters and fan and pump system optimization demonstration projects with the combination of optimized hardware and software for high-efficiency electric motors, drive fans, and pumps. This process increases the efficiency of electric motor systems by 10 to 12 per cent.

Green lighting projects include the certified high-efficiency and energy saving lighting system for the Olympic Games to be held in Beijing in 2008. The green lighting projects increase the market share of fine-quality lighting electric appliances.

Coal-bed gas projects select regions that are rich in the exploration of coal-bed gas to establish two to three projects to enhance the development and utilization of coal-bed gas. This improves the gas utilization ratio and promotes the industrialization of coal-bed gas exploration, development, production and utilization.

Mine gangue utilization projects also construct high-capacity power generation projects to produce new-type building materials for wall and mine re-cultivation projects in key mines.

Mineral resources utilization projects include technologies involved in the ultrafining, brightening and modifying of kaolin. This helps organize a number of projects for the utilization and in-depth processing of bauxite, fire-clay, pyrite, alta-mud and diatomaceous earth. It develops in-depth processing and utilization of coal series paragenous and associated resources. Three regions have been selected to utilize mineral resources – Panzhihua, Baotou and Jinchuan – to extract the minerals schreyerite and magnetite and non-ferrous metals.

Energy-saving and clean enterprise projects select several enterprises in key industries to strengthen administration and create an energy efficient

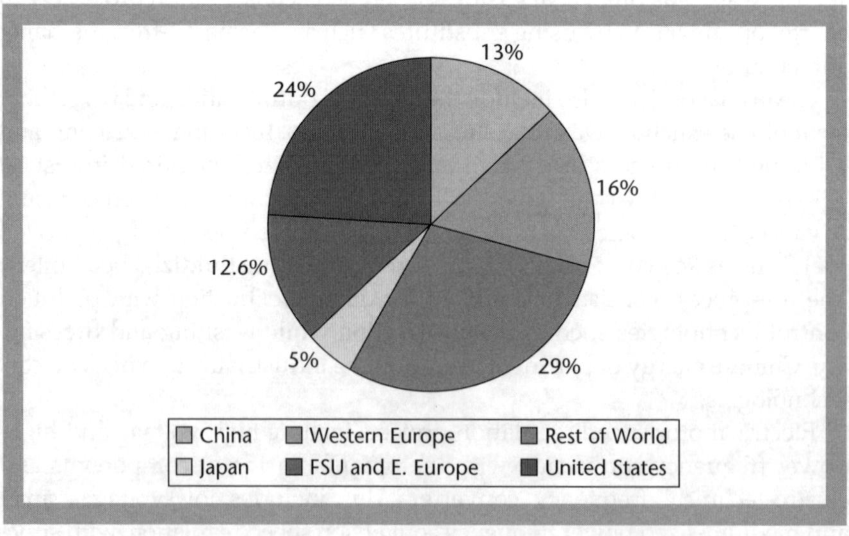

Figure 10.22 Energy-related carbon emissions, 2001

system to reach global competitive standards. Market-oriented energy-saving technical projects are favored in the 9th Five-Year Plan.

With rising consumption come increasing pollution emissions. While Westerners have been the main consumers of energies so far (see Figure 10.22), they are only just now starting to become more environmentally conscious. China is going to become a main consumer and is already re-engineering its energy infrastructures to be less polluting. According to *The Economist* (2007), China and Australia are unveiling new policies with respect to global warming. Makai, the minister who is responsible for energy management and global warming reforms, reports that China is committed to generating 10 per cent of its electricity from renewables by 2010. Much of that will come from hydroelectric power. China also aims to deploy greater use of nuclear power with four new nuclear reactors, as it is more efficient and cleaner. Another measure to reduce pollution is to impose the use of smaller more efficient cars by building vehicles which will travel 100 kilometers on 6.5 liters of fuel or 34 miles per gallon, compared with an average of 9.8 liters per gallon in the United States. Moreover, China intends to reduce carbon dioxide emissions by 330 million tonnes per year. Reforestation has absorbed about 5 billion tonnes of carbon dioxide over the past 25 years. China aims to reduce the ratio of energy consumption to economic output by 20 per cent between 2005 and 2010.

Conclusion

China has made tremendous advances in the construction of infrastructures to enable efficient energy production and emissions reductions, especially since the beginning of this century. Despite the fact that Chinese energy policies are new – initiated only around 50 years ago – the Chinese government has made extensive progress in diversifying into new products, including renewables, multiplying geopolitical partnerships, and developing new performing technologies, especially since 2003. It has also developed substantial strategic alliances to secure its energy reserves and to better manage its own domestic pricing policies. Among those alliances, China counts commercial and corporate partnerships in Africa, South America, the Middle East, Russia and neighboring Asian countries. While its Western counterparts remain heavily dependent on Middle Eastern producers, China is becoming increasingly involved in poorer African and South American countries that have lacked transparencies and governance. Yet, despite these sudden and recent efforts in securing its own strategic energy reserves, producing alternative energies, and inventing still-costly technologies, the effect of globalization on the Chinese population is increasing their energy dependence. At present, China – with about 1.5 billion individuals – is the second largest consumer of oil after the United States, which has a population of only 300 million. As China continues to develop, car ownership – perhaps even two or three cars per household – will become increasingly commonplace and energy needs will become immense.

Since 1993, China has been an exponential importer of oil and its utilization will become increasingly expensive as global oil supplies decrease by 2030. Additionally, the integration of Chinese financial institutions and the introduction of energy derivatives in global energy markets will also have an adverse impact on the pricing of energies. This impact is likely to be an upward inflationary pressure. The tangible collateral assets supporting these upward trending valuations are likely to be physical commodities and energies. Many more private investments and trading opportunities will take place with the introduction of financial engineering and speculative derivatives. Currently, many domestic and international private equity deals and mergers and acquisitions are taking place with Chinese energy corporations in order to profit from lower valuations and the unappreciated foreign exchange rate. Obviously, the rise of energy prices in China will not occur overnight, but it is expected to happen within the next fifty years as Chinese living standards rise. And as the Chinese adopt a Westernized, high-energy consumption life style, so we will see an as yet unquantified impact of the Chinese cultural and economic influence in the global economies.

But in the global competitive race to acquire market shares and achieve economic efficiencies, China remains behind with respect to the energy and commodities sectors. On the one hand, it has remained isolated from financial engineering and speculative derivative trading schemes, but on the other hand, it has benefited from a simplistic and clear view of its energy assets and collateral goods. The absence of a global regulator to monitor corporate governance, transparency, polluting emissions and energy pricing structures will ensure an increasing engagement of Chinese energy corporations with Western style trading entities.

APPENDIX 1

Legal and Regulatory Measures

The Regulatory Assistance Project (2000) recommended the following legal and regulatory measures to deal with the Chinese power system:

- Create a strong central regulatory body with a broad scope charged with implementing reforms to meet China's goals and constraints, and having direct involvement in the restructuring process.
- The regulatory structure should be designed to minimize the possibility of future conflicts between central, regional, and provincial levels of government.
- The scope of the regulatory agency should be broad. It should include oversight of competitive generation markets; anti-monopoly authority; distribution and transmission prices; access; service quality, reliability, and resource planning for captive customers; and environmental performance.
- There are substantial opportunities to reduce the cost of existing generation with or without more extensive restructuring. Dispatch rules and the approach to contracting and privatization provide the main opportunities to reduce costs and prices in the near term.
- Demand response by consumers, distribution companies, and energy service companies should be built directly into the structure of wholesale markets.
- Include some sort of capacity feature in the electricity markets. This is needed to send early price signals for investors to reduce price volatility.
- Regulate transmission and distribution utilities in a way that encourages end-use energy efficiency as well as improvement and expansion of the transmission and distribution system. There are two basic options: price caps and revenue caps. Price caps promote increased electricity sales and discourage utility investment in end-use energy efficiency. Revenue caps encourage cost reductions without giving the incentive to increase sales.

- Incorporate environmental and economic goals in the restructured markets. The study noted: 'Initially, electricity markets in China will be regional in nature and some of these markets may be relatively small. Over time, as the system grows and as transmission expands, markets will combine to form fewer and fewer markets with the distinct possibility that a single national market may ultimately develop. Structuring the regulatory institutions as well as the transmission institutions in ways that minimize the possibility of provincial conflicts between different regions will make the transition to larger and larger markets easier to accomplish.'
- Coordinate with the other closely related laws in China, including the energy conservation law and clean production law.

APPENDIX 2

Chinese Energy Indices

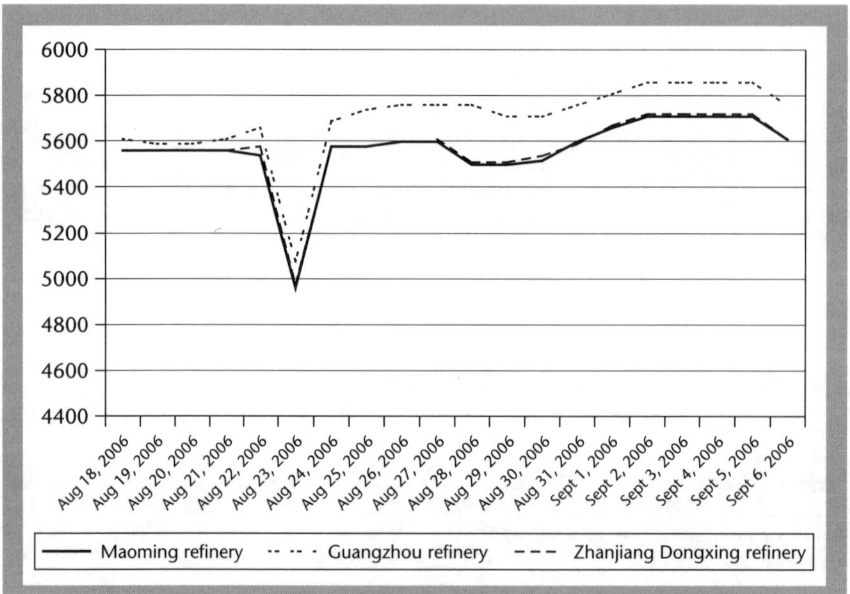

Figure A2.1 South China market refineries Aug–Sept 2006
Source: www.oilgas.com.cn.

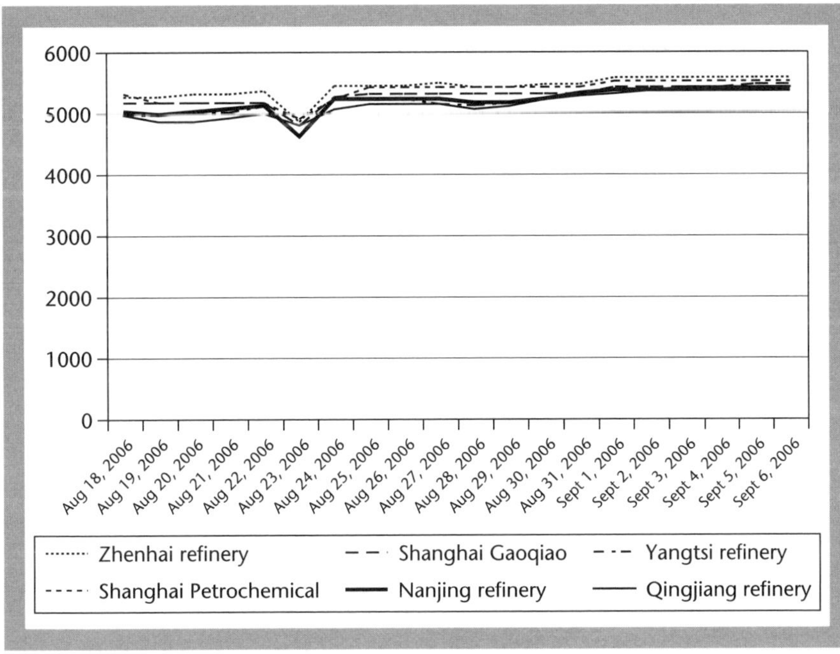

Figure A2.2 East China market refineries Aug–Sept 2006
Source: www.oilgas.com.cn.

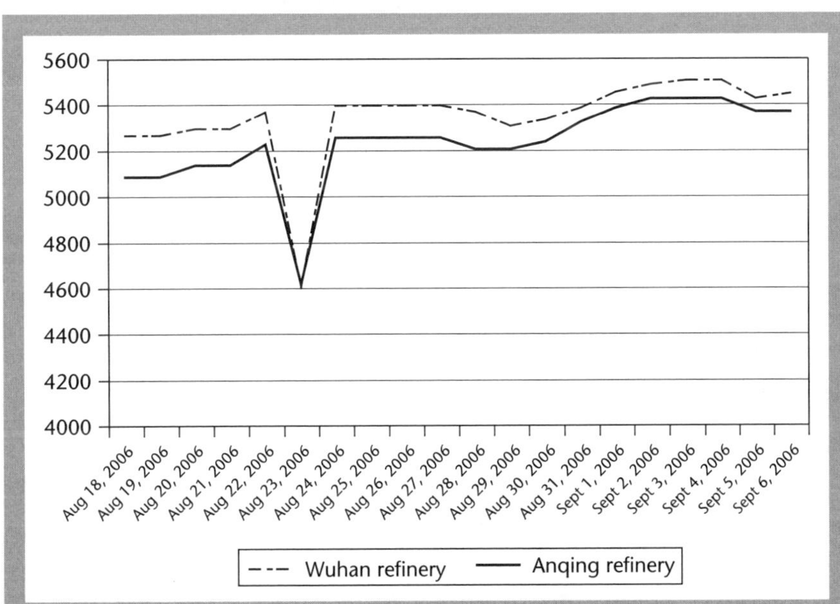

Figure A2.3 Mid and upstream of Yangtse market refineries' pricing index, Aug–Sept 2006
Source: www.oilgas.com.cn.

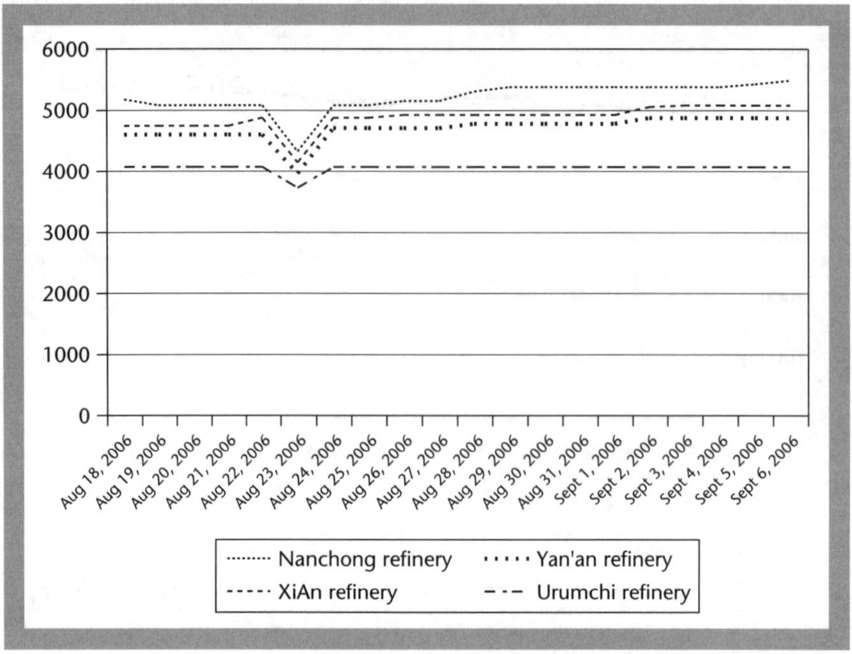

Figure A2.4 Southwest & northwest China market refineries' pricing Aug–Sept 2006
Source: www.oilgas.com.cn.

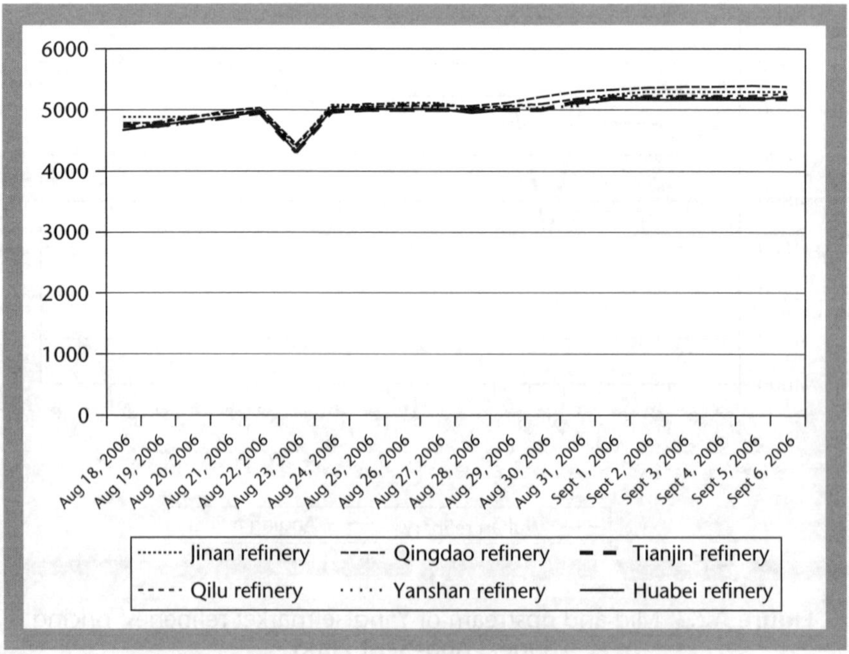

Figure A2.5 Northeastern China market refineries Aug–Sept 2006
Source: www.oilgas.com.cn.

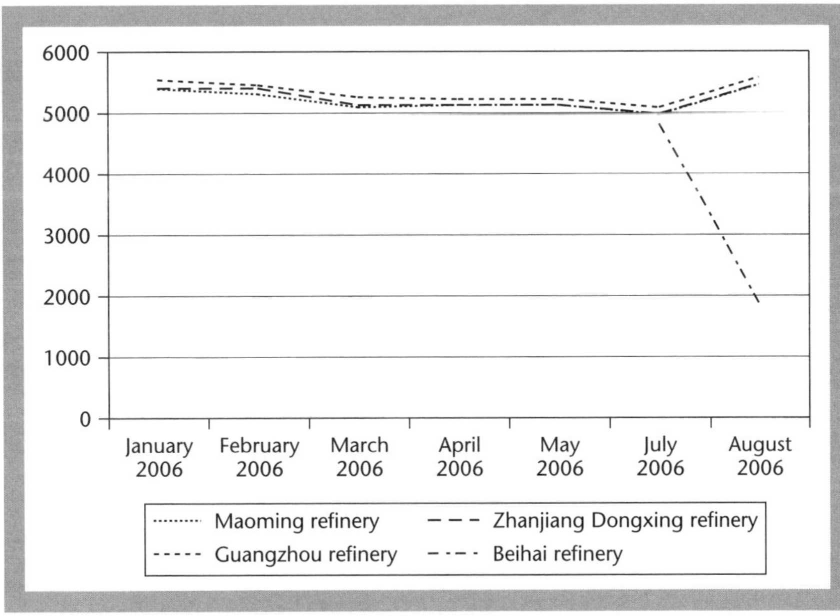

Figure A2.6 South China refinery pricing indices 2006
Source: www.oilgas.com.cn.

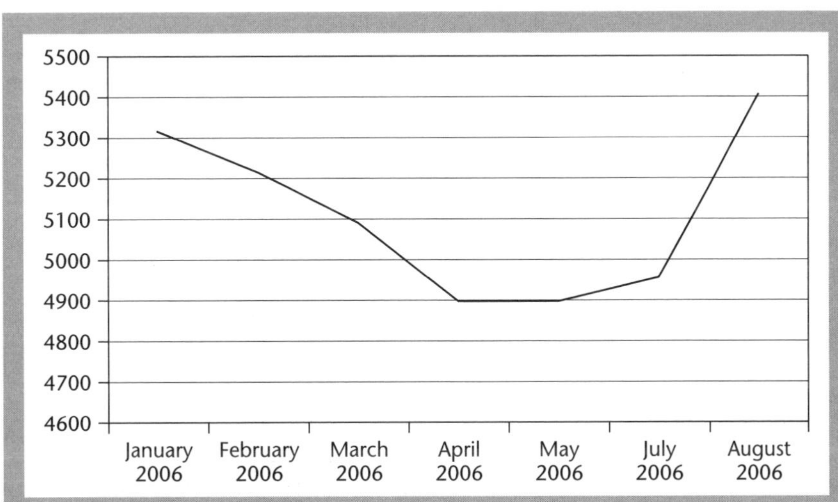

Figure A2.7 Fujian refinery pricing index 2006
Source: www.oilgas.com.cn.

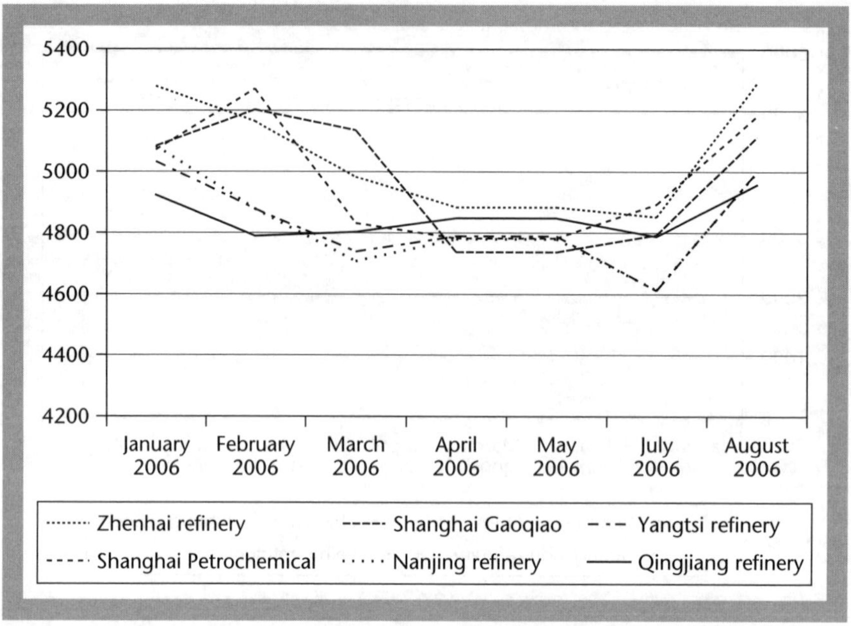

Figure A2.8 Eastern Chinese refineries' pricing index 2006
Source: www.oilgas.com.cn.

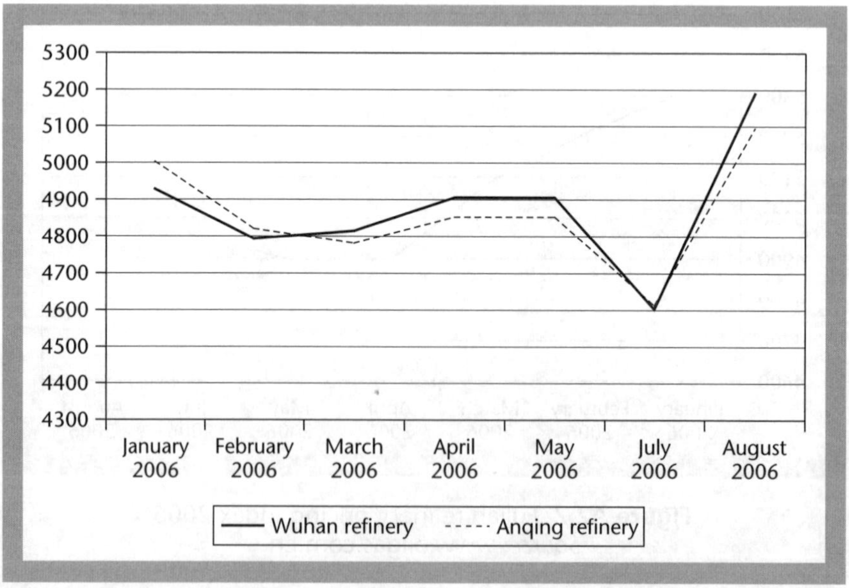

Figure A2.9 Mid and upstream Chinese refineries pricing index 2006
Source: www.oilgas.com.cn.

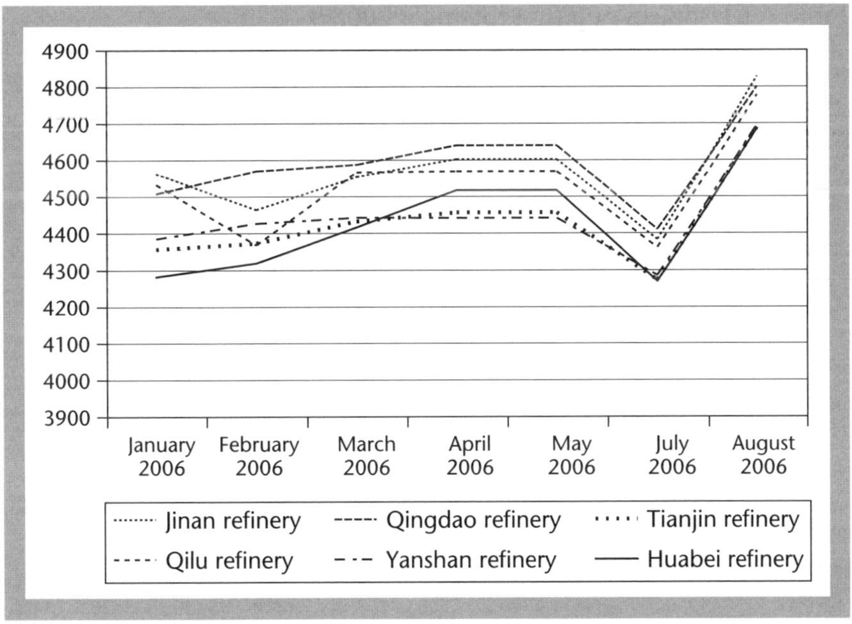

Figure A2.10 North China refineries' pricing indices 2006
Source: www.oilgas.com.cn.

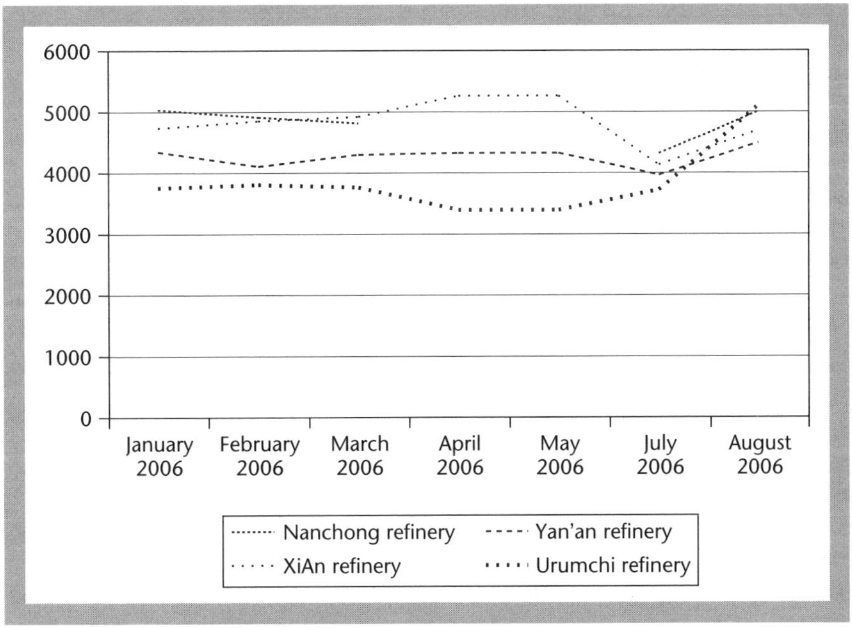

Figure A2.11 Southwestern & northwestern refineries' pricing indices 2006
Source: www.oilgas.com.cn.

Table A2.1 Indices August–September 2006

Refinery	Aug 18 2006	Aug 19 2006	Aug 20 2006	Aug 21 2006	Aug 22 2006	Aug 23 2006	Aug 24 2006	Aug 25 2006	Aug 26 2006
South China market									
Maoming refinery	5550	5550	5550	5550	5530	4952	5570	5570	5590
Guangzhou refinery	5600	5580	5580	5600	5650	5065	5680	5730	5750
Zhanjiang Dongxing refinery	5550			5550	5570	4965	5570	5570	
Beihai refinery						4800			
Fujian market									
Fujian refinery	5500	5400	5400	5400	5450	4953	5550	5550	5550
East China market									
Zhenhai refinery	5250	5250	5300	5300	5350	4849	5430	5430	5430
Shanghai Petrochemical	5300	5150	5150	5150	5150	4890	5210	5400	5400
Shanghai Gaoqiao	5150	5150	5150	5150	5150	4790	5250	5300	5300
Nanjing refinery	5010	4960	5010	5060	5110	4610	5210	5210	5210
Yangtsi refinery	4950	4950	5000	5000	5100	4609	5210	5210	5210
Qingjiang refinery	4950	4850	4850	4900	4980	4785	5050	5130	5130
Mid & upstream of Yangtse River									
Wuhan refinery	5260	5260	5290	5290	5360	4599	5390	5390	5390
Anqing refinery	5080	5080	5130	5130	5220	4610	5250	5250	5250
North China market									
Jinan refinery	4850	4850	4850	4900	4950	4381	5050	5050	5080
Qilu refinery	4720	4770	4870	4900	4950	4359	5000	5060	5060
Qingdao refinery	4750	4750	4840	4940	5000	4409	5010	5010	5030
Yanshan refinery	4680	4690	4750	4850	4950	4284	4970	4990	5000
Tianjin refinery	4700	4700	4770	4830	4910	4271	4930	4970	4970

Aug 27 2006	Aug 28 2006	Aug 29 2006	Aug 30 2006	Aug 31 2006	Sept 1 2006	Sept 2 2006	Sept 3 2006	Sept 4 2006	Sept 5 2006	Sept 6 2006
5590	5490	5490	5510	5590	5650	5700	5700	5700	5700	5600
5750	5750	5700	5700	5750	5800	5850	5850	5850	5850	5750
5600	5500	5500	5530	5580	5660	5710	5710	5710	5710	5600
			5550							
5550	5550		5550	5550	5650	5650	5650	5650	5550	5550
5480	5400	5400	5450	5450	5560	5560	5560	5560	5560	5560
5400	5400	5400	5400	5400	5500	5500	5500	5500	5500	5500
5300	5300	5300	5300	5300	5400	5400	5400	5400	5450	5450
5210	5160	5160	5210	5310	5360	5360	5360	5360	5360	5360
5130	5110	5140	5210	5310	5360	5360	5360	5360	5360	5360
5130	5050	5100	5200	5260	5300	5350	5400	5400	5400	5400
5390	5360	5300	5330	5380	5450	5480	5500	5500	5420	5440
5250	5200	5200	5230	5320	5380	5420	5420	5420	5360	5360
5080	5000	5020	5070	5150	5210	5260	5260	5260	5260	5260
5060	5000	5030	5070	5150	5200	5200	5200	5200	5200	5220
5030	5030	5080	5180	5260	5300	5320	5340	5340	5360	5340
5000	4950	4970	5000	5030	5150	5180	5180	5180	5150	5180
4970	4970	4970	4970	5100	5150	5150	5150	5150	5150	5150

(*Continued*)

Table A2.1 (Continued)

Refinery	Aug 18 2006	Aug 19 2006	Aug 20 2006	Aug 21 2006	Aug 22 2006	Aug 23 2006	Aug 24 2006	Aug 25 2006	Aug 26 2006
Huabei refinery	4630	4700	4780	4850	4930	4267	4960	4980	4980
Northeast China market									
Daqing Petrochemical	4600	4500	4500	4500	4600	3947	4700	4700	4700
Fushun refinery	4784	4200	4784	4784	4784	3966	4784	4784	4784
Jinzhou refinery	4784	4784	4784	4784	4784	4007	4784	4784	4784
Dallan refinery		4716	4716	4716	4780	4036	4780	4880	4880
Southwest & northwest market									
Nanchong refinery	5150	5050	5050	5050	5050	4302	5050	5050	5120
XiAn refinery	4720	4720	4720	4720	4850	4116	4850	4850	4900
Yan'an refinery	4580	4580	4580	4580	4580	3946	4680	4680	4680
Urumchi refinery	4050	4050	4050	4050	4050	3700	4050	4050	4050

Aug 27 2006	Aug 28 2006	Aug 29 2006	Aug 30 2006	Aug 31 2006	Sept 1 2006	Sept 2 2006	Sept 3 2006	Sept 4 2006	Sept 5 2006	Sept 6 2006
4980	4920	4970		5070	5130	5130	5130	5130	5130	5180
4700	4700	4700	4700	4700	4800	4800	4800	4800	5000	5000
4784	4784	4784	4784	4784	4900	4900	4900	4900	4900	4930
5100	4900	4850	4900	4900	5050	5050	5050	5050	4950	5000
4880	4880	4880	4880	4880	4880	4880	4936	5000	5000	5000
5120	5280	5350	5350	5350	5350	5350	5350	5350	5400	5460
4900	4900	4900	4900	4900	4900	5030	5050	5050	5050	5050
4680	4750	4750	4750	4750	4750	4850	4850	4850	4850	4850
4050	4050	4050	4050	4050	4050	4050	4050	4050	4050	4050

Source: www.oilgas.com.cn.

Table A2.2 Indices 2006

	Jan 2006	Feb 2006	Mar 2006	Apr 2006	May 2006	Jul 2006	Aug 2006
South China market							
Maoming refinery	5373	5288	5076	5112	5112	4952	5439
Guangzhou refinery	5525	5437	5241	5204	5204	5065	5555
Zhanjiang Dongxing refinery	5385	5385	5112	5111	5111	4965	5443
Beihai refinery	5393		5550			4800	1850
Fujian market							
Fujian refinery	5313	5212	5088	4895	4895	4953	5403
East China market							
Zhenhai refinery	5276	5162	4980	4880	4880	4849	5287
Shanghai Petrochemical	5068	5268	4828	4780	4780	4890	5179
Shanghai Gaoqiao	5081	5200	5133	4735	4735	4790	5111
Nanjing refinery	5081	4875	4705	4777	4777	4610	4996
Yangtsi refinery	5031	4875	4736	4786	4786	4609	4998
Qingjiang refinery	4920	4787	4800	4845	4845	4785	4954
Yangtse River							
Wuhan refinery	4926	4791	4812	4903	4903	4599	5187
Anqing refinery	5000	4818	4778	4850	4850	4610	5092
North China market							
Jinan refinery	4561	4463	4552	4600	4600	4381	4827

Qilu refinery	4531	4368	4565	4567	4567	4359	4776
Qingdao refinery	4506	4568	4585	4638	4638	4409	4799
Yanshan refinery	4383	4425	4441	4440	4440	4284	4695
Tianjin refinery	4355	4371	4430	4454	4454	4271	4688
Huabei refinery	4280	4317	4414	4515	4515	4267	4684
Northeast China market							
Daqing Petrochemical	3615	3950	3986	4136	4136	3947	4456
Fushun refinery	3765	4068	4052	4191	4191	3966	4548
Jinzhou refinery	4025	3975	4105	4259	4259	4007	4630
Dalian refinery	4250	4250	4247	4305	4305	4036	4583
Southwest & northwest market							
Nanchong refinery	5021	4892	4801			4302	4988
XiAn refinery	4722	4837	4900	5250	5250	4116	4686
Yan'an refinery	4325	4087	4281	4309	4309	3946	4478
Urumchi refinery	3735	3787	3750	3380	3380	3700	5087

Source: www.oilgas.com.cn.

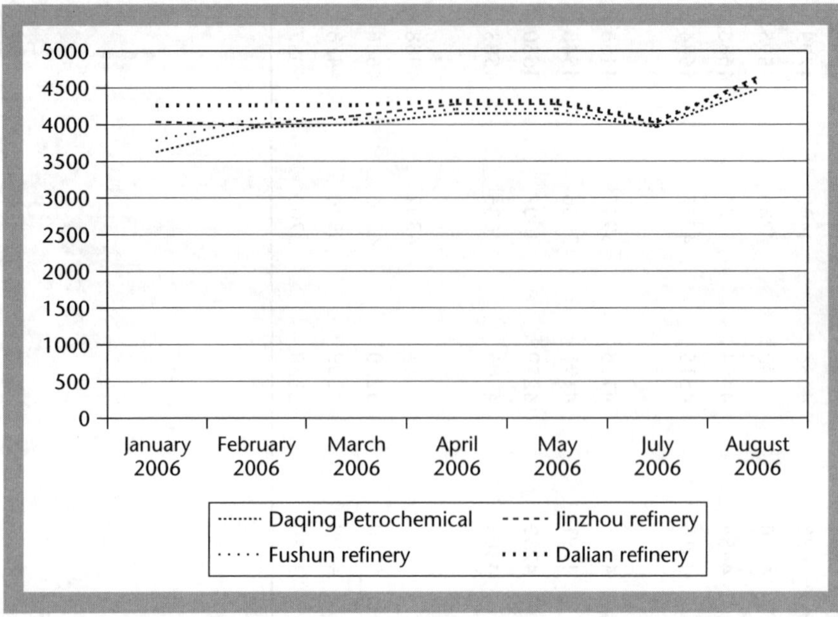

Figure A2.12 Northeastern Chinese refineries' pricing indices 2006
Source: www.oilgas.com.cn.

Bibliography

ABC (1997), 'Annual Report Agricultural Bank of China 1997', Beijing, Internet.

Aizenman, Joshua and Nancy Marion (2004), 'International Reserve Holdings with Sovereign Risk and Costly Tax Collection', *Economic Journal*, 114 (July): 569–91.

Anderson, Jonathan (2003), 'The Complete RMB Handbook', UBS Investment Research, Asian Economic Perspectives, October.

Andrews-Speed, Philip (2004), *Energy Policy and Regulation in the People's Republic of China*, The Hague: Kluwer Law International.

Andrews-Speed, P. (2005), 'China's Energy Needs: Running on Empty', *Far Eastern Economic Review*, June.

Andrews-Speed, P., X. Liao and R. Dannreuther (2002), *The Strategic Implications of China's Energy Need*, Adelphi Papers 346, Oxford: Oxford University Press.

Anil R. Cabraal, Xiaodong Wang, Grayson Heffner and Yabei Zhang (2005), 'Renewable Energy Toolkit Needs Assessment', World Bank, http://wbln0018.worldbank.org/esmap/site.nsf/files/07705 + Final_to_Printshop.pdf/$FILE/077-05 + Final_to_Printshop.pdf.

'Annual Report of the Expert Group on Technology Transfer', (2005) United Nations Framework Convention on Climate Change, 28 November, http://unfccc.int/resource/docs/2005/sbsta/eng/inf10.pdf.

'Aneel Approves Transfer of Coal-Fired Project to Eleja' (2005), *Business News Americas*, (29 April 2005).

APEC (2005), 'Energy Security in APEC', ABARE Research Report 05.2, APEC Energy Working Group, ABARE, Canberra.

APERC (2002), 'APEC Energy Demand and Supply Outlook 2002', Asia Pacific Energy Research Centre, Tokyo.

APERC (2003), 'Energy Security Initiative: Some Aspects of Oil Security', Asia Pacific Energy Research Centre, Tokyo.

APERC (2004), 'Energy in China: Transportation, Electric Power and Fuel Markets', Asia Pacific Energy Research Centre, Tokyo.

Archer et al. (2005), 'Evaluation of Global Wind Power', *Journal of Geophysical Research*, 110, http://www.agu.org/journals/jd/jd0512/2004JD005462/.

Arruda, Michael E. and Ka-Yin Li (2004), 'Framework of Policies, Institutions in Place to Enable China to Meet its Soaring Oil, Gas Demand', *Oil & Gas Journal*, 102(33): 20.

Asia Pulse (2005), 'China to Develop Renewable Energy in West', *Asia Times Online*, 21 April, http://www.atimes.com/atimes/China/GD21Ad02.html.

Atkeson, Andrew and Patrick J. Kehoe (1999), 'Models of Energy Use: Putty-Putty Versus Putty-Clay', *American Economic Review*, 89: 1028–43.

Austin, Angie, Gong Li and Mark Spelman (2005), 'Energy and Power in China, Domestic Regulation and Foreign Policy'.

Avery, Nerys (2006), 'China Investment Jump a Problem', 20 April, Beijing, Bloomberg.

Baily, Martin (2004), 'The US Economic Outlook', Presentation at the Institute for International Economics, Washington, 1 April.

Bank for International Settlements (BIS) (2005), *75th Annual Report*, Chapter 2, http://www.bis.org/publ/arpdf/ar2005e2.htm.

Baputaki, C. (2005), 'Makwinja Takes Over as MCL Board Chair', *Mmegi/The Reporter* (Gaborone) (29 April), www.allafrica.com.

Baum, Ellen and Joe Chaisson (2003), *The Last Straw – Water Use by Power Plants in the Arid West*, Clean Air Task Force, Hewlett Foundation Energy Series, San Francisco: The Energy Foundation.

Barsky, Robert B. and Lutz Kilian (2002), 'Do We Really Know that Oil Caused the Great Stagflation? A Monetary Alternative', in Ben S. Bernanke and Ken Rogoff (eds), *NBER Macroeconomics Annual 2001*, Cambridge, MA: MIT Press, pp. 137–83.

Barsky, Robert B. and Lutz Kilian (2004), 'Oil and the Macroeconomy Since the 1970s', *Journal of Economic Perspectives*, 18(4): 115–34.

Becker, Elizabeth (2003), 'Report Faults China's Policies Since it Joined Trade Group', *New York Times*, 20 December, B3.

Berger, Y. (2004), 'China's Energy Strategy', *Far Eastern Affairs*, 32(3), July–September: 45–65.

Bergesen, C. (2005), 'Platts UDI Country Profile: South Korea', *Platts* (10 February 2005), www.platts.com.

Bergesen, C. (2005) 'Country Profile: China Electric Power Overview' (16 February), www.platts.com/coal/resources.

Bergmann, Ariel, Nick Hanley and Robert Wright (2006), 'Valuing the Attributes of Renewable Energy Investments', *Energy Policy*, 34.

Bernanke, B. (2005), 'Monetary Policy in a World of Mobile Capital', *Cato Journal*, 25(1): 1–12, www.cato.org/pubs/journal/cj25n1/cj25n1-1.pdf.

Bernanke, Ben S., Mark Gertler and Mark Watson (1997), 'Systematic Monetary Policy and the Effects of Oil Price Shocks', *Brookings Papers on Economic Activity*, 1: 91–124.

Bingren, Shao (2005), Vice Chairman of the State Electricity Regulatory Commission, Speech of 25 February, http://www.china.org.cn/e-news/news050225.htm.

Bjorner, T. B. and H. H. Jensen (2002), 'Energy Taxes, Voluntary Agreements and Investment Subsidies – a Micro-panel Analysis of the Effect on Danish Industrial Companies' Energy Demand', *Resource & Energy Economics*, 24(3): 229–49.

BoC (1998), 'Annual Report Bank of China 1998', Beijing.

BoCom (1998), 'Annual Report Bank of Communications 1998', Shanghai.

Bofinger, Peter and Timo Wollmershauser (2000), 'Options for the Exchange Rate Policies in the EU Accession Countries and Other Emerging Market Economies', CEPR Discussion Paper No. 2379, February.

Bohi, Douglas R. (1991), 'On the Macroeconomic Effects of Energy Price Shocks', *Resources and Energy*, 13: 145–62.

Bottelier, P. (2003), 'China's Emerging Debt Markets: Facts and Issues', paper presented at China Economic Policy Reform Conference, Stanford University, 18–20 September, http://scid.stanford.edu/events/China2003/Bottelier.pdf.

Bottelier, Pieter (2004), 'China's Exchange Rate, US-China Economic Relations and Globalization', China Economic Analysis Working Paper, no. 9, Cross-Strait Interflow Prospect Foundation, February.

Boughton, James (2001), 'Silent Revolution: the International Monetary Fund, 1979–1989', International Monetary Fund, Washington.

Bradford, Scott, and Robert Lawrence (2004), 'Has Globalization Gone Far Enough?', Institute for International Economics, Washington.

Brav, Ehren J. (2004), Article 4 of The Law on Energy Conservation of the People's Republic of China, *Reforming China's Energy Conservation Law: Guidelines for Improving Implementation. Report written for the Natural Resources Defense Council in conjunction with the Energy Foundation*, Draft provided by the author.

'Brazilian Mills Plan To Raise Capacity by 30%' (2004), Steel Business Briefing, 21 September, www.steelbb.com.

'Brazil Steel Sector to Invest $8.0 Bln in Capacity Expansion by 2009', *Portuguese News Digest*, 21 August.

Bresnahan, Timothy F. and Valerie A. Ramey (1993), 'Segment Shifts and Capacity Utilization in the U.S. Automobile Industry', *American Economic Review Papers and Proceedings*, 83(2): 213–18.

'The Brightness and Township Electrification Program in China', presentation at the International Conference for Renewable Energies in Bonn, http://www.renewables2004.de.

British Petroleum (2005), *Putting Energy in the Spotlight. BP Statistical Review of World Energy June 2005*, British Petroleum, London.

British Petroleum (2006), *BP Statistical Review of World Energy 2006*, London.

Brooks, Ray and Tao Ram (2003), 'China's Labor Market Performance and Challenges', IMF Working Paper, no. 03/210, International Monetary Fund, Washington.

Brown, Stephen P.A. and Mine K. Yücel (1999), 'Oil Prices and U.S. Aggregate Economic Activity: a Question of Neutrality', *Economic and Financial Review*, Federal Reserve Bank of Dallas (Second Quarter): 16–23.

BTM (2005), 'China's Energy Needs Affecting the Global Economy', *Tokyo-Mitsubishi Review*, Bank of Tokyo Mitsubishi, February.

Burbidge, John and Alan Harrison (1984), 'Testing for the Effects of Oil-Price Rises Using Vector Autoregressions', *International Economic Review*, 25: 459–84.

Burton, David (2004), IMF Survey, International Monetary Fund, Washington, March 29.

Business School of China Science and Technology University and et al. (2000), 'Business Training Materials of Renewable Energy for Enterprisers'.

Bustelo, Pablo (2005), 'China and the Geopolitics of Oil in the Asian Pacific Region', 5 September, Working Paper 38/2005, Real Instituto Elcano, *University of Madrid*.

California Energy Commission (CEC) (2005), 'Integrated Energy Policy Report', http://www.energy.ca.gov/2005publications/CEC-100-2005-007/CEC-100-2005-007-CMF.PDF.

Cambridge Energy Research Associates (2005), 'China Market Commentary, Spring 2005: After the Peak', 15 April, www.cera.com.

Cashin, Paul and C. John McDermott (2003), 'Intertemporal Substitution and Terms of Trade Shocks', *Review of International Economics*, 11 (September): 604–18.

Carruth, Alan A., Mark A. Hooker and Andrew J. Oswald (1998), 'Unemployment Equilibria and Input Prices: Theory and Evidence from the United States', *Review of Economics and Statistics*, 80: 621–8.

CCB (1998), 'Annual Report China Construction Bank 1998', Beijing.

Central Intelligence Agency (CIA) (2001), 'Country Map: China', CIA Maps & Publications, http://www.cia.gov/cia/publications/mapspub/36.shtml.

Central Intelligence Agency (CIA) (2006), 'China', CIA World Factbook, http://www.cia.gov/cia/publications/factbook/geos/ch.html.

Chatterjee, N. (2006), 'World May Turn Back Clock for Liquid Coal Future', Reuters, 21 April, http://today.reuters.com; and Global Insight.

China Annual Energy Review 1997, Department of Resource Conservation and Comprehensive Utilization, SETC.

China Daily, http://www.chinadaily.com.

China Daily (2005), 'Greenpeace Hails China's Energy Law', 3 March, http://www.chinadaily.com.cn/english/doc/2005-03/03/content_421154.htm.

China Daily (2005), 'New Rules to Curb Surging Real Estate Price', 13 May, http://www2.chinadaily.com.cn/english/doc/2005-05/13/content_441874.htm.

China Daily (2006), Sun Min, 'Yuan appreciates to below 8 against greenback', 16 May.

China Energy Research Society (2003), Workshop of Wind Power Development Strategy in China.

'China and India: a Rage for Oil: Beijing has the Upper Hand Now, but Moscow is Old Friends with Delhi' (2005), *BusinessWeek*, 5 September, www.businessweek.com.

China Minsheng Bank (1998), 'Annual Report China Minsheng Banking Corporation 1998'.

China Studies, Department of Sociology, Nanking University, in collaboration with the Social Science Academic Press (China), from 2005.

China's Agenda 21 (1992), White paper on China's Population, Environment and Development in the 21st Century, http://www.acca21.org.cn/indexe5.htm.

'China's Downstream Gas Sector Regulatory Framework', September 2002.

China's Long March into Distributed Energy (2004), *Cogeneration and On-site Power Production*, 4 (Spring), http://www.earthscan.co.uk/news/article/mps/UAN/256/v/4/sp/3328516988 99328391278.

'China's Securities and Futures Markets' (2005), China Securities Regulatory Commission, March 2005.

Chinabidding.com.cn (2005), daily energy reports from January to August 2006.

Chinabidding.com.cn (2007), 'China Adopts New Pricing System for Oil Products', 31 January.

Chinabidding.com.cn (2007), 'Oil Market to Be Open to the World', 2 February.

Chinabidding.com.cn (2007), 'China's 2006, 2007, Oil Demand Forecasts Revised', 12 February.

ChinaHedge.com.cn, daily reports from October 2006 to January 2007.

'China-UK Joint Statement', 10 May 2004, http://www.chineseembassy.org.uk/eng/xwgb/t101947.htm.

Chinn, Menzie and Eswar Prasad (2003), 'Medium-Term Determinants of Current Accounts in Industrial and Developing Countries: an Empirical Exploration', *Journal of International Economics*, 59(1) (January): 47–76.

Chang, F. K. (2001), 'Chinese Energy and Asian Security', *Orbis*, 45(2), Spring: 1–25.

CIEP (2004), *Study on Energy Supply Security and Geopolitics*, Clingendael International Energy Programme, The Hague.

Clarida, Richard, Jordi Galí and Mark Gertler (2000), 'Monetary Policy Rules and Macroeconomic Stability: Evidence and Some Theory', *Quarterly Journal of Economics*, 115: 147–80.

Cleaner Production Promotion Law (2003), http://www.chinacp.com/.

Cline, William R. (2005), 'The United States as a Debtor Nation: Problems, Prospects, and Policies', Washington: Institute for International Economics.

Coal Industry Advisory Board (2004), 'International Coal Markets and Policy Developments in 2003', February.

Cole, B. D. (2003), 'Oil for the Lamps of China – Beijing's 21st-century Search for Energy', The McNair Papers, INSS, National Defense University, Washington.

Constantin, C. (2005), 'China's Conception of Energy Security: Sources and International Impacts', Working Paper, no. 43, Center for International Relations, University of British Columbia.

Crédit Agricole (2005), 'Chine, à l'ombre des terrils', in 'Quelles stratégies energétiques dans le monde?', *CA Perspectives*, 91, July.

Croci, Edoardo (ed.) (2005), *The Handbook of Environmental Voluntary Agreements*, Dordrecht, The Netherlands: Springer.

Crompton, P. and W. Wu (2005), 'Energy Consumption in China: Past Trends and Future Directions', *Energy Economics*, 27(1): 195–208.

Cuñado, Juncal and Fernando Pérez de Gracia (2003), 'Do Oil Price Shocks Matter? Evidence from Some European Countries', *Energy Economics*, 25: 137–54.

'Czech Republic Eyes Brazil Rio Grande do Sul Energy Sector' (2005), *Latin America News Digest* 22 March.

Daly, J.C.K. (2004), 'Energy Concerns and China's Unresolved Territorial Disputes', *China Brief*, 4(24), The Jamestown Foundation, 7 December.

Daniel, Betty C. (1997), 'International Interdependence of National Growth Rates: a Structural Trends Analysis', *Journal of Monetary Economics*, 40: 73–96.

Dannreuther, R. (2003), 'Asian Security and China's Energy Needs', *International Relations of the Asia Pacific*, 3 (2): 197 219.

Database of State Incentives for Renewable Energy (DSIRE) (2006), 'California Incentives for Renewable Energy: Renewable Portfolio Standards', January, http://www.dsireusa.org/library/includes/incentive2.cfm?Incentive_Code=CA25R&state=CA&CurrentPageID=1.

Database of State Incentives for Renewable Energy, (DSIRE) (2006), 'New Jersey Incentives for Renewable Energy: Renewable Portfolio Standards', April, http://www.dsireusa.org/library/includes/incentive2.cfm?Incentive_Code = NJ05R&state=NJ&CurrentPageID=1.

Davis, S.C. and S.W. Diegel (2004), *Transportation Energy Data Book: Edition 24*, ORNL-6973, Oak Ridge, TN: Oak Ridge National Laboratory, December, http://cta.ornl.gov/data/Index.shtml.

Davis, Steven J. (1987a), 'Fluctuations in the Pace of Labor Reallocation', in K. Brunner and A. H. Meltzer (eds), *Empirical Studies of Velocity, Real Exchange Rates, Unemployment and Productivity*, Carnegie-Rochester Conference Series on Public Policy, 24, Amsterdam: North Holland.

Davis, Steven J. (1987b), 'Allocative Disturbances and Specific Capital in Real Business Cycle Theories', *American Economic Review Papers and Proceedings*, 77(2): 326–32.

Davis, Steven J. and John Haltiwanger (2001), 'Sectoral Job Creation and Destruction Responses to Oil Price Changes', *Journal of Monetary Economics*, 48: 465–512.

Dees, Stephane, Filippo di Mauro, M. Hashem Pesaran, and Vanessa Smith (2007), 'Exploring the International Linkages of the Euro Area: a Global VAR Analysis', *Journal of Applied Econometrics*, 22(1): 1–38.

Delmas, M. and A. Terlaak (2000), 'Voluntary Agreements for the Environment: Innovation and Transaction Costs', CAVA Working Paper 00/02/13, February.

Development Research Center (2004), 'National Energy Strategy and Policy Report' (NESP); a condensed and translated version of this report is available at http://www.efchina.org/documents/Draft_Natl_E_Plan0311.pdf.

Dinghuan, Shi (2003) 'Some Thoughts on China's Energy Development Strategy', *Vital Speeches of the Day*, 70(5): 130–3.

'Dominion Announces 7 mt/yr Import Terminal' (2005), McCloskey's Coal Report, No. 25, 9 December.

Dore, John and Yu Xiaogang (2004), 'Yunan Hydropower Expansion: Update on China's Energy Industry Reforms and the Nu, Lancang and Jinsha Hydropower Dams', working paper from Chiang Mai University's Unit for Social and Environmental Research and Green Watershed, http://www.sea-user.org.

Dorian, J.P. (2005), 'Growing Chinese Energy Demand: Dramatic Global Implications', conference on Implications of China's Energy Search, CSIS, Washington DC, 23 March.

Dotsey, Michael and Max Reid (1992), 'Oil Shocks, Monetary Policy, and Economic Activity', *Economic Review of the Federal Reserve Bank of Richmond*, 78/4: 14–27.

Downs, E.S. (2004), 'The Chinese Energy Security Debate', *The China Quarterly*, 177 (March): 21–41.
DRC (2003), 'China's National Comprehensive Energy Strategy and Policy', Development Research Center, State Council, Beijing.
DRC (2004), 'China National Energy Strategy and Policy 2020', Development Research Center, State Council, Beijing.
EBRD (1999), 'Transition Report: Ten Years of Transition', European Bank for Reconstruction and Development, London.
Eckert, V. (2006), 'German Generators Turn to Coal as Gas Prices Soar', Reuters, 6 February, http://today.reuters.com.
Economic and Commercial Section of the Consulate General of the People's Republic of China in Kuching, 'Coal-to-Oil Project To Ease Energy Bottleneck' (November 18, 2005), http://kuching2.mofcom.gov.cn;
The Economist (1997), 'Fragile, Handle with Care: a Survey of Banking in Emerging Markets', Sonderbeitrag, 12 April.
The Economist (2004), 'A Fair Exchange ?' 20 January, http://www.economist.com/finance/displaystory.cfm?Story_id=3219515.
The Economist (2007), 'Apple and the Art of Innovation', 383(8532), 9–15 June: 59–60.
Economist Intelligence Unit (1998), 'Business China', 26 October, http://www.eiu.com.
Economist Intelligence Unit (2006), 'Energy Risk: China and Japan Compete for Gas and Oil', in association with Global Association of Risk Professional (GARP) and Riskcenter.com, 1 September.
'EiEE New/Repowering Generation Project Tracker – September 2005', *Energy in East Europe*, 72 (16 September): 9–39.
Energy Information Administration, *International Energy Annual*, www.eia.doe.gov/iea/, various years.
Energy Information Administration, *International Energy Outlook*, Washington DC, various years, http://www.eia.doe.
Energy Information Administration, US Department of Energy, 'Country Analysis Brief – China', http://www.eia.doe.gov/emeu/cabs/china.html, n.d.
Energy Information Administration, US Department of Energy (2003), 'The Global Liquefied Natural Gas Market: Status and Outlook', December.
Energy Information Administration (2005), Coal Market Module of the National Energy Modeling System, Model Documentation 2005, DOE/EIA-M060(2005).
Energy Information Administration (2006), *Short-Term Energy Outlook*, April, www.eia.doe.gov/emeu/steo/pub/.
Energy Information Administration (EIA) (2006), Monthly Energy Review, February, DOE/EIA-0035 (2006/02), 23 February.
Energy Research Center of the Netherlands (ECN) (2003), 'Renewable Electricity Fact Sheet Germany', http://www.renewable-energy-policy.info/relec/germany/policy/feed-in.html.
Energy Strategy of China Sustainable Development (1998), Chinese Academy of Engineering, December.
ERI-LBNL (2003), 'China's Sustainable Energy Future. Scenarios of Energy and Carbon Emissions', Energy Research Institute (NDRC) – Lawrence Berkeley National Laboratory, October.
ERI-LBNL (2005), 'Evaluation of China's Energy Strategy Options', Lawrence Berkeley National Laboratory, Berkeley, CA.
'Erwin Writes on Power-Generation Challenges' (2006), *Creamer Media's Mining Weekly*, 23 February, www.miningweekly.co.za.
Essays from 'Workshop on China Wind Energy Development in the 21st Century.' Beijing, March 2000.

European Commission (2006), 'Emissions Trading: Commission Sets Out Guidance on National Allocations for 2008–2012', Press Release, 9 January, http://europa.eu.int/press_room/index_en. htm, Reference Number IP/06/9.

ExxonMobil (2004), 'Energy Outlook to 2030', in *Energy Outlook 2004*, ExxonMobil, www.exxonmobil.com.

Facts Inc (2003), *Asia-Pacific Databook 1: Supply, Demand and Prices*, Honolulu, Hawai, Fall, www.factsinc.net/products/databooks.shtml.

Facts Inc. (2005) *GasDatabook*, Honolulu, Hawai.

Facts Inc (2006), 'China's LNG Imports: Delayed Terminal Projects and a Less Bullish Demand Outlook', *China Energy Series: Gas Edition*, 4 (January 2006): 1, 4.

Fan, J. W. Sun and D. Ren (2005), 'Renewable Portfolio Standard and Regional Energy Structure Optimisation in China', *Energy Policy*, 33: 279–87.

Faniso, M. (2005), 'Eskom Signs R1.8bn Deal With Siemens', *Business Report*, 22 July, www.eskom.co.za.

Feller, G. (2005), 'China's Rising Demand for Oil and Pipelines Has Worldwide Implications', *Pipeline & Gas Journal*, May.

Fengqi, Zhou (2003), 'Renewable Energy in China: Status, Goals, and Policy Recommendations', China Energy Laws and Regulation Workshop & The China Sustainable Energy Program Sixth Senior Policy Advisory Council Meeting, National People's Congress, http://www.efchina.org/documents/Briefing_book_031114_EN.pdf.

Ferderer, J. Peter (1996), 'Oil Price Volatility and the Macroeconomy: a Solution to the Asymmetry Puzzle', *Journal of Macroeconomics*, 18: 1–16.

Fernald, John G. and Oliver D. Babson (1999), 'Why has China survived the Asian Crisis so well? What Risks Remain?', Board of Governors of the Federal Reserve System, International Finance Discussion Papers No. 633, February.

Finn, Mary G. (2000), 'Perfect Competition and the Effects of Energy Price Increases on Economic Activity', *Journal of Money, Credit, and Banking*, 32: 400–16.

Folkerts-Landau, David (2004) 'A Small Price to Pay for Growth in Asia', *Financial Times*, 27 October.

Frisch, J.R. (1982), 'Resource Crisis in the Future', China National Energy Strategy and Policy 2020, www.efchina.org.

Ganter, E. (2005), 'Sasol Overview and Strategic Drivers', presentation at the Credit Suisse First Boston Global Oil & Gas Conference, London, 8 June, www. sasol.com.

'Gazprom, Gaz de France Swap Pipeline Gas for LNG Cargo' (2005), *LNG Express*, 15(23) (1 December): 13.

Geller, Howard et al. (2004), 'Policies for Advancing Energy Efficiency and Renewable Energy Use in Brazil', *Energy Policy*, 32: 1437–50.

Gisser, Micha and Thomas H. Goodwin (1986), 'Crude Oil and the Macroeconomy, Tests of Some Popular Notions', *Journal of Money, Credit, and Banking*, 18: 95–103.

Global Insight (2006), 'India: Sasol Targets India's CTL Potential', 7 March, http://myinsight.globalinsight.com.

Global Insight World Markets Research Centre (2006), 'Country Report – Mexico (Energy) Utilities: Electricity and Gas', 9 February, www. worldmarketsanalysis.com.

Gnansounou, E., and J. Dong (2004), 'Opportunity for Inter-regional Integration of Electricity Markets: the Case for Shandong and Shanghai in East China', *Energy Policy*, 32 p. 1737–51.

Goldman-Sachs (2004), 'Charting China, February/March 2004', Goldman-Sachs Asia Economic Research Group, Issue 2004/03, April.

Goldman Sachs (2004), 'The BRICs and Global Markets: Crude, Cars and Capital', Global Economics Paper, no. 118.

Government of Canada (2005), 'Project Green: Moving Forward on Climate Change', Ottawa, Canada, April, www. climatechange.gc.ca/kyoto_commitments/report_e.pdf.

Government of India, Central Electricity Authority, *Monthly Report on Broad Status of Thermal Power Projects in the Country (December 2005)*, www. cea.nic.in.

Government of India, Central Electricity Authority (2006), *11th Plan Shelf of Thermal Power Projects*, web site www.cea.nic.in.

Guizot, Armelle (2007), 'China's Economy: the Financial and Banking System, Foreign Exchange and Interest Rate Risk Policies', *Harriman House*, April.

Haites, Erik (2004), 'Estimating the Market Potential for the Clean Development Mechanism: Review of Models and Lessons Learned, PCFplus Report 19, World Bank Carbon Finance Business PCFplus Research Program, International Energy Agency and International Emissions Trading Association, Washington, DC.

Hamilton, James D. (1983), 'Oil and the Macroeconomy Since World War II', *Journal of Political Economy*, 91: 228–48.

Hamilton, James D. (1985), 'Historical Causes of Postwar Oil Shocks and Recessions', *Energy Journal*, 6: 97–116.

Hamilton, James D. (1988), 'A Neoclassical Model of Unemployment and the Business Cycle', *Journal of Political Economy*, 96: 593–617.

Hamilton, James D. (2003), 'What is an Oil Shock?' *Journal of Econometrics*, 113: 363–98.

Hamilton, James D. and Ana Maria Herrera (2004), 'Oil Shocks and Aggregate Macroeconomic Behavior: the Role of Monetary Policy', *Journal of Money, Credit, and Banking*, 36: 265–86.

Hamilton, James D. (2005), 'Oil and the Macroeconomy', La Jolla, California, University of California, San Diego, http://dss.ucsd.edu/~jhamilto/JDH_palgrave_oil.pdf.

Hansen, Craig and Vince Van Son (2003), 'Renewable Energy Certificates: an Attractive Means for Corporate Customers to Purchase Renewable Energy', World Resource Institute, September, http://pdf.wri.org/gpmdg_corporate_5.pdf.

Harding, James (1999), 'China Throws a lifeline', *Financial Times*, 28 January.

Harrison, S. (ed.) (2005), *Seabed Petroleum in Northeast Asia: Conflict or Cooperation?* Washington DC: Woodrow Wilson International Center for Scholars.

Hazira, Ratnagiri (2006), 'Seek New Supplies: India Struggles with LNG Imports', *LNG Express*, 16(4) (15 February): 3–4.

He, Xin (2003), 'The Development and Status of the Power Grid in China and International Cooperation for Grid Integration in NorthEast Asia', State Grid Cooperation of China, Prepared for the Third Workshop on Power Grid Interconnection in Northeast Asia, Vladivostok, Russia, 30 September – 3 October.

Herberg, M. (2005), 'China's Energy Situation and the Implications for Asia and the US', testimony, Committee on Foreign Relations, US Senate, 7 June.

Hiecker, Nicol-Janin (1999), 'Konkursrechtsgesetzgebung in der Volksrepublik China', Mitteilungen des Instituts fur Asienkunde no. 301, Hamburg.

Holly C. (2005), 'Canada Unveils Kyoto Compliance Plan', *The Energy Daily*, 33(74) (20 April): 1, 4.

Hong Kong and Shanghai Banking Corporation (HSBC) (2005), China Economic Insight, April, www.hsbc.com.cn/cn/common/download/aboutus/200504.pdf.

Hooker, Mark A. (1996), 'What Happened to the Oil Price-Macroeconomy Relationship?', *Journal of Monetary Economics*, 38: 195–213.

Hooker, Mark A. (2002), 'Are Oil Shocks Inflationary? Asymmetric and Nonlinear Specifications versus Changes in Regime', *Journal of Money, Credit and Banking*, 34: 540–61.

Hoover, Kevin D. and Stephen J. Perez (1994), 'Post Hoc Ergo Propter Once More: an Evaluation of "Does Monetary Policy Matter?" in the Spirit of James Tobin', *Journal of Monetary Economics*, 34: 89–99.

Hu, Fred (2004), 'China is Hot, But Not Overheated', *Financial Times*, 12 May.

Hu, Z. and M. Khan (1997), 'Why is China Growing so Fast?' Economic Issues Paper 8, International Monetary Fund, Washington DC, www.imf.org/external/pubs/ft.issues8.

Huaneng Power International Inc. (2005), 'Announcement of Results for 2004', 15 March http://202.66.146.82/listco/hk/huanengpower/annual/2004/res.pdf.

Hurd, D. (2005), 'Global LNG: Key Themes and Choices', 21 April www.db.com.

ICBC (1998), 'Annual Report Industrial and Commercial Bank of China 1998', Beijing.

Institute of Energy Economics of Japan (IEEJ) (2004), 'Asia Outlook of Supply and Demand Trends of Petroleum Products and Crude Oil', Institute of Energy Economics of Japan, Tokyo, July.

'Indonesia: Sasol Eyeing CTL Partnership With Bumi Resources in Indonesia', 22 March 2006, http://myinsight.globalinsight.com.

Information for Marine Energy Section from *China Annual Energy Review 1997*, Department of Resource Conservation and Comprehensive Utilization, SETC.

Institute for Economic System Management, State Council Office for Restructuring Economic System, World Bank, and Private-Public Infrastructure Advisory Facility (PIAFF), Workshop summary.

Institute of Energy Economics (2006), 'Renewables and Regulatory Issues', the Energy and Resources Institute, February.

International Energy Agency (1997), *Voluntary Actions for Energy-Related CO_2 Abatement*, Paris: OECD/IEA.

International Energy Agency (1997), *Voluntary Approaches for Mitigating Greenhouse Gas Emissions*, Conference Proceedings, Bonn 30–31 October, Paris: OECD/IEA.

International Energy Agency (2000), *China's Worldwide Quest for Energy Security*, International Energy Agency, Paris.

International Energy Agency (IEA) (2003), *World Energy Investment Outlook: 2003 Insights*, Paris: OECD/IEA.

International Energy Agency, *World Energy Outlook 2004*, Paris, October, www. worldenergyoutlook.org.

International Energy Agency (2004), 'The Transformation of China's Electric Power Sector: Crisis or Business-as-Usual?', presentation by Jeffrey Logan, China Program Manager, Université Paris Dauphine, 2 June.

International Energy Agency, 'Monthly Oil Market Report', 10 November 2004.

International Energy Agency (2005), 'Oil Market Report', 11 August, International Energy Agency, Paris.

International Energy Agency (2005), *Coal Information 2005*, Paris, August.

International Energy Agency, *Databases for Coal Information 2005*, http://data.iea.org.

International Energy Agency (IEA) (2006), 'Unit Converter', IEA Energy Statistics, http://www.afrepren.org/datahandbook/pdfs/conver.pdf.

International Energy Agency (2006), *Oil Market Report*, Paris, May, http://omrpublic.iea.org/.

International Energy Agency, Renewables Information, various years.

International Energy Administration, *World Energy Outlook*, Yearly Annual Outlook Reports, various years.

International Monetary Fund, *International Financial Statistics Yearbook*, Washington, DC, various years.

International Monetary Fund (2000), 'The Impact of Higher Oil Prices on the Global Economy', IMF Working Paper 01/14, http://www.imf.org/external/pubs/ft/oil/2000/ oilrep.pdf.

International Monetary Fund (2004), 'People's Republic of China: Article IV Consultation – Staff Report', Washington: International Monetary Fund, www.imf.org.

International Monetary Fund's Seminar on China's Foreign Exchange System, Dalian, China, 26–27 May 2004.

International Monetary Fund (2004), *World Economic Outlook*, International Monetary Fund, Washington, April.

Isbell, P. (2005), 'Los precios del petróleo: situación actual y perspectivas futuras', ARI, no. 98/2005, Elcano Royal Institute, 14 July.

Izraelewicz, E. (2005), *Quand la Chine change le monde*, Paris: Grasset.

Jiang, W. (2005), 'Fueling the Dragon. China's Quest for Energy Security and Canada's Opportunities', Asia Pacific Foundation of Canada, Vancouver.

Jianmin, Liu (1999), 'Economic Evaluation of Household Solar Water Heaters', *Solar Energy*.

Jiménez, Viviana (2005), 'World Sales of Solar Jump 32 Percent', Earth Policy Institute, http://www.earth-policy.org/Indicators/2005/indicator12.htm.

Jimenez-Rodriguez, R. and M. Sanchez (2005), 'Oil Price Shocks and Real GDP Growth: Empirical Evidence for some OECD Countries', *Applied Economics*, 37(2): 201–28.

Jing, Fu (2005), 'Renewable Energy Gets Huge Outlay', *China Daily*, 11 November, http://www.chinadaily.com.cn/english/doc/2005-11/08/content_492254.htm.

Jing, Fu (2005), 'China Mulls Deregulating Oil Prices', *China Daily*, 14 November, www.chinadaily.com.

Jhaveri, Rahul (2006), 'Getting Big in Japan', 'The Old and the New', *Asia Risk*: 36–7.

Joskow, P. (2004), 'Regulation and Deregulation after 25 years: Lessons for Research in Industrial Organisation', Economics Department, MIT, http://econ-www.mit.edu/faculty/index.htm?prof_id=pjoskow&type=paper; *Review of Industrial Organization*, 26: 169–93 (2005).

JP Morgan (2004), JP Morgan Real Broad Effective Exchange Rate Indices, http://www2.jpmorgan.com/MarketDataInd/Forex/REXB.xls.

Kaminsky, Graciela, Carmen Reinhart and Carlos Végh, (2004), 'When It Rains, It Pours: Procyclical Capital Flows and Macroeconomic Policies', in *NBER Macro Annual 2004*, ed. Mark Gertler and Ken Rogoff (Cambridge, Massachusetts: National Bureau of Economic Research).

Keane, Michael P. and Eswar Prasad (1996), 'The Employment and Wage Effects of Oil Price Changes: a Sectoral Analysis', *Review of Economics and Statistics*, 78: 389–400.

Kellner, T. (2005), 'La Chine et le pétrole: perceptions sécuritaires et démarche stratégique', in G. Chaliand and A. Jafalian (eds), *La dépendance pétrolière. Mythes et réalités d'un enjeu stratégique*, Paris: Universalis, pp. 71–89.

Kenny, H.J. (2004), 'China and the Competition for Oil and Gas in Asia', *Asia Pacific Review*, 11(2): 36–47.

Kilian, Lutz (2005), 'Exogenous Oil Supply Shocks: How Big Are They and How Much Do They Matter for the U.S. Economy?' Working paper, University of Michigan.

Kim, In-Moo and Prakash Loungani (1992), 'The Role of Energy in Real Business Cycle Models', *Journal of Monetary Economics*, 29(2): 173–89.

'Korea: 5th Long-Term Power Development Plan' (2000), *International Market Insight Reports*, 10 February.

Korea Midland Power Company Ltd (2005), 'Nationwide Construction Status (as of the end of May 2005)', www.komipo.co.kr.

Kose, M. Ayhan, Eswar S. Prasad and Marco E. Terrones (2003), 'Financial Integration and Macroeconomic Volatility', IMF Staff Papers, 50: 119–42.

Kynge, James (1999), 'China: No Devaluation Mantra Dropped', *Financial Times*, 19 July.

KPMG (2005), *Energy Outlook for China*, KPMG, Hong Kong.

'La Meme Plus Grande Centrale Hydro-electrique de Chine Sera Mise en Service en 2007' (2005), 12 March, Xinhua News Release, www.xinhuanet.com.

Lane, Philip R. and Gian Maria Milesi-Ferretti (2001), 'The External Wealth of Nations: Measures of Foreign Assets and Liabilities for Industrial and Developing Nations', *Journal of International Economics*, 55: 263–94.

Lane, Philip and Gian Maria Milesi-Ferretti (2006), 'Exchange Rates and External Adjustment: Does Financial Globalization Matter?', paper provided by Institute of

International Integration Studies in its series The Institute for International Integration Studies Discussion Paper Series with number iiisdp129, 22 May.

Lane, Philip and Gian Maria Milesi-Ferretti (2006), 'Examining Global Imbalances', *Finance and Development Journal*, 43 (1).

Larson, E., Wu Zongxin, P. deLaquil, C. Wenying and G. Pengfei (2003), 'Future Implications of China's Energy Technology Choices', *Energy Policy*, 31: 1189–204.

LBNL (2005), *Evaluation of China's Energy Strategy Options*, Lawrence Berkeley National Laboratory, Berkeley, CA.

Leduc, Sylvain and Keith Sill (2004), 'A Quantitative Analysis of Oil-Price Shocks, Systematic Monetary Policy, and Economic Downturns', *Journal of Monetary Economics*, 51: 781–808.

Lee, Kiseok and Shawn Ni (2002), 'On the Dynamic Effects of Oil Price Shocks: a Study Using Industry Level Data', *Journal of Monetary Economics*, 49: 823–52.

Lee, Kiseok, Shawn Ni and Ronald A. Ratti (1995), 'Oil Shocks and the Macroeconomy: the Role of Price Variability', *Energy Journal*, 16: 39–56.

Lee, Theo L.K. (2005), 'Coal Liquefaction Technologies – Direct and Indirect Routes', presentation at APEC Clean Fossil Energy Technical and Policy Seminar, Cebu City, Philippines, 26–28 January, www. apec-egcfe.org.

Li, Francis (2002), 'Hydropower in China', *Energy Policy* 30(14): 1241–9.

Liu, Yingling (2006), 'Behind the Chilly Air: Impacts of China's New Wind Pricing Regulation', Worldwatch Institute, 30 March, http://www.worldwatch.org/features/chinawatch/stories/20060330-1.

Logan, J. (2005), 'Energy Outlook for China: Focus on Oil and Gas', testimony, Committee on Energy and Natural Resources, US Senate, 3 February.

Loungani, Prakash (1986), 'Oil Price Shocks and the Dispersion Hypothesis', *Review of Economics and Statistics*, 58: 536–9.

LRP (2005), *Crisis on the China Rim. An Economic, Crude Oil and Military Analysis*, Laguna Research Partners, Irvine, CA, April.

Ma, Shenghong (2004), 'Electricity and the Human Prospect', conference hosted by the Stanford Program on Energy and Sustainable Development, 8–9 December, http://iisdb.stanford.edu/evnts/3961/ Rapporteur's_Report.pdf.

Marketplace, American Public Media (2006), 'Special Report on China', 9–20 January, http://marketplace.publicradio.org/features/china2006/.

Martinot et al. (2005), 'Renewable Energy Policies and Markets in the United States', Center for Resource Solutions, http://www.efchina.org/documents/RE_Policies&Markets_US.pdf.

Martinot et al. (2005), Renewable Energy Policy Network, 'Renewables 2005: Global Status Report', Washington, DC: Worldwatch Institute.

Martinot et al. (2002), 'Renewable Energy Markets in Developing Countries', *Annu. Rev. Energy Environ.*, 2002.

Martinot, Eric (2001), 'World Bank Energy Projects in China: Influences on Environmental Protection', *Energy Policy*, 29: 581–94, http://www.martinot.info/Martinot_EP29-8.pdf.

McCloskey's Coal Report (2006), 'At Least Ten New Coal Plants Planned in Germany', 127 (27 January): 22–3.

McKinnon, R. and G. Schnabl (2004), 'The East Asian Dollar Standard, Fear of Floating, and Original Sin', *Review of Development Economics*, 8 (3): 331–60.

McKinnon, R. and G. Schnabl (2003), 'Synchronized Business Cycles in East Asia and Fluctuations in the Yen/Dollar Exchange Rate', *The World Economy*, 26 (8): 1067–88.

Meadows, B. (2005), 'Coal Plants Toast', *The Chronicle Journal* (16 June), www.chronicle-journal.com.

Menkoff, Lukas and Hariolf Teufel (1995), 'Finanzplatz Bangkok, Ausbau zum Regionalen Finanzzentrum', *Zeitschrift fur das gesamte Kreditwesen*, 48(2): 78–87.

Milich, Lenard and Robert G. Varady (1999), 'Openness, Sustainability and Public Participation: New Designs for Transboundary River Basin Institutions', *Journal of Environment & Development*, 8 (3): 258–306.

'Meeting China's Development Challenges Through Energy Efficiency and Renewable Energy', 1999, David and Lucile Packard Foundation, William and Flora Hewlett Foundation, The Energy Foundation, March.

Minchener, A.J. (2004), 'Coal in China', July, www.iea-coal.org.uk.

Mork, Knut A. (1989), 'Oil and the Macroeconomy When Prices Go Up and Down: an Extension of Hamilton's Results', *Journal of Political Economy*, 91: 740–4.

Mork, Knut A., Øystein Olsen and Hans Terje Mysen (1994), 'Macroeconomic Responses to Oil Price Increases and Decreases in Seven OECD Countries', *Energy Journal*, 15 (4): 19–35.

Murray, T. (2006), 'Pemex To Fund Increased Share of Capital Spending from Operating Cashflow', *The Oil Daily*, 56 (100) (25 May): 4.

Mutiga, Amos M. (2001), 'Energy Data Conversions', African Energy Policy Research Network, http://www.afrepren.org/datahandbook/pdfs/conver.pdf.

Nam. H. (2006), 'A New and Improved Coal Development Plan', *Vietnam Economic News*, (23 February), www.ven.org.vn.

National Commission on Energy Policy (2004), *Ending the Energy Stalemate: a Bipartical Strategy to Meet America's Energy Challenges*. Washington, DC: National Commission on Energy Policy.

National Geographic (2004), 'The End of Cheap Oil' (June): 80–109.

National Highway Traffic Safety Administration (2003), *Light Truck Average Fuel Economy Standards, Model Years 2005–2007*, 49 CFR Part 533 (Docket No. 2002-11419; Notice 3), RIN 2127-AI70 (Washington, DC, April), www.nhtsa.dot.gov/cars/rules/rulings/CAFE05-07/Index.html.

National Highway Traffic Safety Administration, Press Release (2006), 'New Light Truck Economy Standards To Save 10.7 Billion Gallons of Fuel, Include Largest SUVs for First Time Ever, Transportation Secretary Mineta Announces', DOT 46-06, 29 March, www.nhtsa.dot.gov.

National Renewable Energy Laboratory (NREL) (2006), 'Photovoltaics', January, http://www.nrel.gov/learning/re_photovoltaics.html.

National Renewable Energy Laboratory (NREL) (2005), 'Renewable Energy in China', http://www.nrel.gov/international/china/pdfs/35786_policy.pdf.

National Renewable Energy Laboratory (NREL) (2004), 'Renewable Energy in China: Township Electrification Program', April, http://www.nrel.gov/docs/fy04osti/35788.pdf.

National Renewable Energy Laboratory (NREL) (2004), 'China's Plan for Renewable Energy', April, http://www.nrel.gov/docs/fy04osti/35787.pdf.

'National Renewable Energy Law' (2005), The People's Republic of China, http://china.lbl.gov/publications/re-law-english.pdf.

NBS (2005), *China's Statistical Yearbook 2004*, National Bureau of Statistics, China's Statistics Press, Beijing.

NEAEF (2005), *Promoting a Northeast Asian Energy Community*, Northeast Asia Economic Forum, Honolulu.

The News Hour with Jim Lehrer (2005), 'Special Report: China Rising', 4 October, www.pbs.org/newshour/bb/asia/july-dec05/china_10-04.html.

Nyakazeya, P. (2005), 'Rio Tinto Seeks Partners for Sengwa Coal Project', *The Herald* (Harare) (5 April), www.allafrica.com.

Ogden, Doug (2005), China Sustainable Energy Program, April.

Oil & Gas Journal (2005), 'Worldwide Look at Reserves and Production', 103 (47) (19 December): 24–5.

Ontario Ministry of Energy (2006), 'Electricity Information', www.energy.gov.on.ca.

Organization of Petroleum Exporting Countries (2005), *Annual Statistical Bulletin 2004*, Vienna.

Ostry, Jonathan D. and Carmen M. Reinhart (1992), 'Private Savings and Terms of Trade Shocks: Evidence from Developing Countries', IMF Staff Papers, Vol. 39 (September): 495–517.

Oud, Engelberus (2002), 'The Evolving Context for Hydropower Development', *Energy Policy*, 30: 1219.

Pacala, S. and R. Socolow (2004), 'Stabilization Wedges: Solving the Climate Problem for the Next 50 Years with Current Technologies', *Science*, 13 August, 305.

Painuly, J. (2001), 'Barriers to Renewable Energy Penetration: a Framework for Analysis', *Renewable Energy*, 24: 73–89.

Pichel, Jesse (2006), 'China's Solar Push More Than Just Low-Cost?' Renewable Energy Access, 27 March, http://www.renewableenergyaccess.com/rea/news/story?id=44457.

People's Bank of China (2004), Report on the Implementation of Monetary Policy in 2004Q3, www.pbc.gov.cn.

People's Bank of China (2004), Monetary Policy Report for 2003, People's Bank of China, Beijing, March.

People's Bank of China (2005), Table of the Exchange Rate and Gold and Foreign Exchange Reserves, www.pbc.gov.cn.

People's Daily (2005), 'China Won't Be Pushed Into Yuan Reform', 11 May, http://english.people.com.cn/200505/11/eng20050511_184541.html.

People's Daily (2005), 'Central Bank: US Pressure not Favoring Reform on RMB Exchange Rate', 12 May, http://english.people.com.cn/eng20050512_184687.html.

People's Daily (2005), 'China Never Yields to Outside Pressure on RMB Exchange Rate: Premier Wen, 13 May, http://english.people.com.cn/eng20050516_185298.html.

Perkins, D. (1994), 'Completing China's Move to the Market', *Journal of Economic Perspectives*, 8 (2): 23–46.

Pesaran, M. Hashem, Til Schuermann and Scott Weiner (2004), 'Modeling Regional Interdependencies Using a Global Error-Correcting Macroeconometric Mode', *Journal of Business and Economics Statistics*, 22: 129–62.

Petroleum Intelligence Weekly (2004), No. 5 (July), www.EnergyIntel.com.

Platts International Gas Report (2005), 'Northern South America Takes First Step Towards Gas Pipeline Integration', No. 535 (4 November).

Platts, Power in Asia (2006), 'Nam Ngum-II Project Advances', No. 447 (16 February).

Platts, Power in Asia (2006), 'Song Da Advances Projects', No. 447 (16 February).

Platts, Power in Asia (2006), 'Bhel Wins Hydro Project', No. 446 (2 February).

Power Engineering International, 'South Korea's Power Industry Has Been Restructured', 13 (3) (1 March).

Prasad, Eswar and Thomas Rumbaugh (2004), 'Overview', in Eswar Prasad (editor), 'China's Growth and Integration into the World Economy: Prospects and Challenges', MF Occasional Paper, No. 232, IMF, Washington.

Prasad, E. and S. Wei (2005), 'The Chinese Approach to Capital Inflows: Patterns and Possible Explanations', IMF Working Paper (WP/05/79).

Prasad, Eswar (editor), Steven Barnett, Nicolas Blancher, Ray Brooks, Annalisa Fedelino, Tarhan Feyzioglu, Thomas Rumbaugh, Raju Jan Singh and Tao Wang (2004), 'China's Growth and Integration into the World Economy: Prospects and Challenges', IMF Occasional Paper No. 232 (Washington: International Monetary Fund).

Prasad, Eswar, Kenneth Rogoff, Shang-Jin Wei and M. Ayhan Kose (2003), 'The Effects of Financial Globalization on Developing Countries: Some Empirical Evidence', IMF Occasional Paper No. 220 (Washington: International Monetary Fund).

Prasad, E., T. Rumbaugh and Q. Wang (2005), 'Putting the Cart before the Horse? Capital Account Liberalization and Exchange Rate Flexibility in China, IMF Policy Discussion Paper (PDF/05/01), www.imf.org/external/pubs/ft/pdp/2005/pdp01.pdf.

Preeg, Ernest (2003), 'Exchange Rate Manipulation to Gain an Unfair Competitive Advantage: the Case against Japan and China', in C. Fred Bergsten and John Williamson (eds), *Dollar Overvaluation and the World Economy*, Washington: Institute for International Economics.

'Program Updates April 2005', *China Clippings*, The China Sustainable Energy Program, Issue 17, April, http://www.efchina.org/documents/CSEP_Update_Apr_05.pdf.

Qiumao, Niu et al. (1999), 'High Efficient Urban Waste Power Generation by Heat-Electricity of Alkali Metals', Proceedings on Workshop and Process Exhibition on Clean Energy Technology in China, July.

Quaisser, Wolfang (1995), 'Vergleich der Wirtschaftlichen Transformation in Osteuropa und Ostasien', *Osteuropa-Wirtschaft*, 40(4): 279–98.

Rajan, Raghuram and Arvind Subramanian (2004), 'Exchange Rate Flexibility is in Asia's Interest', *Financial Times*, 26 September.

Rader, Nancy and Scott Hempling (2001), 'The Renewables Porfolio Standard: a Practical Guide', National Association of Regulatory Utility Commissioners, February.

Rasche, R.H. and J.A. Tatom (1977), 'Energy Resources and Potential GNP', *Federal Reserve Bank of St Louis Review*, 59 (June): 10–24.

Rasche, R.H. and J.A. Tatom (1981), 'Energy Price Shocks, Aggregate Supply, and Monetary Policy: the Theory and International Evidence', in K. Brunner and A.H. Meltzer (eds), *Supply Shocks, Incentives, and National Wealth*, Carnegie-Rochester Conference Series on Public Policy, vol. 14, Amsterdam: North-Holland.

Raufer et al. (2002), 'Wind Resource Concession Approach in China', IEEE Power Engineering Review, September.

Raymond, Jennie E. and Robert W. Rich (1997), 'Oil and the Macroeconomy: a Markov State-Switching Approach', *Journal of Money, Credit and Banking*, 29 (May): 193–213; Erratum 29 (November, Part 1), p. 555.

Regulatory Assistance Project (2000), 'Regulation or Competition: the California Experience Shows Both Are Needed', Comments at the International Symposium on Restructuring and Regulation of China's Electric Industry, hosted by the State Council Office for Restructuring Economic Systems (SCORES), http://www.raponline.org/Pubs/China/RegOrCmp.pdf.

'Renewable Electricity Systems into a Mandated Market Share Mechanism in China' (2003), *ECN*, October.

'Renewable Energy Policy in Germany: Pioneering and Exemplary Regulations', *Energy for Sustainable Development*, VIII (I): 47–57.

Renewable Energy Policy Project (REPP) (2004), 'RPS Texas: Renewable Portfolio Standards – State Summary', July, http://www.crest.org/articles/static/1/Texas_RPS.html.

Renewable Energy Policy Project (REPP) (2004), 'RPS California: Renewable Portfolio Standards – State Summary', July, http://www.crest.org/articles/static/1/California_RPS.html.

Renewable Energy Policy Project (REPP) (2004), 'RPS Maine: Renewable Portfolio Standards – State Summary', July, http://www.crest.org/articles/static/1/Maine_RPS.html.

Reserve Bank of Australia (2004), 'Oil Market Developments and Macroeconomic Implications', Bulletin, October.

'Richards Bay To grow to 92 mt/yr' (2005), *McCloskey's Coal Report*, No. 123 (November).

'Rio Grande do Sul To Hold Referendum on 650MW Coal Project' (2004), *Business News Americas*, 7 December.

'Rio Grande do Sul: US$5bn To Double Capacity by 2010' (2004), *Business News Americas*, 24 May.

Roach, Stephen (2004), 'Global Rebalancing and the China Play', Morgan Stanley, Economic Trends, March.

Rogers, Jim (2005), 'Expect Oil at $150 by 2015', IncisiveMedia Conference 2005, *Hedge Fund Review*, July, www.incisivemedia.com.

Rotemberg, Julio J. and Michael Woodford (1996), 'Imperfect Competition and the Effects of Energy Price Increases', *Journal of Money, Credit, and Banking*, 28 (part 1): 549–77.

Roubini, N. and J. Setser (2005), 'Will the Bretton Woods 2 Regime Unravel Soon? The Risk of a Hard Landing in 2005–2006', http://ideas.repec.org/a/fip/fedfpr/y2005ifebx13.html.

Rural Energy Development Study for China (draft) (1999), Asian Development Bank.

'S. Korea Posco To Liquidate Posenergy Unit', *Dow Jones International News*, 27 May, Korea South-East Power Company Ltd, www.kosep.co.kr.

Salameh, M.G. (2003), 'Quest for Middle East Oil: the US versus the Asia Pacific Region', *Energy Policy*, 31 (11): 1085–91.

Sanghoon, Ahn (n.d.), 'Competition, Innovation and Productivity Growth: a Review of Theory and Evidence', Economics Department, OECD, Working Papers, No. 317.

Santini, Danilo J. (1985), 'The Energy-Squeeze Model: Energy Price Dynamics in U.S. Business Cycles', *International Journal of Energy Systems*, 5: 18–25.

Santini, Danilo J. (1992), 'Energy and the Macroeconomy: Capital Spending after an Energy Cost Shock', in J. Moroney (ed.), *Advances in the Economics of Energy and Resources*, vol. 7, Greenwich, CN: JAI Press.

Schiermeier, Q. (2006), 'Europe Fails on Carbon Limits: Industry Emitted Less Than Allowed in 2005; But That Isn't Good News', *news@nature.com* (15 May), www.nature.com/news/.

Schlotthauer, Nicolas (1999), 'Will the People's Republic of China Trigger off the Next Asian Crisis', *Intereconomics*, 34 (3): 124–35.

Schlotthauer, Nicolas (2000), 'Einige Notwendige Reformen fur eine erfolgreiche Abwicklung des chinesischen WTO-Beitritts', *GieBener Beitrage zur Wirtschafts und Gesellschaftspolitik Chinas*, No. 26.

Schmoker, J.W. and T.R. Klett (n.d.), 'Chapter RG: Estimating Potential Reserve Growth of Known (Discovered) Fields: a Component of the USGS World Petroleum Assessment 2000', in *U.S. Geological Survey Digital Data Series 60*, http://energy.cr.usgs.gov/WEcont/chaps/RG.pdf.

Sequeira, Sabine (2006), 'Renewable Portfolio Standards, Feed-in Tariffs, and Tendering: Instituting Effective Mandated Market Policies in China', unpublished paper, Woodrow Wilson School, Denver, Colorado.

Shimbun, Yomiuri (2005), 'China Gorging and Japan-China Resource and Energy Conflicts', *Asia Times*, 1 July.

Shiu, A. and Pun-lee Lam (2004), 'Electricity Consumption and Economic Growth in China', *Energy Policy*, 32: 47–54.

Siegel, Jack (2004), 'Foreign Involvement in Combined Heat and Power in China: Policies and Successes', ERI-3049-0401, Energy Resources International, Washington DC.

Sijm, J.P.M. (2002), 'The Performance of Feed-In Tariffs to Promote Renewable Electricity Generation in European Countries', Energy Research Center of the Netherlands (ECN), http://www.ecn.nl/docs/library/report/2002/c02083.pdf.

Sinochem Corporation, 'Company Information', http://www.sinochem.com/en/.

Sinton, Jonathan, Rachel Stern, Nathaniel Aden and Mark Levine (2005), 'Evaluation of China's Energy Strategy Options', May, sponsored by Laurence Berkeley National Laboratory, and Jean Y. Ku National Renewable Energy Laboratory; http://china.lbl.gov/publications/nesp.pdf.

'South Africa: Eskom Recommissions Mothballed Power Plants To Boost Supplies' (2005), May, www.worldmarketsanalysis.com.

South China Morning Post (2004), 'Guangdong Orders Closure of Hydropower Plants to Save Water', 23 November.

State Economic and Trade Commission (SETC) (2000), 'The 10th Five-Year Plan for Industrial Water Saving', Beijing: SETC.

State Statistical Bureau (1995), *China Statistical Yearbook 1995*, Beijing: China Statistical Publishing House.

Stern, Rachel (2001), 'North American Development Bank and the Border Environment Cooperation Commission', see Rachel E. Stern (2001), 'Addressing Cross-Boundary Air Pollution: A Comparative Case Study of the US-Mexico Border and The Hong Kong-Guangdong Border', Hong Kong: Civic Exchange, http://www.civic-exchange.org/publications/2001/airCrossborder.pdf.

Study on the Survey and Forecast of China's Coal Consumption (1999), SDPC and MOCI Joint Survey Team.

'Swaziland: Plans To Expand Power Production' (2005), UN Integrated Regional Information Networks, 4 May, www.allafrica.com.

SSY Consultancy and Research Ltd (2005), *SSY's Coal Trade Forecast*, 14(3) (London, September).

The State of Queensland, Department of Natural Resources and Mines (2005), *Queensland's Ports*, October, www.nrm.qld.gov.au/mines/coal/rail_ports.html.

Sukhanov, Alexander (2005), 'Caspian Oil Heading East', *Asia Times*, 9 February.

'Tanzania Now To Pay $2.5m Monthly to Songas' (2004), LiquidAfrica Holding Limited, 11 June.

TERI (2006), 'Renewables and Regulatory Issues', The Energy and Resources Institute, February.

Tian, M. (2004), 'Renewable Energy Law under Consideration', *China Daily*, 26 November, http://www.chinadaily.com.cn/english/.

Torcellini, P., N. Long and R. Judkoff (2003), *Consumptive Water Use for U.S. Power Production*, Golden, CO: National Renewable Energy Laboratory.

Tseng, Wanda and Harm Zebregs (2002), 'FDI in China: Lessons for Other Countries', IMF Policy Discussion Paper 02/3, International Monetary Fund, Washington.

Umbach, F. (2004), 'Future Impacts of Chinese and Asian Dependency upon Energy from the Middle East and Central Asia', DGAP, Berlin, mimeo.

USCC (2004), 'China's Energy Needs and Strategies', *Annual Report 2004*, US-China Security and Economic Commission, USCC, Washington DC.

van de Venter, E. (2005), 'Sasol Coal-to-Liquids Developments', Presentation at Gasification Technologies Council Conference, San Francisco, 10–12 October, www.gasification.org.

'Venezuela Moves on Socuy and Cape Port' (2005), *McCloskey's Coal Report*, Issue 117, 19 August.

Wall Street Journal (Eastern Edition) (2005), 'Chinese Juggernaut hits Europe Head On; Strong Euro Fuels Trade Gap, Adds to Continent's Economic Malaise, 17 May.

Wang, Changgui et al. (2003), 'New Energy Generation Technology', China Electricity Power Press.

Wang, Qian (2006), 'China Maps out Natural Gas Pipelines', 23 June, China.org.ch.

Wang, Sicheng (2003), 'Solar Voltaic Generation', China/Spanish Renewable Energy Technology Symposium, December.

Wang, Tao (2004), 'Exchange Rate Dynamics', in Eswar Prasad (ed.), 'China's Growth and Integration into the World Economy: Prospects and Challenges', IMF Occasional Paper, No. 232, International Monetary Fund, Washington.

Warnock, Francis E. and Veronica Cacdac Warnock (2006), 'International Capital Flows and U.S. Interest Rates', http://www.faculty.darden.virginia.edu/warnockf/research.htm.

Wei, Shang-Jin (2000), 'How Taxing is Corruption on International Investors?', *Review of Economics and Statistics*, February, 82 (1): 1–11.

Weidou, Ni and T. Johansson (2004), 'Energy for Sustainable Development in China', *Energy Policy*, 32: 1225–9.

Weiss, J. and S. Gao (2003), 'People's Republic of China's Export Threat to ASEAN: Competition in the US and Japanese Markets, Asian Development Bank Institute Discussion Paper No. 2, http://www.adbi.org/discussion-paper/2003/01/02/86.prc.export.threat.

Wiser, Ryan and Ole Langniss (2001), 'The Renewables Portfolio Standard in Texas: an Early Assessment', Lawrence Berkeley National Laboratory, November, http://eetd.lbl.gov/ea/EMS/reports/49107.pdf.

Wiser, Ryan, Jan Hamrin and Meridith Wingate (2002), 'Renewable Energy Policy Options for China: a Comparison of Renewables Portfolio Standards, Feed-In Tariffs, and Tendering Policies', Center for Resource Solutions, June.

Wiser, Ryan, Kevin Porter and Robert Grace (2005), 'Evaluating Experience with Renewable Portfolio Standards in the United States', *Mitigation and Adaptation Strategies for Global Change*, 10: 237–63.

Wolak, F. (2004), 'Reforming the Indian Electricity Supply Industry', Department of Economics, Stanford University, Preliminary Draft May 2004, http://www.stanford.edu/~wolak.

Woodman, Sophia (2000), 'China, Resettlement and the International Review of Big Dams', *China Rights Forum*, Spring: 16.

World Bank (2002), 'Establishment of a State Electricity Regulatory Commission in China: a Suggested "Roadmap"', Working Paper, http://www.worldbank.org.cn/English/content/roadmap.pdf.

World Bank (2005), *The Little Green Data Book 2005*, World Bank, Washington DC.

World Commission on Dams (2000), *Dams and Development: a New Framework for Decision Making*, Cape Town: World Commission on Dams.

World Energy Council (2004), *2004 Survey of Energy Sources*, 20th Edition, London, December.

World Energy Outlook, 1998, US Energy Information Agency.

World Market Research Centre (2004), 'Automotive Sector Analysis: India and China', 18 October, www.wmrc.com.

World Market Research Centre (2005), 'China: Urban Centers in China To Face Power Crunch', 7 April, www.worldmarketsanalysis.com.

World Nuclear Association (2004), 'Nuclear Power in China', http://www.worldnuclear.org/info/inf63.htm.

World Oil (2005), 226 (9) (September).

World Summit on Sustainable Development (2002), 'Facts About … Energy', Johannesburg, South Africa, 26 Aug–4 Sep.

World Trade Organisation (1998), 'Energy Services Background Note by the Secretariat', Council for Trade in Services, 9 September, 98–3480.

'Worldwide Look at Reserves and Production' (2005), *Oil & Gas Journal*, 103(47): 19 December, 24–5.

Wu, K. and F. Fesharaki (2005), 'Natural Gas Pipelines and LNG Terminals in China: an Update', *Facts Gas Insights*, 46 (March), www.factsinc.net.

Wu, K., L. Wang and F. Fesharaki (2006), 'China's LNG Imports: Delayed Terminal Projects and a Less Bullish Demand Outlook', *FACTS Gas Insights*, 4 (January), www.factsinc.net.

Wu, L. (2004), 'China's Oil Security Challenges and its Countermeasures', *Geopolitics of Energy*, 26 (11) (November): 2–5.

Wu, Yanrui (2003), 'Deregulation and Growth in China's Energy Sector: a Review of Recent Development', *Energy Policy*, 31: 1417–25.

Xu, Y. (2000), 'China's Exchange Rate Policy', *China Economic Review*, 11 (3): 262–77.

Yamagata H. (2005), 'Japan's Climate Change Policy and Kyoto Mechanisms', presentation at the Climate Policy in the US and Japan: Prospects in 2005 and Beyond Workshop, Tokyo, Japan, 11–12 May), sponsored by Resources for the Future, www.rff.org/rff/Events/conferences.

Yao et al. (2005), 'Energy Policy and Standard for Built Environment in China', *Renewable Energy*, 30: 1973–88.

Yixin, Dai et al. (1999), 'Technical Innovation of the Chinese Photovoltaic Industry', Center for the Integrated Study of the Human Dimensions of Global Change, http://www.andrew.cmu.edu/user/kf0f/Solar_PV_Report_8-99.pdf.

Yu, Yongding (2004), 'China's Macroeconomic Situation Since 2003', *China and the World Economy*, 12 (4).

Yunxian, Chen (2001), 'China Stock Market and Challenge of Globalization', Speech notes in Harvard China Review Annual Conference held on 6 May.

Zhang Guobao (2004), Speech at the International Conference for Renewable Energies in Bonn, http://www.renewables2004.de.

Zhang Wendan, Zhou Shangqing and Zhu Yantao (2006), 'Welcoming a New Era in the Development of China's Forex Market', online.

Zhang Zhengming (1999), 'Impact Evaluation of Renewable Energy Technology on Society, Resources, and the Environment', November.

Zhang Zhengming (1999), 'Commercialization Strategies for Biomass Technologies in China', China Environmental Publishing House.

Zhang Zhengming, Deng Keyun and Ralph Overend (1999), 'Strategy Design for the Commercialization of China Biomass Energy Technology', China Environmental Science Publishing House, January.

Zhang Zhengming et al. (2006), 'Renewable Energy Development in China: the Potential and the Challenges', The Center for Resource Solutions, http://www.efchina.org/documents/China_RE_Report_EN.pdf.

Zhang Zhengmin et al. (2003), 'Economic Incentive Policy Research of Wind Power Generation in China', China Environment Science Press.

Zhang Zhengming et al. (2005), 'China National Energy Strategy and Policy 2020', Renewable Energy Strategy and Policy, The Development Research Center of the State Council of the People's Republic of China, http://www.efchina.org/documents/8_Renewable_energy.pdf.

Zhao, Jiarong (ed.) (1998), 'Economic Incentives Policy Study for China Renewable Energy Development', China Environment Science Publishing House, October.

Zhao, Jimin (n.d.), 'Reform of China's Energy Institutions and Policies: Historical Evolution and Current Challenges', Energy Technology Innovation Project Belfer Centre for Science and International Affairs.

Zhipeng, Liang (2003), 'Targets and Suggested Policy Framework for Renewable Energy Development in China', China's National Energy Strategy and Reform, China Development Forum, http://www.efchina.org/documents/Briefing_book_031115_EN.pdf.

'Zhongguo jiangshe fubuji nengyuandanwei choubeigongzuo jin chongciqi; Preparatory Work Enters Urgent Phase for China's New Vice Ministerial Energy Agency' April 2005, http://www.phoenixtv.com/phoenixtv/.

Zhou, Dadi (2003), 'China's Sustainable Energy Senarios in 2020', China Environment Science Press.

Zhou Fengqi et al. (2002), '50 Years of China Energy', China Electricity Power Press.

Zhou Nan, Chris, Marnay et al. (2004), 'The Potential for Distributed Generation in Japanese Prototype Buildings: A DER-CAM Analysis of Policy, Tariff Design, Building

Energy Use, and Technology Development', LBNL-56359, Lawrence Berkeley National Laboratory, Berkeley, CA.

OTHER SOURCES

China Renewable Energy Information Network
http://www.crein.org.cn
Center for Energy Efficiency & Renewable Technologies
http://www.ceert.org
A collaboration between major environmental organizations, public interest groups and clean technology companies working to achieve a more sustainable energy future. CEERT brings a unique alliance to the fight for cheaper, cleaner power. There are many articles about clean power, and there is clean power journal.
Energy Technology Data Exchange
http://www.etde.org
ETDE is a program of the International Energy Agency (IEA). Its mission is to collect, and exchange energy research and technology information, which it does through ETDE's Energy Database, but only ETDE member countries have access to the more than 3.8 million scientific and technical citations in the Energy Database.
National Renewable Energy Laboratory
http://www.nrel.gov
The nation's primary federal laboratory for renewable energy research. Performs functions assigned by DOE in research, development and testing.
The Energy Outlet
http://www.energyoutlet.com
A resource center promoting electrical energy conservation, its new location is filled with interesting displays that show energy saving products and techniques, and there is much energy efficiency information for homeowners.

WEBSITES

www.csrc.gov.cn
www.chinaclear.com.cn
www.sac.net.cn
www.cfachina.org
www.cnoocltd.com/ and http://www.cnooc.com.cn/english/companyoverview/dashi.html
www.iht.com
www.sse.com.cn
www.szse.cn
www.shfe.com.cn
www.dce.com.cn
www.czce.com.cn
http://scic.cec.eu.int/china/events/presen_gra/moftec/moftec.html
www.moftec.gov.cn
www.chinabidding.com (daily news from July to August 2006)
www.chinadaily.com (daily news from July to August 2006)
www.safe.gov.cn State Administration of Foreign Exchange

www.sohu.org
www.efchina.org
http://www.sznews.com/szdaily/
http://www.cbiz.cn/NEWS/
http://www.nadbank.org/english/general/general_frame.htm
www.worldsteel.org

Index